TOURISM AND HOSPITALITY DEVELOPMENT AND MANAGEMENT

RESEARCH STUDIES ON TOURISM AND ENVIRONMENT

JOSE MONDEJAR-JIMENEZ
GUIDO FERRARI
AND
MANUEL VARGAS-VARGAS
EDITORS

Nova Science Publishers, Inc.
New York

Copyright © 2012 by Nova Science Publishers, Inc.

All rights reserved. No part of this book may be reproduced, stored in a retrieval system or transmitted in any form or by any means: electronic, electrostatic, magnetic, tape, mechanical photocopying, recording or otherwise without the written permission of the Publisher.

For permission to use material from this book please contact us:
Telephone 631-231-7269; Fax 631-231-8175
Web Site: http://www.novapublishers.com

NOTICE TO THE READER

The Publisher has taken reasonable care in the preparation of this book, but makes no expressed or implied warranty of any kind and assumes no responsibility for any errors or omissions. No liability is assumed for incidental or consequential damages in connection with or arising out of information contained in this book. The Publisher shall not be liable for any special, consequential, or exemplary damages resulting, in whole or in part, from the readers' use of, or reliance upon, this material. Any parts of this book based on government reports are so indicated and copyright is claimed for those parts to the extent applicable to compilations of such works.

Independent verification should be sought for any data, advice or recommendations contained in this book. In addition, no responsibility is assumed by the publisher for any injury and/or damage to persons or property arising from any methods, products, instructions, ideas or otherwise contained in this publication.

This publication is designed to provide accurate and authoritative information with regard to the subject matter covered herein. It is sold with the clear understanding that the Publisher is not engaged in rendering legal or any other professional services. If legal or any other expert assistance is required, the services of a competent person should be sought. FROM A DECLARATION OF PARTICIPANTS JOINTLY ADOPTED BY A COMMITTEE OF THE AMERICAN BAR ASSOCIATION AND A COMMITTEE OF PUBLISHERS.

Additional color graphics may be available in the e-book version of this book.

Library of Congress Cataloging-in-Publication Data

Research studies on tourism and environment / editors, Jose Mondejar-Jimenez, Guido Ferrari, Manuel Vargas-Vargas.
    p. cm.
 Includes index.
 ISBN 978-1-61209-946-0 (hardcover)
 1. Tourism--Environmental aspects. 2. Tourism--Research. 3. Environmental management. 4. Environmental engineering. 5. Sustainable development. I. Mondejar-Jimenez, Jose. II. Ferrari, Guido. III. Vargas Vargas, Manuel.
 G156.5.E58R47 2011
 338.4'791--dc22
                                        2011008479

Published by Nova Science Publishers, Inc. † New York

CONTENTS

Preface		ix
Chapter 1	Application of Ultrasonic Wave Irradiation for Environmental Management Emphasis on Cost Benefit in Dairy Wastewater Treatment Plants *N. Merdadi, A. Zahedi, A. Aghajani Yasini, A. R. Karbassi and A. Mohamadi Aghdam*	1
Chapter 2	Environmental Certification as a Tool to Measure and Promote Eco-innovation in the Tourist Sector *Lluís Miret-Pastor, María-del-Val Segarra-Oña and Ángel Peiró-Signes*	13
Chapter 3	Analysis of the Spending of People Attending a Cultural Festival: The Cuenca Religious Music Week *Juan-Antonio Mondéjar-Jiménez, María Cordente-Rodríguez, María-Leticia Meseguer-Santamaría, José Mondéjar-Jiménez, and Manuel Vargas-Vargas*	27
Chapter 4	Looking for a Northern Region: Attributes of Choice *Normand Bourgault, Patrice Leblanc, Judy-Ann Connelly and Ann Gervais*	43
Chapter 5	Visitors Profile of a Mediterranean Green Corridor: Case Study of the Parc Fluvial del Túria (Valencia, Spain) *María José Viñals, Zeina Halasa and Mireia Alonso-Monasterio*	57
Chapter 6	Segmentation of Tourists and Day-trippers in Castilla-La Mancha (Spain): Main Differences and Similarities *Águeda Esteban-Talaya, Carlota Lorenzo-Romero and María-del-Carmen Alarcón-del-Amo*	69
Chapter 7	Touristic Sector Companies at the Warsaw Stock Exchange *Krzysztof Kompa and Dorota Witkowska*	89
Chapter 8	Overbooking as a Double-selling Practice in the Tourist Industry: A Brief View from Spanish Jurisprudence *María Valmaña Ochaíta*	111

Chapter 9	Hunting and the Environment *Pilar Domínguez Martínez*	119
Chapter 10	Monumental Tree Heritage as Urban Tourist Attraction *Francisca Ramón-Fernández, Lourdes Canós-Darós and Cristina Santandreu-Mascarell*	133
Chapter 11	Modeling Dimensions of the Tourist Satisfaction: The Cultural/Heritage Case of Toledo *Gema Fernández-Avilés, José-María Montero Lorenzo and Jean Pierre Lévy-Mangin*	145
Chapter 12	Luxury Resorts and Sustainable Tourism in the Maldives *Blanca de-Miguel-Molina, María de-Miguel-Molina and Mariela Rumiche-Sosa*	157
Chapter 13	Towards an Explanatory Model of Innovation in the Cultural Tourism Districts *Pedro M. García-Villaverde, Dioni Elche-Hortelano, María J. Ruiz-Ortega, Gloria Parra-Requena, Pilar Valencia-De Lara, Ángela Martínez-Pérez, Job Rodrigo-Alarcón and Miguel Toledo-Picazo*	169
Chapter 14	You Are an Order Citizen, Are you? Then, why don't you Separate Selectively your Trash? An Experimental Test *Marta Magadán and Jesús Rivas*	189
Chapter 15	Advertising, Unfair Commercial Practices and Consumer Protection in Spain *María Ángeles Zurilla Cariñana*	201
Chapter 16	Some Issues about the Inviolability of the Hotel Establishment in Spanish Law *Silvia Valmaña Ocháita*	211
Chapter 17	A Historical View about the Liability by Safekeeping in the Parking Business *Alicia Valmaña Ocháita*	223
Chapter 18	Ecotourism Capability in Sensitive Wetland Conservation, Case Study: Cheqakhor Wetland, Central Iran *Homa Irani Behbahani, Hassan Darabi and Zhale Shokouhi*	237
Chapter 19	Vineyard Site Qualification as a Strategy for Value Creation in Wine Tourism: The Case of "D.O. de Pago" Producers in the Spanish Region of Castilla-La Mancha *Ricardo Martínez Cañas and Pablo Ruiz Palomino*	253
Chapter 20	The Uniform System of Accounts for the Lodging Industry and XBRL: Development and Exchange of Homogeneous Information *Adelaida Ciudad Gómez*	269

Chapter 21	Tourism: Accessible Destination? *Isabel Mª. Ferrandiz Vindel, Jose Luis González Geraldo and Ana Mª. Bordallo Jaén*	**281**
Chapter 22	The Teaching of Spanish as a Sustainable Resource for Tourism *Pilar Taboada-de-Zúñiga Romero*	**289**
Chapter 23	Connecting Urban and Rural Areas through a Green Corridor: Case Study of Turia Fluvial Park (Valencia, Spain) *María José Viñals, Maryland Morant and Pau Alonso-Monasterio*	**313**
Chapter 24	Sustainable Forest Management in the Law 3/2008 of Castilla la Mancha and Adjustment to State Law and the European Union on the Matter *Jose Antonio Moreno Molina*	**325**
Chapter 25	Recognition of the Responsibility of the Local Spanish in Cases of Noise *A. Patricia Domínguez Alonso*	**335**
Index		**343**

PREFACE

The potential of environmental resources as tourist resources and the impact of tourism on these resources are an open research area and, because of its social and economic impact, arouse great interest among the social stakeholders. It is necessary to understand and measure their mutual influence in order to achieve positive mutual links between tourism and the environment. This new book addresses the interaction between tourism and the environment through several disciplines, a multidisciplinary perspective, and different theoretical and methodological approaches. In addition, this book presents a wide range of current research and promotes debate and analysis on this research.

Chapter 1 - Dairy industries generated large amount of wastewater in comparison to other industries. For example in Tehran Pegah Dairy Complex (TPDC) as the biggest Iranian dairy producer, around 3500 cube meter of wastewater generated for processing of 1000 tons milk, per day. Biological treatment like activated sludge is common technology for achieving discharge standard but producing large amount of biosolids as a by-product. Biosolids needs additional processing for disposal that allocating up to 60% of capital cost and more than 50% of operational cost in wastewater treatment plants (WTPs). This is more expensive for such industry. Therefore in modern WTPs reducing the volume of waste biosolids is main concern. This biosolids contain high amount of nutrients such as nitrogen and phosphors that could use for nutrients balancing in biological treatment. Ultrasonic wave irradiation (UWI) is one of the new methodologies in this field for release nutrients from solids phase to soluble phase which easily degraded by micro organism in biological treatment. In this paper effect of UWI on nitrogen and phosphors content of waste biosolids in TPDC was evaluated and possibility to applying large-scale UWI for reducing the cost of WTPs in different operational condition estimated. The equipment was used for generating ultrasonic wave is Scientz-IID with 20 kHz operational frequency which manufactured by Xinzhi Bio-Instrument LTD. Experimental investigation was carried out at 60,120,240,480 s, contact times and in output power 100,300,500 W. The immersion depth of probe (transferring the ultrasonic wave to biosolids) was 20 mm. and the volume of biosolids with exposures to ultrasonic wave was 300 ml.

Chapter 2 - The tourist industry is evolving towards a new model where it will be necessary to combine profitability with sustainability. Adapting to this new tourism model requires companies and tourist destinations innovate in the field of environmental awareness. In this new framework, eco-innovation appears as a novel and important concept, but in any case, eco-innovation still presents difficulties in its conceptualization and measurement. This chapter reviews the literature on eco-innovation and proposes the use of three environmental certifications (ISO 14001, EMAS and Eco-label) as tools to measure and promote eco-innovation in the tourist sector.

Chapter 3 - Cultural events and festivals have become a booming attraction, complementing the investment made in alternative commercial channels, because of the fidelity these periodic events generate among their public. In addition, holding them generates important economic effects in the economic activity of the region, both directly and indirectly in other economic sectors. So, aware that the investments made are justified by the good results, public institutions favour, encourage and even support the holding of such festivals, considering them an advanced factor in tourism competitiveness. This is the case of Cuenca's *Semana de Música Religiosa* (SMR), Religous Music Week, the 49th edition of which was held in 2010, consolidating it in the Spanish and international music world. The purpose of this study is to analyse the spending of the people attending the *Semana de Música Religiosa de Cuenca*, and how it is distributed among the different expenditure items analysed. The purpose of this is to identify and differences and specific needs according to the characteristics of the people attending the festival, and so be able to modify the strategy followed by the event's organizers in terms of price, product and promotion.

Chapter 4 - It is not an easy task to attract tourists and new residents north of the 48th parallel in a country. This research was conducted in two phases. In the first phase, a qualitative approach was used to identify broader themes of concern to people when choosing not only a holiday destination, but also a place to live. Four focus groups generated some overall themes that form the consideration set of attracting variables. With this as a substrate, a quantitative approach was then applied in the second phase. Fourteen items were developed. Importance and presence were measured in a statistically representative sample of 306 respondents from different regions of Québec Province, in Canada. The analysis showed that nearness to wildlife in safe conditions is the most strongly present factor of attraction, but people who are approachable and very hospitable toward travelers is quite a close second. Eleven other attributes of choice were tested. An advertising campaign theme, as revealed by the research, is presented.

Chapter 5 - The *Parc Fluvial del Túria*, Valencia (Spain) is a green corridor with a riverside shared-use 30km path along the towns of Vilamarxant, Benaguasil, Riba-roja de Túria, L'Eliana, Paterna, Manises, Quart de Poblet, Mislata and the city of Valencia. A multi-disciplinary team of researchers from the Universidad Politécnica de Valencia has carried out a survey between March and April 2009, before the official inauguration of the Park. This campaign has been done for monitoring the visitors' behaviors and attitudes, determining the visitor's profile of this corridor and unfolding people's thoughts and perceptions, in order to design the signage of the Park as well as the etiquette and the code of ethics. Basic descriptive statistics has been applied over the 1,824 visitor respondent forms. Additionally, attention has been given to the interpretation of the richness of the data obtained. The findings of this survey show that the prevailing visitors' profile is dominated by a slight majority of male visitors'(63.3%), age ranges between 36-50 years old, working in the service sector, and coming from the residential adjacent areas. Pedestrian activities with 82.7% (including walking, walking the dog and jogging) are the preferable activities developed in the site. Cycling activities (64.2%) are also very popular. Other activities such as observing nature, going for a picnic and horse riding were also practiced in a regular manner. The results show a high level of satisfaction related with the greenway setting and the most significant problem detected was the conflict between different users.

Chapter 6 - According to tourist official data, the region of Castilla-La Mancha in Spain ranks in the fifth position as favourite destination for Spanish nationals tourists. Due to the importance of tourist sector in Spain, and specifically in this autonomous region, this chapter carries out a comparative analysis of the different groups of tourists who visited Castilla-La Mancha during 2008, taking into consideration the period of travel (Easter Holidays vs. weekends). In order to achieve this objective, this research has used a cluster analysis which has allowed us to classify tourists and day-trippers in different groups according to two variables: kind of tourism (cultural, rural, nature, etc.), and activities carried out by the tourists (rest, gastronomy, nightlife, etc.). The association between the group to which a tourist belongs and these two variables have also been taken into consideration in this analysis, as well as other characteristics like age, marital status, and autonomous region of origin, which have been analyzed separately for each of the two types of visitors, tourists and day-trippers, in the wider context of the separation between Easter holiday travel and weekend journey. The results of this research show the differences and similarities between the visitors analyzed, which should be of interest for the tourist industry, especially when planning marketing strategies adapted to each of the tourist profiles described here.

Chapter 7 - The paper is divided into two main parts. The first one describes the selected aspects of touristic sector in Poland. The aim of the research, discussed in the second part, is to present and analyze the companies belonging to that sector and being listed at the Warsaw Stock Exchange. Research is conducted applying comparative analysis, trend function and taxonomic measures based on real data.

Chapter 8 - Overbooking is a commercial strategy that has becamed very relevant in the last years in the tourist industry. Although overbooking is well known in sectors like air transport overbooking is also used by other kind of companies like hotels in order to optimize resources and infrastructures. Despite overbooking being regulated in the European Union by an specific regulation in order to protect consumers there are still many questions debatable from economic as well as legal point of view.

Chapter 9 - Hunting plays an important role in the conservation of species, and the inter-relation between fauna and flora in the management of natural heritage is aimed at maintaining biodiversity. For this reason the need for protection of wild fauna requires to make the existence of fences compatible with the protection demands on fauna mobility with a view to ensure its conservation and biodiversity. Not surprisingly, in compliance with EIA legislation, conditions relating the so-called impact statements have been imposed on fencing. According to hunting legislation in Spain, fence installation is not mandatory, moreover, a prior authorization is required before installing fencing, being the exclusive obligation of the owner of the hunting ground to place signposts in the property. The aim of this paper is to analyze whether the absence of fencing can be determinant of careless conduct for the purposes of attribution of liability to the holder of the game preserve for damages resulting from traffic accidents caused by collision with animals from the said farm.

Chapter 10 - When authors think about a city's tourist attraction, we always focus on its architecture and historical heritage. We forget that there is also a living heritage, composed of trees with environmental and historical characteristics, which form the ornamentation of towns and cities, and that, in the case of the region of Valencia (Spain) is especially taken into consideration and legislated. This chapter attempts to demonstrate the importance of trees as living monuments, the monumental tree heritage, in sum, the city's tourist attraction, taking into account its scenic and visual value for potential visitors. This cultural heritage consists in outstanding, monumental or singular trees that are a potential for the city. Their regulation by Law 4/2006 of monumental tree heritage, is the most important tool for their conservation, protection and cataloguing. As authors will see, this protection is completely justified by their characteristic exceptions of historical, cultural, scientific and recreational value, since these specimens represent a singular part of the people's environmental and cultural heritage, and, thus, an enormous potential as instrument of reputation and identity of the city as tourist destination.

Chapter 11 - Because of people's inclination to look for new attractive activities including traditional cultures, the heritage tourism has become a major "new" area of tourism demand and almost all policy–makers are now aware of and anxious to develop. Despite the emergence of new strategies (those based on equity, norm, or perceived overall performance, those where emotions play an important role in satisfaction-formation, and those with a cognitive affective view) the models of expectation are still very popular in tourist research. In this context, authors pursuit to measure tourist satisfaction through three first-order factor variables: accommodation, offered services and tourist attractiveness by using a causal model among other three second-order factor variables: importance given to tourist attributes, satisfaction and expectations. This research relies on the information provided by 1,500 respondents who were given a small questionnaire specially designed to measure tourist satisfaction in the emblematic part of Toledo, Spain (a UNESCO World Heritage City).

Chapter 12 - Small islands are favoured locations for luxury resorts. However, this type of tourism, which aids economic development in some countries, can cause a threat to these islands' local environment. In the case of the Maldives, the country's government has established specific regulations to prevent this from happening but these are not sufficient to ensure the future conservation of local ecosystems. Therefore, this chapter examines, from a conceptual sustainable tourism framework, whether luxury resorts in the Maldives have implemented measures to ensure sustainable tourism and whether they have communicated them to their visitors in order to establish whether luxury and sustainable tourism are compatible. To carry out our study, authors analysed around 40 luxury resorts in the Maldives using a content analysis methodology and we examined those which were supposed to be more eco-friendly.

Chapter 13 - In the last decade there has been an increasing political, economic and scientific interest for the dynamic of the urban and cultural tourism as a basis for sustainable local development, especially in small and medium size historical cities. Cultural touristic cluster remain common features of industrial districts: interdependence between firms, flexible boundaries of companies, partnerships and competition, confidence through sustained collaboration among stakeholders and a cultural community with the support of public policies. We try to advance into this field to analyse in depth the heterogeneity in the behaviour and innovation of firms belonging to cultural tourism districts. We propose an integrative model to analyse the role that two key factors, such as relationships and a higher level of knowledge, has on the innovation developed by firms in the cultural tourism districts. In this sense, authors have established several theoretical propositions based on different approaches to the literature. We have improved the previous models by introducing an intermediary effect, which provides a better explanation of the causal process that leads to the generation of a greater performance of innovation. Thus, authors highlight the role of knowledge acquisition, as a factor that mediates between the combination of internal and external relationships and the development of innovation by firms in the cultural tourism districts. This research contributes to integrate three theoretical perspectives such as network and relationships approach, knowledge-based approach and territorial perspective to the study of innovation.

Chapter 14 - Yellow, green, blue and black trash cubes can be found by any street, in front of each block of houses. Usually, the authors' municipalities are which determine the day of the week for throw away plastics, glasses, cardboard and paper. All in favor of recycling. All for recycling. But, How many people believe in recycling? More even, how many people are recycling practicants? To answer these questions authors need to dive into the most unknown motivations of human behavior. Not many people face voluntarily to a recycling routine. Then, which are the variables that give us a good explanation to understand a non recycling attitude? Our experiment tries to throw some light to this respect.

Chapter 15 - Law 29/2009 of 30 December, amending the rules on unfair competition and advertising in order to improve the protection of consumers, transposes Community Directive 29/2005 to Spanish law and amends two very important Spanish laws: Law 3/1991 on Unfair Competition and General Law 34/1988 on Advertising. It establishes a unitary legal framework on unfairness in misleading and aggressive practices, imposing the same standards regardless of whether the addressees are consumers or businesses. The traditional distinction between unfair practices and the regulation of unfair or misleading advertising is set aside. It is worth noting that the legislators have opted to maintain the General Advertising Law and that the concept of unlawful advertising in the framework thereof has been preserved, safeguarding actions and remedies allowing that practice to be combated. In this paper authors will examine the most notable aspects of Law 29/2009, with special attention to those concerning the regulation of unlawful advertising and the mechanisms to protect consumers from it.

Chapter 16 - The protection granted by the legal system to privacy in Spanish law is not in the majority of cases applicable to the hotels, hostels and other hotelier establishments. Casuistry is so large that it is necessary to distinguish the different nature of the object of protection. This protection would depend on the room concerned for the invasion of the privacy, the person who commits the intrusion and the moment of the entry. This chapter tries to systematize and clarify some essential points about the treatment given by the Spanish law, especially the criminal law, to these issues, distinguishing the application of the concept of domicile, and therefore the application of the crime of unlawful entry of dwelling-houses, from other unlawful entries in hotel establishments open to the public.

Chapter 17 - The contract of parking has been regulated a few years ago in the Spanish Law, in particular in Act 40/2002, of 14th November, which tried to arrange a business activity more and more frequent from mid of the last century. With the Act, was solved the existing scientific and jurisprudencial discussion about the juridical nature of the parking - deposit or rent-, and was provided a duty of vigilance and safekeeping connected to the obligation of restitution of the vehicle; however, the Act creates problems on its own scope and the extension of the liability. In this chapter this question will be approached from the perspective of the Law in force and from the protection that the Roman Law gave to similar situations to the present ones, saved the distances, as it were the case of *stabularii*, owners of stables in which the cavalries could be left by a period of more or less long time, for reasons of terrestrial transport.

Chapter 18 - Wetlands environmental functions in semi-dry and dry regions are very important, sensitive and fragile at the same time. In order to compensate deficit, wetlands marginal rural communities exploit of wetland resources directly such as; wetland plants collection for feeding livestock, hunting of migratory birds and the like. These actions damage ecosystem seriously and put sensitive ecological systems at risk. On the other hand, wetlands have considerable characteristics and landscape attractions that are introduced them as tourism attractions. Communities-based ecotourism provides alternative income possibility for rural households. Ecotourism provide training and participation of indigenous people chance. These opportunities would cause destructive agents change into the environmental protections and their interests will be depended on the protection of wetland resources. This chapter addresses the needs for conservation endangered sensitive ecosystems and ecotourism capability to protect wetland ecosystems. An example of community-based ecotourism in CheqaKhor Wetland, Chahar Mahal Bakhtiari province, central Iran as a sensitive wetland will be highlighted.

Chapter 19 - Wine regions in Spain are classified according to the quality wine they produce. The best quality of wine denomination comes from "*Denominación de Origen de Pago*", or "D.O. de Pago", that is the highest vineyard site legal qualification. This qualification is a very special guarantee of origin under which it is attempted to protect certain singular wines that are produced through a close relationship between vineyard and winery (like French *Chateaus*). D.O. de Pago wine has exceptional quality recognised both by specialists and by consumers. As approved plots, legislation allows the possibility of a guarantee of origin consisting of a name of a wine growing place that, because of its natural characteristics and the varieties of vines and growing systems, produces grapes from which distinctive wines are produced by means of specific preparation methods. In this paper authors analyze the effect of this recognised quality trademark on wine tourism. Wine tourism in Pago's vineyards is a new and emergent industry that combines the synergic effect of commercial and tourism strategy from wine producers. To illustrate this phenomenon authors analyze the case of wineries under that recognition that have emerged in the Spanish region of Castilla-La Mancha.

Chapter 20 - The internationalization of markets and users, new management tools using information based on a homogeneous mass, or the use of the Internet for exchange, has made as a whole that standardized information available internationally, not only in development but also in its publication and exchange, should become a requirement. Therefore, this paper aims, first, to describe the possibilities for the lodging sector to enable it to develop, present and exchange information uniform, focusing both on financial and environmental information. On the one hand, the *Uniform System of Accounts for the Lodging Industry (USALI)*, whose great contribution is to facilitate homogenization in the presentation of information worldwide, allowing comparison between hotels and build statistical databases from figures uniform and standardized, including providing segment information according to the requirements of *International Financial Reporting Standards* -IFRS 8 and *Statement of Financial Accounting Standards*-SFAS 131, and on the other, to allow that information to be transmitted electronically in a prompt, transparent, accurate, quality, implementation of the XBRL standard. The second objective authors aim to propose is those issues which should be addressed in a near future.

Chapter 21 - The link between Accessibility and Tourism has been one of the greatest challenges in recent years, and the reality has proved its possibilities. Today, one of the market sectors with more relevance is the one which takes care of people with reduced mobility (PRM) and/or disabilities. Its relevance is justified, mainly, by two reasons: the vast number of people who can be numbered within this group as well as their growing participation in the tourism market. Accessibility in tourism should not only be understood as requirement that a specific service or destination must accomplish in order to foster access to people with reduced mobility or disabilities. On the contrary, accessibility in tourism has turned out to be one of the intrinsic factors of touristic quality because there cannot be real quality in tourism if the destination is not reachable for everyone. None should be excluded from tourism for any reason or circumstance. In this chapter, authors would like to increase the awareness of the institutions responsible for tourism, as well as the general public, about the relevance of tourism with quality; a tourism which everyone can join and enjoy. Having said that, authors will approach this issue from the perspective of those collectives which traditionally speaking had problems with tourism accessibility.

Chapter 22 - The promotion and development of Cultural Tourism has become one of the primary objectives of tourism policy for many countries, attempting to bring about less invasive tourism, with less marked seasonal trends, targeting a public with higher cultural awareness and consequently more respectful of the environment. In a globalized world the learning of languages gains more importance every day, not only amongst young people as part of their education, but also amongst adults. This desire to learn other languages has brought about the concept of 'linguistic tourism', consisting in the provision of organised trips, where the primary objective is the learning or perfecting of a language, usually in another country, combined with tourism activities. Linguistic tourism has a long tradition in countries like the United Kingdom (UK) and France. In Spain, however, it is a more recent phenomenon. Nevertheless, there is significant willingness on the part of those bodies responsible for the promotion and commercialisation of cultural tourism to develop this sector, whilst maintaining quality. This study presents two well defined sections. In the first section, authors analyse the great potential for growth in the learning of Spanish as a foreign language, and how this contributes to the development of linguistic tourism in Spain. In the second, the study focuses both on the particular case of Santiago de Compostela as well as using evidence from Salamanca to make a comparison of two similar cities differently developed in terms of linguistic tourism. The demand for Linguistic Tourism in Santiago de Compostela is demonstrated through a survey

Chapter 23 - This chapter analyzes the connectivity effects afforded by a green corridor that links urban environments with natural and rural areas. This is the case of the Turia Fluvial Park in Valencia (Spain), a strategic project of a green corridor promoted by the Jucar River Basin Agency (*Confederación Hidrográfica del Júcar*). This corridor encompasses the Turia River as it flows through seven villages to the City of Valencia. The Turia River Valley and the natural corridor have become a particular focus of recreation in the rural-urban fringe, positively affecting more than 1 million people. This green corridor includes a shared-use trail, providing a breathing space visually pleasurable, situated in a low-stress environment which is suitable for bicycling, horse riding and walking. It is an alternative setting that is separate from road traffic and concreted, where people can feel at ease and indulge their leisure needs in a natural area. The most important outcome of this work is to highlight the vision of how this corridor can make this urban-rural fringe more attractive, accessible, and multi-functional, and to state the benefits of creating a connection both in people's mind as well as in the functioning of the environmental systems.

Chapter 24 - The work carried out an analysis of several of the recent forest legislation of Castilla-La Mancha, in line with international forestry law, the law of the European Union and the State Basic Law on the matter, taking into account the many uses that the authors' society demands of the mountain today, but above all, is based on the prioritization of environmental protection of the forest, in the protection of the biological dimensions of the forest. The object of study in particular the modern concept of forest as forest ecosystem.

Chapter 25 - The paper analyzes the requirements for recognition by the administrative litigation law the liability of local authorities in cases of failing to address noise pollution. It is a jurisprudence that has opened in recent years an important means of struggle of citizens against the serious problem of noise in Spain.

In: Research Studies on Tourism and Environment
Editors: J. Mondejar-Jimenez et al.
ISBN: 978-1-61209-946-0
© 2012 Nova Science Publishers, Inc.

Chapter 1

APPLICATION OF ULTRASONIC WAVE IRRADIATION FOR ENVIRONMENTAL MANAGEMENT EMPHASIS ON COST BENEFIT IN DAIRY WASTEWATER TREATMENT PLANTS

N. Merdadi, A. Zahedi[*], A. Aghajani Yasini, A. R. Karbassi and A. Mohamadi Aghdam

University of Tehran, Iran

ABSTRACT

Dairy industries generated large amount of wastewater in comparison to other industries. For example in Tehran Pegah Dairy Complex (TPDC) as the biggest Iranian dairy producer, around 3500 cube meter of wastewater generated for processing of 1000 tons milk, per day. Biological treatment like activated sludge is common technology for achieving discharge standard but producing large amount of biosolids as a by-product. Biosolids needs additional processing for disposal that allocating up to 60% of capital cost and more than 50% of operational cost in wastewater treatment plants (WTPs). This is more expensive for such industry. Therefore in modern WTPs reducing the volume of waste biosolids is main concern. This biosolids contain high amount of nutrients such as nitrogen and phosphors that could use for nutrients balancing in biological treatment. Ultrasonic wave irradiation (UWI) is one of the new methodologies in this field for release nutrients from solids phase to soluble phase which easily degraded by micro organism in biological treatment. In this paper effect of UWI on nitrogen and phosphors content of waste biosolids in TPDC was evaluated and possibility to applying large-scale UWI for reducing the cost of WTPs in different operational condition estimated. The equipment was used for generating ultrasonic wave is Scientz-IID with 20 kHz operational frequency which manufactured by Xinzhi Bio-Instrument LTD. Experimental investigation was carried out at 60,120,240,480 s, contact times and in output power 100,300,500 W. The immersion depth of probe (transferring the ultrasonic wave to

---

[*] Contact author: ali_1980_8@yahoo.com

biosolids) was 20 mm. and the volume of biosolids with exposures to ultrasonic wave was 300 ml.

Keywords: Environmental management, Ultrasonic Wave Irradiation, Release of nutrients, Dairy biosolids, Tehran Pegah Dairy Complex

1. INTRODUCTION

Dairy industries generated large amount of wastewater in comparison to other industries. Tehran Pegah Dairy Complex (TPDC) as the biggest Iranian dairy producer, around 3500 cube meter of wastewater generated for processing of 1000 tons milk, per day. Biological wastewater treatment involves the transformation of dissolved and suspended organic contaminants to biomass and evolved gases (CO_2, CH_4, N_2 and SO_2) [1]. The activated sludge process is the most widely used biological wastewater treatment for both domestic and industrial plants in the world[2]. One of the most serious challenges in biological wastewater treatment is production large amount of biosolids(or sludge) as a by-product. Biosolids needs additional processing for disposal that allocating up to 60% of capital cost and more than 50% of operational cost in wastewater treatment plants (WTPs)[3,4]. This is more expensive for such industry. Therefore in modern WTPs reducing the volume of waste biosolids is main concern. This biosolids contain high amount of nutrients such as nitrogen and phosphors that could use for nutrients balancing in biological treatment. Ultrasonic wave irradiation (UWI) is one of the new methodologies in this field for release nutrients from solids phase to soluble phase which easily degraded by micro organism in biological treatment [5]. The application of ultrasounds to sludge processing is attractive because of the possibility for obtaining significant effects without the necessity to apply additional components, e.g. chemical agents, simplicity of the construction and operation of reactors and air tightening of the installation [6].

Ultrasound includes a wide range of frequencies between 20 kHz and 10 MHz [7–10]. When acoustic energy is supplied to a liquid, gas bubbles are formed and grow by absorbing gas and vapor from the liquid [9–12]. As a result of alternating expansion and compression cycles, these bubbles can implode, locally resulting in very extreme conditions of temperature (~5000 K) and pressure (~500 bar), whereas the bulk solution remains near ambient conditions. This phenomenon is called cavitation and is encountered only at frequencies ~ 1 MHz. Higher frequencies will cause stable, oscillating bubbles that will not implode, as discussed in the above-cited references. Cavitation will result in (1) the promotion of chemical reactions as a result of a locally high temperature and pressure; (2) extreme shear forces in the liquid, thereby mechanically attacking components; (3) the formation of highly reactive radicals (H°and OH°), which can again facilitate chemical reactions for destroying organic contaminants to take place; and (4) the additional destruction of specific compounds, given that cavitation bubbles are surrounded by a liquid hydrophobic boundary layer that preferentially permeates volatile and hydrophobic substances, subsequently reacting in the gas bubble. Sonochemic reactions are commonly applied at frequencies between 10 and 1000 kHz. A detailed discussion of the cavitation phenomenon is beyond the scope of the present paper and the reader is referred to more general literature [11, 13, 14]. The radius of the bubbles is related to the frequency of the generating ultrasound and in water at ambient conditions this relationship is represented by the following equation [15]:

$$R \approx \frac{3.28}{v}$$

Eq.(1)

Where R is the radius of the bubble (m) and v is the frequency of the ultrasound wave (Hz = s^{-1}).

Lower frequencies thus generate larger bubbles. It is expected that the energy released by the implosion of the bubble is a function of the bubble size, and is higher at lower frequencies. Tiehm et al. [15] confirm that the rate of sludge disintegration increases proportionally with the bubble size (as calculated by Eq. 1). Laborde et al. [9] noticed that the implosion of bubbles was more pronounced at low frequencies (20 kHz) than at higher frequencies (500 kHz).

Another mechanism that occurs when sludge is sonicated is acoustic streaming. Acoustic streaming has been studied since 1831 (Faraday) and occurs at the solid/liquid (sludge) interface when the solid interface experiences harmonic vibrations. The main benefit of streaming in sludge processing is mixing, which facilitates uniform distribution of ultrasound energy within the sludge mass, convection of the liquid and distribution of any heating that occurs. Overall there are three regions of acoustic streaming. The largest region, Eckart streaming, (Region I as shown in Figure 1), is the furthest from the vibrating tool and has circulating currents that are defined by the shape of the container and the size of the wavelength of the acoustic wave in the liquid. The region near the tooling (Region II) has circulating currents and its size and shape are primarily defined by the acoustic tooling. These circulations are typically called Rayleigh streaming, with much longer wavelengths than that of the acoustic wave in the liquid. The region nearest to the tool (Region III), called Schlichting streaming, is adjacent to the fluid acoustic boundary layer. This is a region where the tangential fluid velocity is near the velocity of the horn face. This layer is relatively thin. For example, at 20 kHz, the acoustic boundary layer for water at 20° C is less than 4 μm [16]. All three regions play a critical role in mixing of the fluid.

2. METHODOLOGY

The settled waste activated sludge (WAS) or return sludge was collected from Tehran Pegah Dairy Complex on a simple basis and analyzed immediately for TKN, TP, COD, SCOD and soluble part of TKN and TP. The sludge sample was sorted at 4 °C storage room to prevent biodegradation and transferred to lab for applying ultrasonic wave. All the analyses were done according to standard methods.

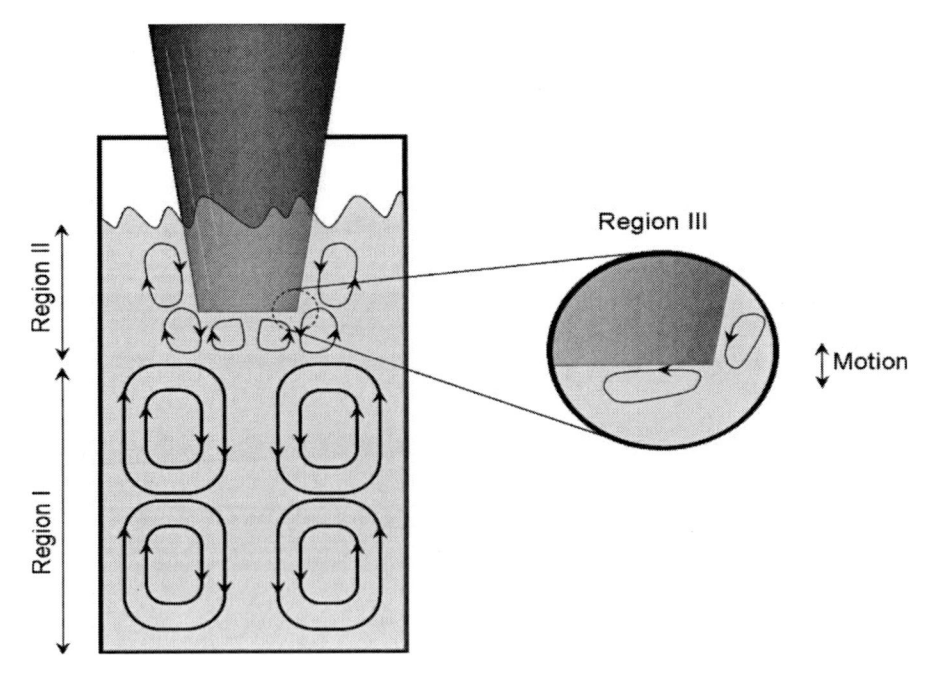

Figure1. Regions of acoustic streaming.

The most important role in the cost optimization of the ultrasonic wave irradiation process in operating conditions will be played by an appropriate matching of the output parameters of the ultrasonic system with sludge characteristics.

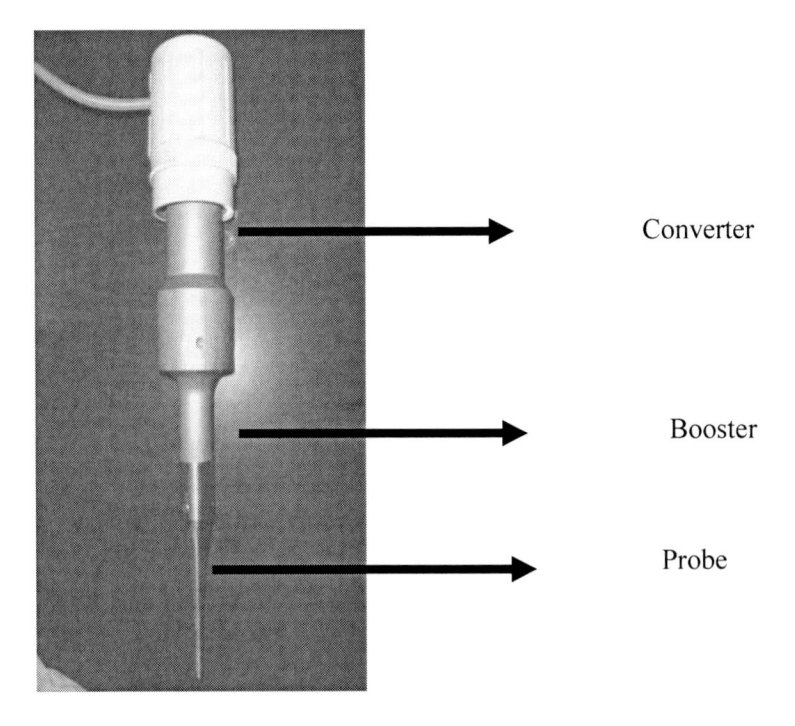

Figure 2. Component of sonotrode.

Table 1. Details specification of ultrasonic equipment

Scientz-IID sonificator system	
Maximum energy out put (W)	950 W
Duty ratio	0.1-99.9%
Amplitude pole	Φ6
Electricity supply	220 V, 50 Hz.
Frequency(KHz.)	20
Dimensions of sound box proof	220x400x290 mm
cable	UNF 1/2 –in.-20
Working time	Adjustable (1 s-99h99min)

2.1. Ultrasonic Equipment

The WAS samples was sonificated using Scientz-IID system (manufactured by Ningbo Scientz Biotechnology Co. Ltd.). The ultrasound unit consist of the converter (or transducer), booster and sonotrode. A converter basically converts electrical energy into ultrasound energy (or vibration). The booster is a mechanical amplifier that promote higher amplitude which generated by the converter. The sonotrode is a specially designed tool that contributes the ultrasonic energy to the sludge. The component of sonotrode is shown in the Figure 1.

The full specification of the apparatus is present in table 1. The operational frequency is 20 kHz. The maximal energy output is 950 W. Figure 2 shows the ultrasonic equipment.

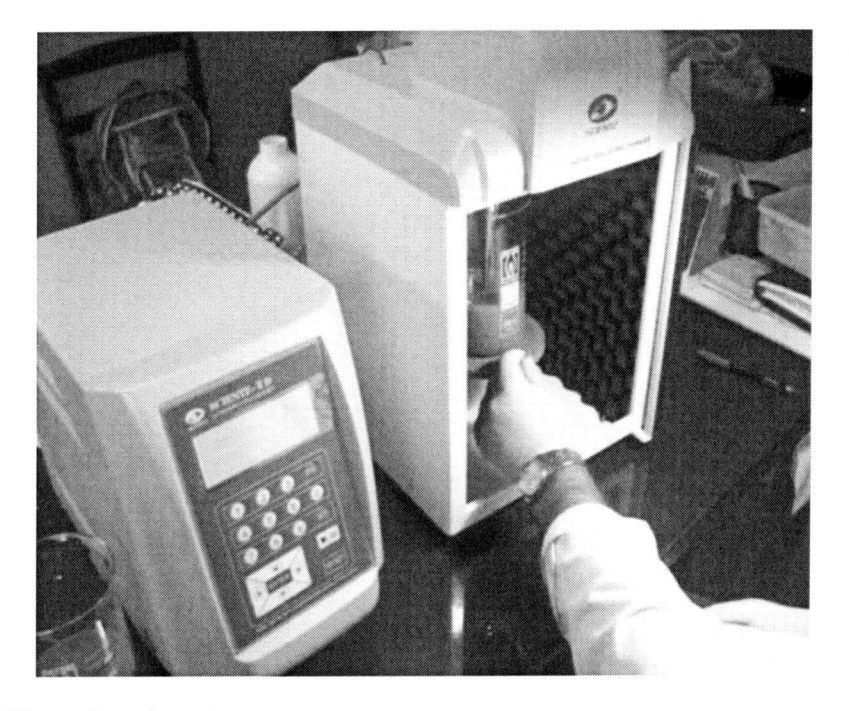

Figure 3. Ultrasonic equipment.

Table2. Characteristics of waste activated sludge from Tehran Pegah Dairy Complex

parameter	Unite	Number of repetition	Rage of data	Mean	Standard deviation
COD	mg/l	3	19580-21211	20297	833.15
SCOD	mg/l	3	784-820	800	18.23
TKN	mg/l	3	1095-1170	1130	37.74
Soluble TKN	mg/l	3	334-379	360	23.30
TP	mg/l	3	253-276.5	265.5	11.82
Soluble TP	mg/l	3	16.5-22	19.6	2.84

2.2. Operational Condition of Sonification

All experiment carried out with fixed volume of dairy WAS (300 ml) and at ambient temperature in all sonification operation. And also same dairy WAS applied for all sonification tests. Different contact times of 60,120,240,480 seconds were selected for this investigation. The ultrasonic energy outputs set at 100, 300,500 W. Small probe with diameter of 6 mm in all operational condition applied and immersed at the depth 20 mm into sludge during ultrasonic irradiation. Figure 3 shows the ultrasonic equipment.

2.3. Selection of Waste Activated Sludge Container

300 ml sample of the sludge was taken into the container with a volume of 700 ml and internal diameter of 7.2 cm.

3. RESULTS AND DISCUSSION

Characteristic of waste activated sludge from Tehran Pegah Dairy Complex present in table 2. The mean of soluble TKN is 360 mg/l and the mean of soluble TP is about 19.6 mg/l.

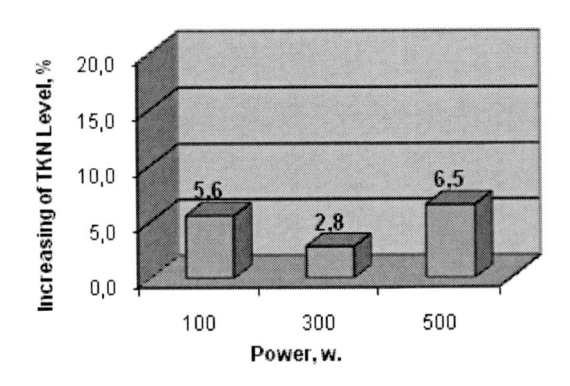

Figure 4. Released TKN from Waste activated sludge at 60 seconds sonification time versus different powers.

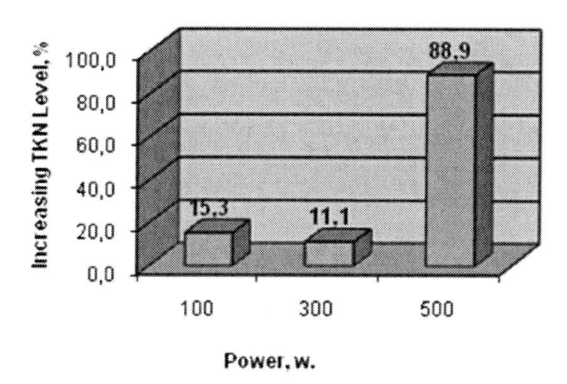

Figure 5. Released TKN from Waste activated sludge at 120 seconds sonification time versus different powers.

3.1. Evaluation of Effect of Ultrasonic Wave on Release TKN Level

According to characteristic of the waste activated sludge (see table2) more than 96 % of COD and 68% of TKN are in insoluble form. Micro organisms in biological treatment are able to degrading only soluble form of substrates and if substrates is in insoluble form micro organisms forced to hydrolysis the substrates, which is limiting step in biological treatment [15]. Beside that, balancing of nitrogen in biological unit is very important to achieve high rate of the removal. Figure 4-7 illustrate the percentage of increase of TKN level in different contacts times of 60,120,240 and 480 seconds respectively. The maximum release of TKN was found around 170.8 percent which observed in power 100w and contact time of 480 seconds. In all different powers of 100,300 and 500w with increasing sonification times, hydrolyses of TKN shows increasing trends. Although in power 300w achieving values is relatively lower which observed in power 100w. This could be linked to consumption some part of release TKN because of domination free active radicals during sonification with power 300w. But in power 500w as comparison to power 300w high tends of increasing TKN level observed which could related to nature of nitrogen compounds in dairy waste activated sludge that break down in this power and going to soluble form.

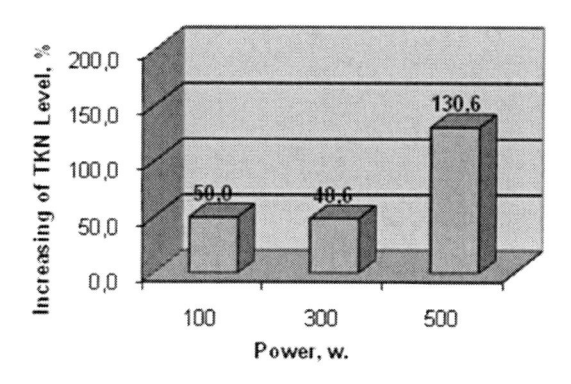

Figure 6. released TKN from Waste activated sludge at 240 seconds sonification time versus different powers.

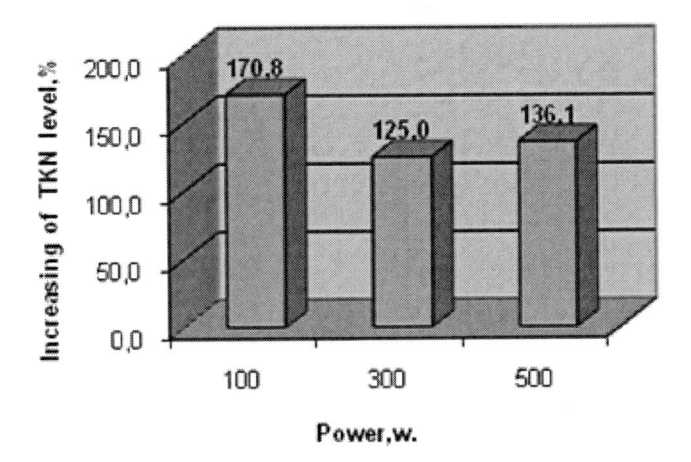

Figure 7. Released TKN from Waste activated sludge at 480 seconds sonification time versus different powers.

3.2. Evaluation of Effect of Ultrasonic Wave on Release TP Level

According to characteristic of the waste activated sludge more than 96 % of COD and 92% of TP are in insoluble form. Phosphor is one of the initial substrates for growth of microbial activity in biological treatment. Tehran Pegah Dairy Complex suffering from balancing phosphor in biological treatment. As shown in the waste activated sludge characteristic (see table2), only small part of TP existing in soluble form (about 8%). Therefore release of TP from sludge for balancing phosphor is very important in this plant.

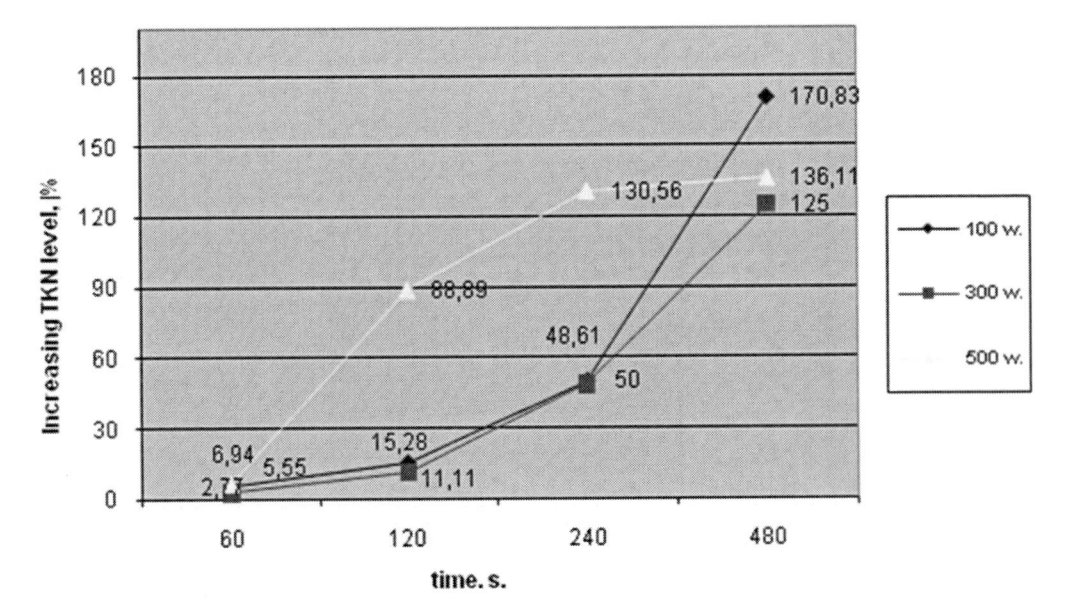

Figure 8. Comparison of release TKN level from Waste activated sludgeat different sonification times versus powers.

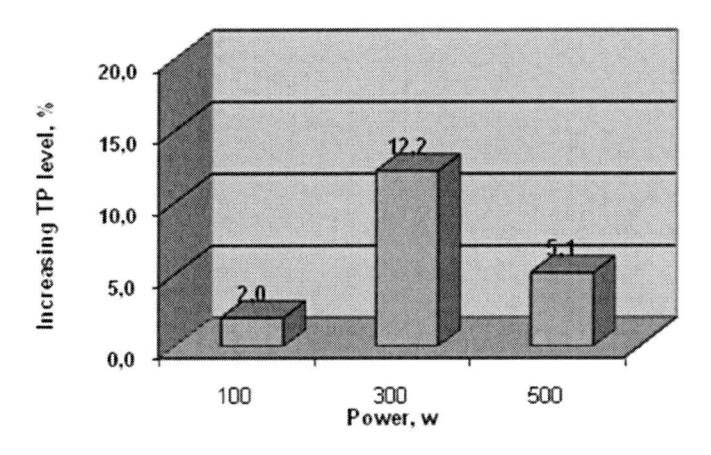

Figure 9. released TP from Waste activated sludge at 60 seconds sonification time versus different powers.

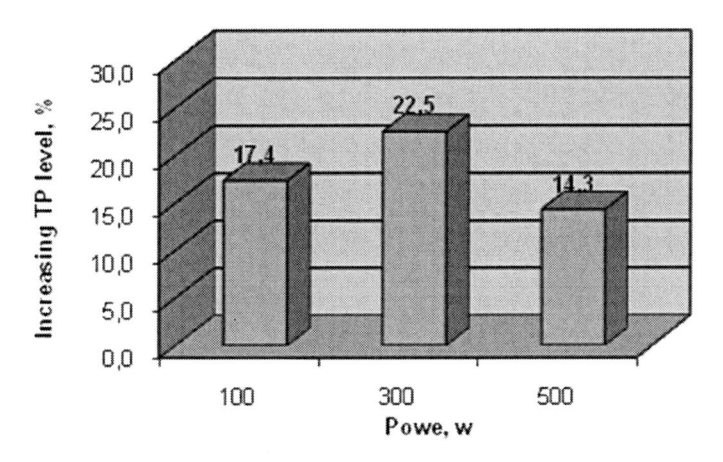

Figure 10. released TP from Waste activated sludge at 120 seconds sonification time versus different powers.

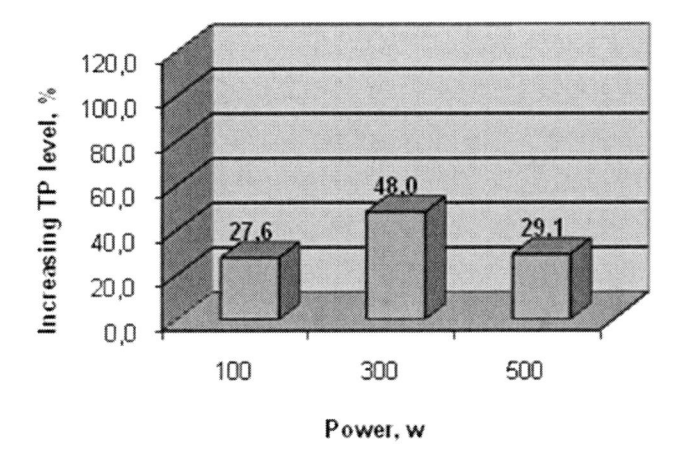

Figure 11. released TP from Waste activated sludge at 240 seconds sonification time versus different powers.

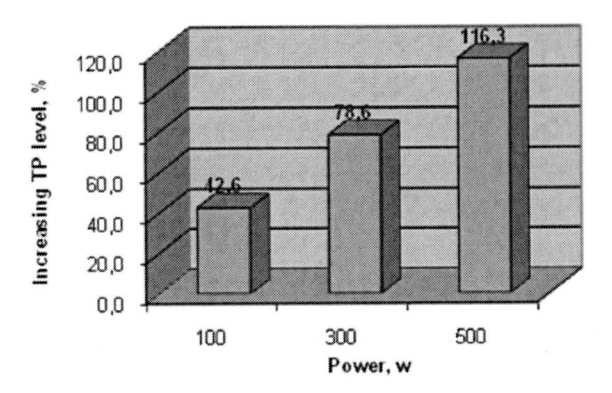

Figure 12. released TP from Waste activated sludge at 480 seconds sonification time versus different powers.

Figures 9-12 illustrate the percentage of increase of TP level in different sonification contacts times of 60,120,240 and 480 seconds respectively. The maximum of release TP observed at power 500 w and in contact time of 480 seconds. In all of the different powers, releasing TP shows increasing trend. All the obtained value in power 500 w as comparison to power 300 shows less increasing ,expect the contact time of 480 seconds. This could be related to domination of free active radicals activities and consummating some part of release TP in power 500w during sonification. Although at the contact time of 480 seconds with rising power obtained value for TP show increasing trend. This could be related to nature of phosphoric compounds in dairy waste activated sludge which breakdown in this power (500w) and going in soluble form. Therefore if solubilisation of phosphoric compounds is considered, it is better to used low power (e.g 100 and 300w) and long sonification time (e.g. 120 and 480s). And whenever removed of some part of soluble TP is considered, high power levels (e.g. 500w) cloud be applying but contact times in high power levels is very important because some phosphoric compounds maybe breakdown at longer sonification times. Figure 13 comparisons different contacts times versus different applied powers.

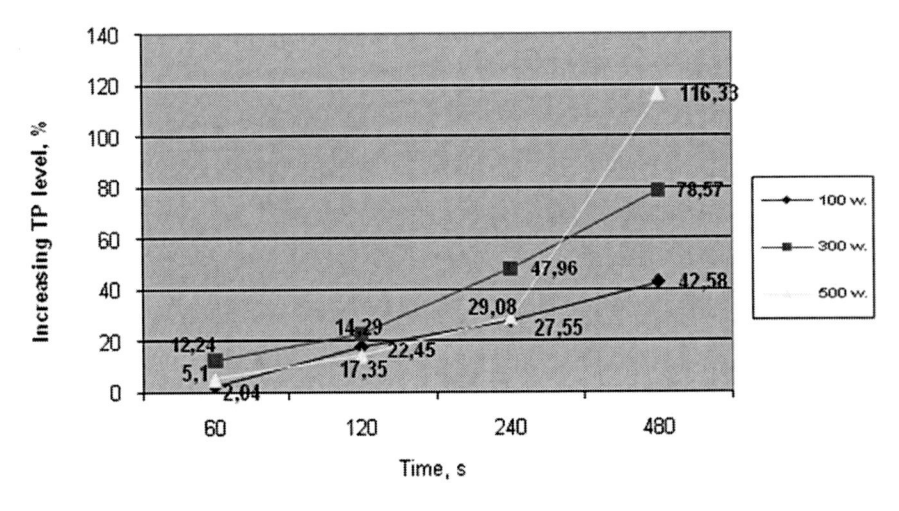

Figure 13. Comparison of release TKN level from Waste activated sludge at different sonification times versus powers.

CONCLUSION

The capital costs for building sludge processing in Tehran Pegah Dairy Complex was about $45,000 (in the year of 2000) and operational cost for waste activated sludge processing is about $22,000 per month which it is expensive for such as industries. Waste activated sludge (or biosolids) contain high amount of nutrients such as nitrogen and phosphors that could use for nutrients balancing in biological wastewater treatment. Micro organisms in biological treatment are able to degrading only soluble form of substrates and if substrates is in insoluble form micro organisms forced to hydrolysis the substrates, which is limiting step in biological treatment [17]. The application of ultrasounds to sludge processing is attractive because of the possibility for obtaining significant effects for release nutrient to soluble form without the necessity to apply additional components, e.g. chemical agents, simplicity of the construction and operation of reactors and air tightening of the installation [6]. According to characteristic of the waste activated sludge (see table2) more than 68% of TKN and 92% of TP are in insoluble form. The maximum release of TKN was found around 170.8 percent which observed in power 100 w and contact time of 480 seconds. Nitrogen compounds in dairy waste activated sludge shows different behavior break down and going to soluble form. In all different powers of 100,300 and 500w with increasing sonification times, hydrolyses of TKN shows increasing trends. The maximum of release TP observed at power 500 w and in contact time of 480 seconds. In all of the different powers, releasing TP shows increasing trend. Phosphoric compounds in dairy waste activated sludge shows different breakdown behavior. If solubilisation of phosphoric compounds is considered, it is better to used low power (e.g100 and 300 w) and long sonification time (e.g. 120 and 480 s). And whenever removed of some part of soluble TP is considered, high power levels (e.g. 500w) cloud be applying but contact times in high power levels is very important because some phosphoric compounds maybe breakdown at longer sonification times.

In general the capital cost for an ultrasonic sludge treatment process is roughly $30,000 per kW. One kW of the ultrasound process treats approximately ten thousand population equivalents. Operation and maintenance costs are minimal, which include the replacement of probe once in every 1.5-2 years. The pay back periods for ultrasound treatment plants are reported to be in the range of 8 months to 3 years [18]. In terms of environmental concerns and safety, ultrasound treatment is superior to ultraviolet and ozone treatment [19].

The relative cost for ultrasound treatment can be reduced by developing better technology and increasing the efficiency of electrical to acoustic energy conversion. Future studies can be directed towards the development of efficient transducers, reduction of energy loss in the ultrasonic system and design of economically feasible full scale ultrasonic systems.

REFERENCES

[1] Bartholomew R. Ultrasound disintegration of sewage sludge: an innovative wastewater treatment technology. PA Department of environmental protection, Bureau of Water Supply and Wastewater Management, Division of Municipal Financial Assistance, Innovative Technology Section;2002.Available at http://www.dep.state.pa.us/DEP/

DEPutate/Watermgt/WSM/WSM_TAO/InnovTech/ProjReviewes/ultrasound-Disintegr. htm. Accessed on October 22, 2007.

[2] Chiu, Y.C., Chang, C.N., Lin, J.G., & Huang, S.J. (1997). Alkaline and ultrasonic pre-treatment of sludge before anaerobic digestion, *Water Science and Technology,* 36, 155–162.

[3] Chowdhury P., Viraraghavan T., 2008, Sonochemichal degradation of chlorinated organic compounds, phenolic compounds and organic dyes- A review. *J.scitotenv.*2008.12.031.

[4] Clark, P.P., & Nujjoo, I. (1998). Ultrasonic sludge pre-treatment for enhanced sludge digestion, Cambridge,MA: *Proceedings of Conference on Innovation* 2000.

[5] Davis RD, Hall JE. Production, treatment and disposal of wastewater sludge in Europe from a UK perspective. *Eur Water Pollut Control* 1997; 7(2):9–17.

[6] Graff, K. Lecture Notes. WE 795, *Independent Study on High Power Ultrasonic*, The Ohio State University, Columbus, OH, 1988.

[7] Laborde, J.L., Bouyer, C., Caltagirone, J.P., & Ge´rard, A. (1998a). Acoustic bubble cavitation at low frequencies, *Ultrasonics*, 36, 589–594.

[8] Laborde, J.L., Bouyer, C., Caltagirone, J.P., & Ge´rard, A. (1998b). Acoustic cavitation field prediction at low and high frequency ultrasounds, *Ultrasonics*, 36, 581–587.

[9] Lauterborn, W., & Ohl, C.D. (1997). Cavitation bubble dynamics, *Ultrasonics Sonochemistry,* 4, 65-75.

[10] Lin,J.g., C.N., Chang, S.C., 1997. Enhancement of anaerobic digestion of waste activated sludge by alkaline solubilization. *Bioresour. Technol.* 62, 85-90.

[11] Low EW, Chase HA. Reducing production of excess biomass during wastewater treatment *Water Res* 1999; 33(5):1119–32.

[12] Mehrdadi N., (2006). Pakistan journal of 6 sciences, *Petrochemical wastewater treatment using an anaerobic hybrid reactor.*

[13] Neis U. (2002). *Intensification of biological and chemical processes by ultrasound* (Volume 35), Hamburg, Germany: Technical University Hamburg–Harburg Reports on Sanitary Engineering.

[14] Neis, U. (2000). Ultrasound in water, wastewater and sludge treatment, W*ater,* 21, 36–39.

[15] Portenlanger, G. (1999). Mechanical and radical effects of ultrasound, *Environmental Engineering,* 25, 11–22.

[16] Spellman FR. *Wastewater biosolids to compost.* Lancaster, PA, USA: Technomic Publishing Company: 1997. p. 223–35.

[17] Tiehm, A., Nickel, K., Zellhorn, M., & Neis, U. (2001). Ultrasonic waste activated sludge disintegration for improving anaerobic stabilization, *Water Research,* 35, 2003–2009.

[18] Zahedi A. (2009), Application of ultrasonic wave irradiation in sludge processing, *Master thesis at university of Tehran.*

[19] Zielewicz E. and Sorys P. The comparison of ultrasonic disintegration in laboratory and technical scale disintegrators. *Eur. Phys. J. special Topics* 154, P. 289-294 (2008).

In: Research Studies on Tourism and Environment
Editors: J. Mondejar-Jimenez et al.
ISBN: 978-1-61209-946-0
© 2012 Nova Science Publishers, Inc.

Chapter 2

ENVIRONMENTAL CERTIFICATION AS A TOOL TO MEASURE AND PROMOTE ECO-INNOVATION IN THE TOURIST SECTOR

Lluís Miret-Pastor[*], *María-del-Val Segarra-Oña* and *Ángel Peiró-Signes*
Universidad Politécnica de Valencia, Spain

ABSTRACT

The tourist industry is evolving towards a new model where it will be necessary to combine profitability with sustainability. Adapting to this new tourism model requires companies and tourist destinations innovate in the field of environmental awareness. In this new framework, eco-innovation appears as a novel and important concept, but in any case, eco-innovation still presents difficulties in its conceptualization and measurement. This chapter reviews the literature on eco-innovation and proposes the use of three environmental certifications (ISO 14001, EMAS and Eco-label) as tools to measure and promote eco-innovation in the tourist sector.

1. INTRODUCTION

The concept of eco-innovation is very recent and suggestive. Eco-innovation should play a key role in the way towards a sustainable economy and society, since it connects both pillars of sustainability: environmental quality and economic well-being. Eco-innovation allows to increase the value of producers and consumers, and at the same time reduces environmental impacts.

Both the limited work on eco-innovation and the many studies on innovation, often focus on manufacture industries. However, the statistics warn about the outsourcing of the economy of developed countries, therefore, special attention should be given to eco-innovation in the

---
[*] Contact author: luimipas@esp.upv.es

service sector. Tourism appears as the main industry of the service sector, for the generation of employment as well as for the effects on the development of regions and countries (Holjevac, 2003).

The study of the eco-innovation in the tourism industry seems particularly important. The irruption of new technologies and the changes in the consumption patterns are producing important changes in the tourist industry. Some authors like Ioannides and Debbage (1997) refer to a "post-Ford tourist industry". Following the classification by Knowles and Curtis (1999), the sun and beach tourism would correspond to the so-called second generation tourism, characterized by the homogenization of tourist packages and the search for scale economies. This model is considered by many authors as unsustainable and outdated (Agarwal, 2002; Aguilo, Alegre and Sard, 2005; Miret, Segarra and Hervás, 2010). These same authors argue the appearance of a more demanding third generation of tourism, an intensive use of new technologies and a product based on quality criteria and respect for the environment. The adaptation to this new tourism, in most cases, will not constitute a radical breaking-off, but an evolution of the existing model. In this way towards a new tourist model, eco-innovation shall play a determining role.

Tourism is an intensive sector in resources that leaves an important mark on the environment (Hunter and Shaw, 2007). As pointed out in a recent study of the Spanish Ministry of Environment: "the maintenance of the dynamism of the tourist sector can only be assured if instruments are designed and applied to incorporate processes of technological eco-innovation, which include areas ranging from energy saving, to environmental protection, as well as the creation of new products. (Network of environmental studies, 2007).

Eco-innovation appears as one of the great challenges of the European tourist industry, since environmental quality is essential for the survival of tourism (Tzschentke et al, 2008). However, eco-innovation is a complex and difficult concept to measure. After this first introductory section, the second and third section describes a theoretical framework where the concept of eco-innovation is delimited, and the difficulties for its measurement and its relation with the tourism are analyzed. The fourth section proposes the use of the standards of environmental quality as a method not only for the promotion and certification of tourist eco-innovation, but also as an indicator for its measurement and study. The fifth section analyses environmental quality standards (EMAS, Eco-label and ISO 14001); finally, the last section presents the conclusions of the research work.

2. ECO-INNOVATION: CONCEPT AND MEASUREMENT

There are many definitions for the concept of eco-innovation (Fussler and James, 1996, OECD, 2008, Hupes et al, 2008). Anyway, eco-innovation can be defined "as the production, assimilation or exploitation of a product, production process, service or management or business method that is novel to the organisation (developing or adopting it) and which results, throughout its life cycle, in a reduction of environmental risk, pollution and other negative impacts of resources use (including energy use) compared to relevant alternatives" (Kemp and Pearson, 2008).

The concept implies that eco-innovation can be given at any stage of the lifecycle of the service or product considered it is not a simple "curative technology" (Wuppertal institute,

2009). It is a wide concept that includes different types of innovations (process, product, or system). According to Kemp and Pearson (2007), eco-innovation should not be limited to environmental products or services or those innovations that have had a merely environmental motivation. The term eco-innovation should be expanded to any innovation that involves an environmental benefit, whether searched for or not. The term eco-innovation is closely related to concepts such as eco-efficiency, or eco-industry, but the differences should be specified.

Eco-efficiency is a philosophy of management to orient and measure companies and other development agents in the environmental performance. Eco-efficiency measures the environmental impact of a product or service. It is a concept aimed at obtaining the greatest value with the smallest environmental impact, combining thus environmental and economic benefits. The underlying idea in eco-efficiency is to dissociate growth and environmental pressure (WBCSD, 2000).

Eco-industry is defined as "The environmental goods and services industry which produces goods and services to measure, prevent, limit, minimise or correct environmental damage to water, air and soil, as well as problems related to waste, noise and eco-systems. This includes cleaner technologies and products and services that reduce environmental risks and diminish pollution and the use of resources" (EUROSTAT/OECD, 1999).

Eco-innovation and eco-efficiency are related but are not identical concepts. Although both try to measure sustainability, eco-efficiency offers a static measurement, while eco-innovation is based on a dynamic measurement. On the other hand, eco-innovation goes beyond eco-industry. There is eco-innovation in the eco-industry, but there is also eco-innovation in a wide range of sectors that cannot be called eco-industrial. One of the major problems in the design and implementation of policies that promote eco-innovation is the lack of data and indicators.

Therefore, a great deal of studies on eco-innovation focus on searching for indicators that are effective and feasible to measure it. Kemp and Pearson (2008) or Arundel and Kemp (2009) have centred their methods on the analysis of surveys, patents and variable documentation; on the other hand, Foxton, Pearson and Speirs (2008) have tried to adapt the theory of innovation systems and their indicators in accordance with eco-innovation, while Huppes et al (2008) try to establish a model framework for the analysis and indicators of eco-innovation.

The measurement of eco-innovation is a complex problem that entails the same problems that make difficult the measurement of innovation in general. In practice, there are all sorts of indicators, which Kemp and Pearson (2008) classify in four groups:

Table 1. Eco-innovation measurement indicators

• The inputs measures (R&D expenditure, research personnel, etc)
• The output measures (for example, the number of patents)
• Direct measures (number of innovations, sales of new products, etc.)
• Indirect measures (changes in the efficiency and productivity through the analysis of added data).

Source: Own elaboration from Kemp and Pearson (2008).

Table 2. Eco-innovation analysis methods

• The analysis of patents: Used in the works by Lanjouw and Mody (1996), Popp (2006) or Nameroff et al (2004).
• The analysis of surveys. Used in the works by Horbach (2006) or Rehfeld, Rennings and Ziegler (2007).
• The analysis of digital document and resources. Used in the works by Huber (2004) or Newell et al (1999).

Source: Own elaboration from Kemp and Pearson (2008).

As far as analysis methods, the same work by Kemp and Pearson (2008) emphasize three main methods shown in Table 2.

In any case, in addition to the difficulties to measure innovation it should be added that the measurement of eco-innovation involves finding indicators that include economic as well environmental aspects. The works on the subject (Kemp and Pearson, 2008; Arundel and Kemp, 2009, Huppes et al, 2008) provide different methods and indicators, but they admit that all of them are insufficient. In addition to the problems mentioned are the difficulties of the fact that eco-innovation is not an official sector. EUROSTAT has been working in the elaboration of an Environmental Goods and Services Sector (EGSS), but this sector would identify more with eco-industries than with eco-innovation. In this same sense, the next Survey of Community Innovation (CIS, 2008) will have an eco-innovation track, that will be available throughout 2010.

3. ECO-INNOVATION AND TOURISM

Tourist eco-innovation is the application of the concept of eco-innovation to the tourist sector. It is necessary to distinguish tourist eco-innovation from eco-tourism.

Eco-tourism is "an activity where the authorities, the tourist industry and the local population cooperates to allow tourists to travel to genuine zones to admire, study and enjoy nature and culture without exploiting the resources and contributing to a sustainable development (Bjork, 2000). It is a concept that appeals to conservation and development through ecological maintenance, with a low environmental impact and a little use of the local resources (Stem, 2003).

We can understand eco-tourism as eco-innovation; however tourist eco-innovation goes beyond eco-tourism, since it is not limited to this sub-sector. The definitions of the term eco-innovation entail the possible existence of eco-innovations (searched for or not) in other tourist subsectors such as cultural tourism, adventure, nature and, even, in large scale tourism.

Eco-innovation appears as a key competitive source for sustainable development. People in charge of the economy in general and tourism in particular are starting to show a great interest in this concept which translates in lines of financing, legislative initiatives and academic works. Eco-innovation is due to mark not only innovating policies, but also future tourist policies. The interest of the European Union takes shape through the Competitiveness and Innovation Framework Programme (CIP) and the European Commission's Environmental Technologies Action Plan (ETAP), that allows financing eco-innovation projects, among which can be sustainable tourism innovating projects which stimulate the economic growth and competitiveness. On the other hand, the Spanish Ministry of Environment, through the Network of Environmental Studies (2007) proposes eco-innovating actions and initiatives in the tourist sector from two approaches: tourist products and destinations.

Regarding tourist products, it proposes:

1. The reform and promotion of already existing product.
2. The application of new technologies in tourist products.
3. The creation and promotion of marks for eco-innovating tourist products.
4. A greater regional coordination for the development of innovating tourist activities.
5. Environmental communication.

As far as tourist destinations, the initiatives would materialize in:

1. Actions for the contribution of tourism to the improvement of cultural and natural heritage.
2. Actions to make tourist destinations more sustainable.
3. Develop and promote environmental certifications of tourist destinations.

As seen above, the Spanish Ministry proposes the use of marks and environmental certifications as an instrument for eco-innovation of tourist destinations and products; on the other hand, it was also mentioned that a great deal of literature (Foxton, Pearson and Speirs, 2008; Kemp and Pearson, 2008) proposed the use of these same certifications as indicators of eco-innovation. The review of this theoretical framework of eco-innovation leads us to propose, in the following section, a methodology based on the systems of environmental certification as indicators of tourist eco-innovation.

4. METHODOLOGY: ENVIRONMENTAL CERTIFICATION SYSTEMS AS INDICATORS OF TOURIST ECO-INNOVATION

Certification can be defined as a voluntary process that verifies audits and ensures in writing that a process, product or service meets specific standards (Bien, 2003). These standards, concerning environmental management, serve as a guide to establish a System of Environmental Management

A system of environmental management (SEM) is a continued cycle of planning, action, revision and improvement of the environmental actions of the company in relation to the nature, size, and environmental impacts of its activities, products and services. A SEM is based on the accomplishment of an environmental diagnosis to identify, value, reduce and prevent environmental impacts in the environment, assuming a commitment of continuous improvement of the organization's behaviour with the environment.

Since the 1980's, all types of certifications and eco-labels have appeared in the tourist sector, currently there are more than 100 different certifications (Font, 2002) from which at least 50 are located in Europe (the majority in Germany, Spain and Italy), including a great variety of tourist offer, from lodgings, beaches, restaurants, golf courses, ports, tourist municipalities, holiday packages, etc. The criteria, contents, objectives and quality of these certifications are very heterogeneous, which cause confusion among the potential clients (Lubbert, 2001), this was one of the reasons that led the EU to present its own "European Ecological Label". In spite of the large number of certifications and more than 4000 tourist lodgings with some type of environmental certification, this amount is hardly 1% of the total of the existing lodgings in the European Union.

The environmental certification systems most used are 14001 ISO and EcoManagement and Audit Scheme (EMAS). These two systems are used in the hotel industry and are mentioned in the works by Pearson et al (2008) and Kemp and Pearson (2008) as indicators of eco-innovation. The problem is that these two systems present levels of exigency and costs that, sometimes, make them nonviable for small companies, which causes that a large part of the tourist industry has developed less demanding systems of environmental certification (Synergy, 2000). In representation of these eco-labels, this work will introduce in its analysis the study of the "European Ecological Label".

These standards been have used for the study of different aspects of the environmental or business management of the tourist industry (Bracke, and Albrecht, 2007; Bonilla and Avilés, 2008)

Kemp and Pearson (2008) as well as Pearson et al (2008), propose the environmental certification as one of the indicators to use for the measurement of eco-innovation in companies, adapting what is already indicated in the OECD report OECD (Governance of Innovation Systems) by Remøe et al (2005) for the use of certification in the studies on innovation in general.

The advantage of this indicator is twofold, since the certification is not only considered relevant for the study of eco-innovation, but also that is reasonably practical to measure. In any case, certification would not be indicative of radical eco-innovations, but of the so-called incremental eco-innovations. Next, we validate the suitability of each one of these three environmental standards for their use as indicator of eco-innovation in the tourist sector; we will use different official reports, academic works, statistics and surveys.

5. RESULTS. ANALYSIS OF THREE ENVIRONMENTAL CERTIFICATES AS INDICATORS OF ECO-INNOVATION

5.1. Eco-labels

Eco-labels are the European official seal for the products that ensures an adequate environmental behaviour. It is based on rigorous and trustworthy criteria considering the complete lifecycle of products. With the support of the European Commission, a Board formed by national associations of labelling, representatives of consumers and environmental NGOs, unions and companies in charge of their management.

Eco-labels are closely related to Environmental Management Systems and they can be to be considered as point of reference in relation to the ecological excellence at the European level.

Table 3. Restricting, Reducing and Promoting

Restricting energy consumption.
Restricting water consumption.
Reducing waste.
Promoting the use of natural resources and less harmful substances with the environment.
Promoting education and environmental communication.

Source: Own elaboration from European Commission, 2009.

There are 22 types of products which can be awarded eco-labels. The "tourist lodging services" sector is the largest, since it represents 34% of the total of eco-labels granted by the European Union. The popularity of eco-labels has grown exponentially during the last years, as it has increased from 95 eco-labels in 2001 to 839 in 2009. Per countries, Italy with 265 and France with 153, are the countries with the largest number of eco-labels. A large part of this Italian leadership is explained by the strong implantation of eco-labels in hotel establishments. If the growth of the number of eco-labels has increased (it has doubled from the end of 2006 to the end of 2009), the growth of eco-labels in the tourist sector has been still greater (in this same period it has multiplied by three).

The award of eco-labels to a tourist lodging involves:

A great deal of these requirements is related to eco-innovation. The EU relates eco-labels with eco-efficiency and eco-innovation when stating that "the majority of the environmental measures requires a certain amount of time and effort, but on the other hand, offer an added value. They help to discover the ecological and economic weak points in the company and promote innovations. An example is the reduction in the consumption of natural resources, such as energy or water that involve a reduction of costs" (European Commission, 2009). The World Tourism Organization estimates in 20%, with respect to the average of the sector, the savings in water and energy consumption per stay in European hotels with eco-label (WTO, 2002)

5.2. ISO 14001

ISO 14001 is a Standard elaborated by the International Organization for Standardization whose aim is to provide companies with an Environmental Management System (EMS) effective, contrasted and integrated in the rest of the productive activities. ISO (1996)

It is the most used environmental management standard. From 1996 to 2006 it has increased from 1491 to 129,199 certificates at a global level. 40% of these certificates take place in the European Union, being Spain and Italy the European countries with the highest number of ISO 14001 certifications. The majority of ISO 14001 certifications belong to the industry, although the certifications in the services sector are experiencing the highest growth globally (especially in the European Union).

There is a wide range of literature that analyzes the economic impact of the companies that establish an ISO 14001 system. Melnyk et al (2003), Kelly et al (2007) and Montabon et al (2002) find empirical evidences to state that ISO 14001 improves the environmental as well as the economic results. These latter are perceived by the managers through the reduction of costs, the improvement of quality or the improvement of the reputation. These improvements are justified since ISO 14001 contributes to generate valuable resources to maintain or create competitive advantages (Cañón and Garcés, 2006). Other authors such as Russo and Fouts (1997) or Melnyk et al (2002) maintain that environmental management systems provide the company with an information system that not only also reduces pollution but also contribute to the improvement of the corporate results. These evidences serve us to validate the existing relation between ISO 14001 and different aspects of eco-innovation.

In the last years, the literature has confirmed this positive relation between the enterprise establishment of an ISO 14001 system and business results within the hotel sector (Claver-Cortés et al, 2007, Tarí et al, 2009, Molina-Azorín et al, 2009).

5.3. EMAS

EMAS is a voluntary EU standard that acknowledges those organizations that have established a EMS (Environmental Management System) and have committed to a continuous improvement, verified by means of independent audits

The EMAS complies with an EU regulation and, therefore, it has a legal status, regulation 761/2001. Its implantation is lower than ISO 14001. Heras et al (2008) indicates three main reasons: it is more expensive, less recognized world-wide and the regulating pressure is greater.

Sectorially, the industrial sectors should be noted, although the hotel industry already involves more than 5% of the certifications. In fact, between 2001 and 2007, the greatest growth has taken place in the activities of "public cleaning" and "hotel trade". In any case, it should be emphasized the similarity between the sectorial distribution of ISO 14001 and EMAS.

Table 4 shows the evolution during the last few years:

It should be noted that although most tourist companies with EMAS are hotel establishments, this certification is reaching to other services of the tourist industry such as campings, travel agencies, natural parks and even theatres.

The establishment of the EMAS system per countries is very irregular, thus Germany, Italy and Spain groups almost 80% of the total of certifications (European Commission, 2009b).

Table 4. Evolution During Past Years

	Registered Organizations	Hotel Organizations
2005	3126	200
2006	3225	217
2007	3935	233
2008	3967	236
2009	4347	241

Source: Own elaboration from EU reports.

Organizations with EMAS certificate

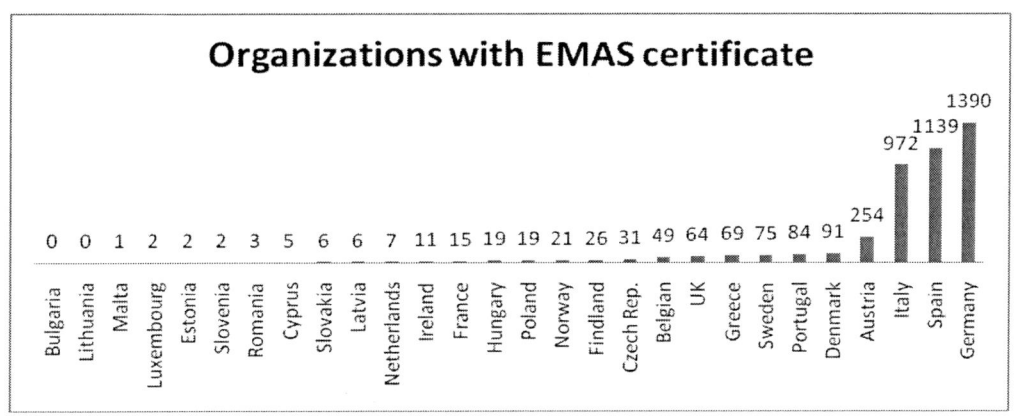

Country	Count
Bulgaria	0
Lithuania	0
Malta	1
Luxembourg	2
Estonia	2
Slovenia	2
Romania	3
Cyprus	5
Slovakia	6
Latvia	6
Netherlands	7
Ireland	11
France	15
Hungary	19
Poland	19
Norway	21
Findland	26
Czech Rep.	31
Belgian	49
UK	64
Greece	69
Sweden	75
Portugal	84
Denmark	91
Austria	254
Italy	972
Spain	1139
Germany	1390

Source: Own elaboration from European Commission reports (2009b).

Figure 1. Organizaitons with EMAS Certificate.

In the case of the certifications in the hotel sector, the concentration is still greater since in this case Spain groups 80% of the certifications.

The relation between the EMAS certification system and eco-innovation is patent throughout different works. The objective of EMAS is to increase the organizations' awareness and commitment in adopting measures that improve continuously their behaviour with the environment (Carmona, 1996), but keeping in mind that the main reason used to establish an Environmental Management System is the attainment of competitive advantages (Spanish Ministry of Environment, 2002; Ayuso and Fullana, 2000)

In this line, the results of the survey carried out by the Spanish Ministry of Environment (2006) are significant; Its "Questionnaire of Participation and Consultation of Regulation 761/2001" on 583 organizations where it verifies that companies perceive the following main advantages of the implantation of a EMAS:

1. The reduction of the risk of failure to comply with the environmental legislation (11%)
2. The enhancement of image (10%)
3. The improvement in the process control (7%)
4. The improvement in the management of maintenance activities of facilities and equipment (7%)
5. The reduction of costs related to energy consumption (7%)
6. The improvement in the relation with the administrations and the control institutions (7%)

Many of these advantages also appear in the study by the European Commission (2005) and confirm that entrepreneurs relate the establishment of an environmental certification system like the EMAS to environmental as well as economic advantages. In sum, with different aspects related to eco-innovation.

EMAS certificates in hotels

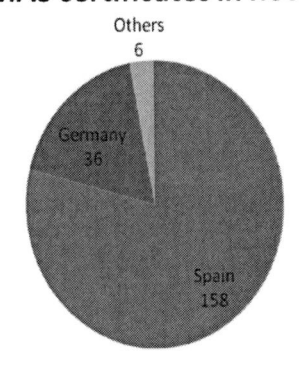

- Others: 6
- Germany: 36
- Spain: 158

Source: Own elaboration from European Commission data (2005).

Figure 2. EMAS Certificate in Hotels.

CONCLUSION

The tourist sector is undergoing a period of deep changes that will transform the current touristic model. Different works insist on one of the core characteristics of the new tourism, the growing demand for greater environmental quality.

Eco-innovation appears as a new and important concept. Hotels seeking to adapt to new competitive framework must innovate in environmental matters. Anyway, this eco-innovation does not usually respond to radical changes in management or truly innovative products, but it usually refers to changes in management models, in the procedures or the technology used. Businesses must manage better their resources, achieve energy savings, add environmental value to the brand, etc. All of these environmental improvements are accredited through the environmental certification. This chapter has discussed the three most popular environmental certificates in the European market (the ISO 14001, EMAS and the Eco-Label). The conclusion matching different essays, opinion polls and official reports is that there is a direct relationship between environmental certification and environmental and economic improvements.

Environmental certification appears as a simple and useful tool not only to promote eco-innovation in the tourist sector, but also to measure it. The analysis of the three certificates agrees that the hospitality industry is one of the most established areas where the environmental certification exists and where it is growing faster in recent years.

In any case, it should be noted that the difference in geographical location of the certificates and the wide variety of models, difficult to use as an indicator of eco tourism. In the next years it will be necessary to work on further standardization of environmental certification, as well as a detailed analysis of how environmentally proactive companies achieve to increase their income or reduce costs.

Acknowledgment

The authors acknowledge the financial aid received from the Spanish Ministry of Science and Innovation through research project reference no. EC02008-05895-C02-01/ECON, submitted to the 2008 Call for Grants for R&D projects.

The translation of this study was funded by the Universidad Politécnica de Valencia.

References

Aguiló, E., J. Alegre, and M. Sard (2005): "The Persistence of the Sun and Sand Tourism Model". *Tourism Management*. 26. pp 219–231.

Argawal, S. (2002): Restructuring Seaside Tourism: The Resort Lifecycle. *Annals of Tourism Research* 29:pp 25–55.

Arundel A. y Kemp R. (2009): "Measuring Eco-innovation". *Working paper series*. United Nations University.

Ayuso, S y Fullana P (2000): *Report for the EMAS Comission Eco-Management and Audit-Scheme (EMAS)- Applicability for Sustainable Tourism Network?*

Björk, P. (2000). Ecotourism from a conceptual perspective, an extended definition of a unique tourism form. *International Journal of Tourism Research, 2*(3), pp 189-202.

Bonilla M.J; Avilés, C (2008): "Analysis of Environmental Statements Issued by EMAS-Certified Spanish Hotels" *Cornell Hospitality Quartely* 49(4), pp 381-394.

Bracke, R y Albrecht, D (2007): "Competing Environmental Management Standards: How *ISO* 14001 Outnumbered *EMAS* in Germany, the UK, France, and Sweden" *Environment and Planning C: Government and Policy* v. 25 (4), pp. 611-27.

Cañón J y Garcés C (2006): "Repercusión económica de la certificación medioambiental ISO 140011". *Cuadernos de Gestión* Vol. 6. (1) (Año 2006), pp. 45-62.

Carmona Ibáñez (1996): "La Unión Europea ante el reto ambiental: El Reglamento de Ecogestión y Ecoauditoría". *Actualidad Financiera*. Nº 23. Pp 231-245.

Ernst & Young (2006): *European Commission DG Environment Eco-industry, its size, employment, perspectives and barriers to growth inan enlarged EU.*

European Commission (2004,a): *Stimulating Technologies for Sustainable Development: An Environmental Technologies Action Plan for the European Union.*

European Commission (2004b): *Facing the challenge. The Lisbon strategy for growth and employment* (Kok Report), Available in: http://ec.europa.eugrowthandjobs/pdf/kok_report_en.pdf)

European Commission (2005): *Evaluation of EMAS and Eco-Label for their Revision.* Available in: http://ec.europa.eu/environment/emas/pdf/eversummary.pdf

European Commission (2006). *The Competitiveness and Innovation Framework Program (2007-2013).* Available in: http://europa.eu/legislation_summaries/energy/european_energy_policy/n26104_en.htm

European Commission (2009). *Eco-labels.* Available in: http://ec.europa.eu/environment/ecolabel/about_ecolabel/facts_and_figures_en.htm

European Commission (2009b). *EMAS Newsletters.* Available in: http://ec.europa.eu/environment/emas/documents/brochure_en.htm

Font, X (2002). "Environmental Certification in Tourism and Hospitality: progress, process and prospects". *Tourism Management.* Vol 23, issue 3. Pp 197-205.

Fussler, C. y James P, (1996); *Driving Eco-Innovation: A Breakthrough Discipline for Innovation and Sustainability*, Pitman Publishing: London.

Heras, I; Arana, G y Molina J.F. (2008): "EMAS versus ISO 14001. Un análisis de su incidencia en la UE y España". *Boletín Económico del ICE.* N° 2936. Pp 49-63.

Holjevac, I. A. (2003). A vision of tourism and the hotel industry in the 21st century. *International Journal of Hospitality Management*, 22. Pp129-134.

Horbach, J. (2008): Determinants of Environmental Innovation – New Evidence from German Panel Data Sources. *Research Policy* 37. Pp 163-173.

Huber, J. (2004) *New Technologies and Environmental Innovation.* MPG Books Ltd, Bodmin, Cornwall.

Hunter, Colin, Shaw, Jon (2007): "The ecological footprint as a key indicator of sustainable tourism". *Tourism Management*, Vol 28, No 1. Pp 46-57.

Huppes G.; Kleijn R.; Huele R.; Ekins P.; Shaw B.; Schaltegger S. & Esders M. (2008): *Measuring eco-Innovation: Framework and typology of indicators based on Casual Chains.* Ecodrive Project.

International Organization for Standardization (1996): "Norma ISO 14001. Environmental Management System. Specification with Guidance for Use". Geneve.

Ioannides, D. y Debbage, K. (1997): "Post-Fordism and Flexibility: The travel industry polyglot" *Tourism Management*, Vol 18, n° 4.

Kelly, J; Haider, W; Williams, P. y Englund, K. (2007) Stated Preferences for EcoEefficient Destination Planning Options. *Tourism Management* Vol 28, No 2, pp 377-390.

Kemp R. y Pearson P. (eds.), 2008; Final report of the project 'Measuring Eco-Innovation' (MEI) . Available in: (http://www.merit.unu.edu/MEI/index.php).

Knowles, T., and S. Curtis (1999) "The Market Viability of European Mass Tourist Destinations: A Post-Stagnation Life-Cycle Analysis". *International Journal of Tourism Research.* 1: pp 87–96.

Lanjouw, J.O., and A. Mody (1996) 'Innovation and the International Diffusion of Environmentally Responsive Technology', *Research Policy* 25: pp 549-571.

Lubbert, C. (2001). "Tourism ecolabels market research in Germany". En X. Font, & R. Buckley (Eds.), *Tourism ecolabelling: certification and promotion of sustainable management.* Wallingford: CAB International.

Melynk, S.; Sproufe, R. y Calantone, R. (2003): "Model of Site-Specific Antecedents of ISO 14001". *Certification Production and Operations Management.* Vol 12, No 3, pp 369-387.

Melnyk, S.; Sroufe, R. y Calantone, R. (2002): "Assessing the impact of environmental management systems on corporate and environmental performance", *Journal of Operations Management*, Vol. 336, pp. 1-23.

Ministerio de Medio Ambiente (2002): Guía EMAS para PYMES. Guía Interactiva de Aplicación del Reglamento Europeo EMAS en PYMES. Ministerio de Medio Ambiente. Madrid.

Ministerio de Medio Ambiente de España (2006): "Estudio de opinión relativo al proceso de revisión del Reglamento (CE) no 761/2001 (EMAS). Dirección General de Calidad y Evaluación Ambiental". Ministerio de Medio Ambiente.

Miret-Pastor, L; Segarra-Oña, M. y Hervás-Oliver, J.L. (2010): "Nuevas Medidas de Concentración Espacial Aplicadas al Sector Turístico: El Papel de las Externalidades en el Turismo de la Comunidad Valenciana". in: Ferrari, G; Montero, J.M., Mondéjar, J, y Vargas, M (eds). *Investigaciones, Métodos y Análisis del Turismo*. Ed. SEPTEM. Oviedo. Pp 269-280.

Motabon, F.; Melnyk, S.; Sroofe, R. y Calantone, R. (2000): «ISO 14000: Assessing its perceived impact on corporate purchasing performance», *The Journal of Supply Chain Management*, Vol. 36, No. 2, pp. 4-15.

Nameroff T.J., Garant R.J., Albert M.B. (2004) 'Adoption of green chemistry: an analysis based on US patents', *Research Policy*, Vol.33, pp. 959-974.

Newell, R. G., Jaffe, A. B. and Stavins, R. N. (1999) The Induced Innovation Hypothesis andEnergy-saving Technological Change. *Quarterly Journal of Economics*, 114: pp 941-975

OCDE (2008), Sustainable manufacturing and eco-innovation: First steps in building a common analytical framework. DSTI/IND(2008)16/REV1, OECD, Paris.

Popp, D. (2006), 'International Innovation and Diffusion of Air Pollution Control Technologies: The Effects of NOX and SO2 Regulation in the U.S., Japan and Germany', *Journal of Environmental Economics and Management*, 51, pp 46-71.

Rehfeld K.-M., Rennings K., Ziegler A. (2007) "Integrated product policy and environmental product innovations: An empirical analysis", *Ecological Economics* 61: pp 91-100

Remøe, S., (2005). *Governance of innovation systems*. Paris, OECD

Russo, M. y Fouts, P. (1997): «A Resource-Based perspective on corporate environmental performance and profitability», *Academy of Management Journal*, Vol. 40, No. 3, pp. 534-559

Stem, C. J., Lassoie, J. P., Lee, D. R., Deshler, D. D., and Schelhas, J. W. (2003). Community participation in ecotourism benefits: The link to conservation practices and perspectives. *Society and Natural Resources, 16*, pp 387-413

Synergy Ltd (2000). *Tourism certification: an analysis of Green Globe 21and other certification programs*. Godalming: WWF UK.

Technopolis Group (2008) Eco-innovation. Final report for sectoral innovation watch, (May), Europe-INNOVA (Alasdair Reid, Michal Miedzinski), (http://www.technopolis-group.com/resources/downloads/661_report_final.pdf); Brussels.

Tzschentke, Nadia, A.; Kirk, David and Lynch, Paul A. (2008): "Going Green: Decisional Factors in small hospitality operations". *International Journal of Hospitality Management*, Vol 27, No 1, pp 126-13

WBCSD (2000). *Measuring Eco-Efficiency: A guide to reporting company performance*.

WTO (2002). "Voluntary Iniciative for Sustainable Tourism". World Tourism Organisation. Madrid.

Wuppertal Institute (2009) *Eco-innovation : Putting the EU on the path to a resource and energy efficient economy*. Blischwitz, R; Giljun, S; Kuhndt, M; Scmidt- Bleek et al. Bruselas.

In: Research Studies on Tourism and Environment
Editors: J. Mondejar-Jimenez et al.
ISBN: 978-1-61209-946-0
© 2012 Nova Science Publishers, Inc.

Chapter 3

ANALYSIS OF THE SPENDING OF PEOPLE ATTENDING A CULTURAL FESTIVAL: THE CUENCA RELIGIOUS MUSIC WEEK

Juan-Antonio Mondéjar-Jiménez[*], María Cordente-Rodríguez, María-Leticia Meseguer-Santamaría, José Mondéjar-Jiménez and Manuel Vargas-Vargas

University of Castilla-La Mancha, Spain

ABSTRACT

Cultural events and festivals have become a booming attraction, complementing the investment made in alternative commercial channels, because of the fidelity these periodic events generate among their public.

In addition, holding them generates important economic effects in the economic activity of the region, both directly and indirectly in other economic sectors.

So, aware that the investments made are justified by the good results, public institutions favour, encourage and even support the holding of such festivals, considering them an advanced factor in tourism competitiveness.

This is the case of Cuenca's *Semana de Música Religiosa* (SMR), Religous Music Week, the 49th edition of which was held in 2010, consolidating it in the Spanish and international music world.

The purpose of this study is to analyse the spending of the people attending the *Semana de Música Religiosa de Cuenca*, and how it is distributed among the different expenditure items analysed. The purpose of this is to identify and differences and specific needs according to the characteristics of the people attending the festival, and so be able to modify the strategy followed by the event's organizers in terms of price, product and promotion.

---

[*] Contact author: JuanAntonio.Mondejar@uclm.es

Table 1. Cultural activities according to its nature

Painting	**Plastic arts**	Movie	**Visual arts**
Sculpture		Discs	
Photography		Videos	
Architecture		Television	
Theatre	**Performing arts**	Radio	
Dance		Books	**Graphic arts**
Opera		Design	**Applied arts**
Music (instrumental and vocal)		Handicraft	

Source: Cuadrado & Berenguer (2002).

1. INTRODUCTION

Culture nowadays has two dimensions: on one hand, it represents the support for society's collective memory and identification, and on the other, it is a source of wealth and generation of economic activities [Herrero, 2004] allowing urban regeneration and social cohesion.

According to their nature, cultural activities are classified in the way shown in Table 1.

Fine art and the performing arts are considered cultural services, because of their low level of reproduction, while audiovisual, graphic and applied arts are considered cultural goods, because of their high level of reproduction [Cuadrado & Berenguer, 2002].

Cultural activities generate employment and added value directly, but they also contribute to the development of other sectors, for example, tourism, transport, etc., because of the effect these cultural activities cause on the rest of the economic system in a given geographical area in terms of production, income and employment.

Among cultural activities, holding festivals has an important effect on the economic activity in a particular geographical area, generating an additional economic impact to that produced by other sectors of the economy [González & Ramírez, 2008].

A *festival* is an event or celebration, generally held by a local community or municipality, which centres on a particular subject or a particular unique aspect of the community, which serves as platform for cultural consumption.

In the social sciences, a scientific definition of festival is "a periodic celebration held with multiple ritual forms and events which directly or indirectly affect all the members of a community and which explicitly or implicitly show the basic values, ideology, world view shared by members of the community and which are the basis of its social identity" [Falassi, 1997].

Another, later definition is offered by Sassatelli [Sassatelli, 2008], who considers festivals "social happenings held periodically in which, through multiple forms, levels and a series of coordinated events, all the members of a community take part directly or indirectly."

There is a wide variety of types of festival: ethnic, arts, film, music, theatre, literary, food, hippy, Renaissance, alternative, experimental, summer, and sports festivals, etc.

Cultural festivals are the cultural object they provide to a geographical area [González & Ramírez, 2008]:

- Mass attraction of visitors, with their corresponding spending.
- Creation of an image.
- Creativity and social progress factor.

As a result, festivalization is a strategy for socioeconomic development, territorial development, urban development and maintenance of a place's position, so holding large events is today a strategic option for cities on which they can base their *city-marketing*.

The event analysed for this study is the *Semana de Música Religiosa de Cuenca* or Cuenca's Religious Music Week, the fourth-oldest music festival in Spain, the 49th edition of which was held in 2010.

This festival has been declared a Festival of International Tourist Interest, and belongs to the European Festival Association (EFA).

This music festival is held from the last Friday in Lent until Easter Sunday, in parallel with Cuenca's Holy Week celebrations, which have been declared of Cultural Interest.

The purpose of this study is to find out the flows of activity and revenues that holding the Cuenca Religious Music Week brings into the city, as well as the consumption of the people attending the festival, so that marketing and communication strategies suitable for the profile and needs of that public can be developed.

2. METHODOLOGY

To achieve the proposed objectives, we have information about the actual public attending the Semana de Música Religiosa de Cuenca, obtained by direct interview with those attending the concerts, randomly selected to try to cover the widest possible spectrum [Mondéjar, Cordente, Gázquez, Pérez & Milanés, 2010].

Table 2. Weighting of the allocation of expenditure by groups in Spain

Percentage (%)	2008	2007	2006	2005	2004	2003	2002
General	100,00	100,00	100,00	100,00	100,00	100,00	100,00
Food and beverages	20,28	22,06	22,28	22,60	22,60	21,93	21,86
Alcohol and tobacco	2,67	2,82	3,07	3,17	3,17	3,18	3,22
Clothing and footwear	8,81	9,03	9,25	9,73	9,73	9,90	9,93
Housing	10,26	10,36	10,71	10,69	10,69	10,68	11,03
Household goods	6,67	6,15	6,17	6,41	6,41%	6,41	6,36
Medicine	3,04	2,83	2,72	2,68	2,68	2,75	2,81
Transport	15,20	14,89	14,91	14,40	14,40	15,32	15,58
Communications	3,68	3,58	3,28	2,99	2,99	2,73	2,57
Leisure and culture	7,50	7,11	6,78	6,77	6,77	6,83	6,73
Education	1,47	1,60	1,68	1,67	1,67	1,67	1,74
Hotels, cafes and restaurants	11,87	11,55	11,45	11,23	11,23	11,18	11,27
Other goods and services	8,57	8,02	7,72	7,64	7,64	7,39	6,91

As a result 180 valid questionnaires were obtained, divided into four sections:

- Block 1: includes the *sociodemographic characteristics* of the interviewees.
- Block 2: questions about their *attendance at the SMR*.
- Block 3: includes questions related with the *spending structure* of those attending the festival.
- Block 4: contains questions about the *impact Cuenca's Religious Music Week* has had on the interviewee, in terms of returning to and recommending the festival.

When the information had been collected, the data obtained were analysed using the software SPSS, version 19.0, using the following techniques:

- *Descriptive analysis* of variables.
- *Contingency analysis* to find out the relationships between variables.

3. SPENDING BY THE FESTIVAL PUBLIC

3.1. Total Budget

Over the course of time, changes occur to the distribution of the total spending of the homes assigned to each spending group (Table 2); so spending on leisure and culture, which represents a very important part of the total mean spending per household, rose from 7.11% in 2007, to 7.50% in 2008.

An increase in spending on hotels, cafeterias and restaurants can also be seen, rising from 11.50% in 2007 to 11.87% in 2008 [National Statistics Institute, 2010].

Examining the total budget of the individuals attending the Religious Music Week during their stay in Cuenca (Figure 1), and considering all the costs of their stay (transport, accommodation, meals, tickets, etc.), a mean figure of 301.15 € is obtained; considering that the mean length of the stay is 3.02 gives a high daily spending figure of 99.72 € a day per person.

The most frequent spending bands are 150-300 € and that of over 500 €, each with 26.44%. The next spending band is 300-500 €, with 21.84%. The least frequent spending bands are 50-150 € (18.97%) and that of up to 50 € (6.32%).

The people attending the festival from abroad have a total mean budget higher than that of the Spanish public, 480 € compared with 290.24 €; so for the international public the most frequent spending band (Table 3) is that of over 500 € (80%), followed by 300-500 € (20%); so being in the highest spending bands. Whereas Spaniards on average have a high percentage in the band of over 500 € (22.35%), but the most frequent band for the total budget is that of 150-300 € (27.06%). This relationship is significant to 99%.

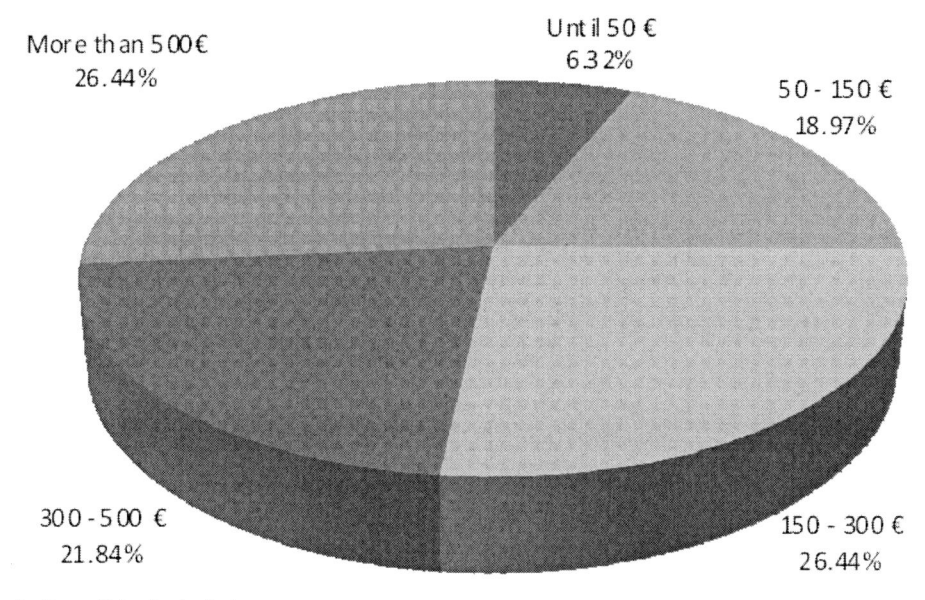

Figure 1. Overall budget of stay.

Analysing the difference in total budget between those attending the festival who live in Cuenca (city or province) and those who come from elsewhere, the budget including all the costs of their stay (transport, accommodation, meals, tickets, etc.), comes to 362.71 € for those from outside Cuenca; considering that the average length of stay is 3.02 days (for those coming from outside Cuenca), this gives a high daily spending figure, of 120.10 € per day per person. And 171.43 € total budget for their stay for those who live in Cuenca. However, this relationship is not significant.

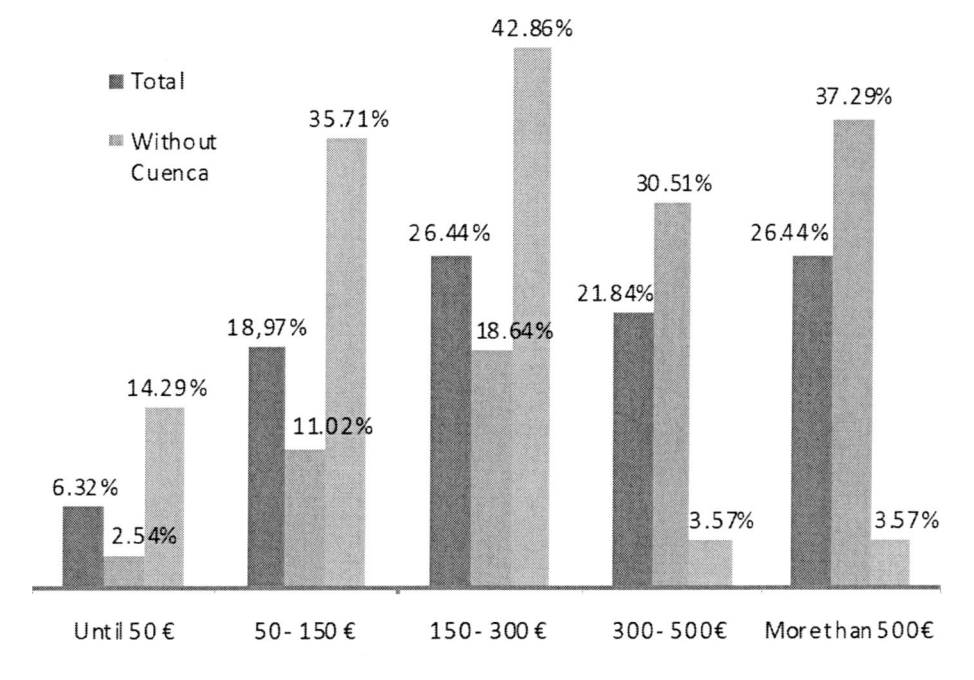

Figure 2. Comparison overall budget of stay according to place of residence.

Table 3. Relationship between overall budget and nationality's attendees

Percentage (%)		Overall budget (€)						Total
		Until 50	50-150	150-300	300-500	More than 500	DK/-NA	
Natio-nality	Spanish	6.47	19.41	27.06	21.18	22.35	3.53	100.00
	International	0.00	0.00	0.00	20.00	80.00	0.00	100.00

If we compare the distribution of the total budget (Figure 2), separating total visitors from the visitors who live in Cuenca, we note that the lower spending bands, up to 300 €, are more frequent among individuals who live in Cuenca, while the spending bands of more than 300 € for the whole stay are more frequent among the public which lives outside the province of Cuenca.

Analysing the total budget according to the main activity the concertgoers carry out during their stay in Cuenca (Table 4), differences are observed in this relationship which are significant to 99%. The mean total budget is thus higher for those whose main activity is visiting family and friends (433.33 €), followed by those whose activity is attending the Religious Music Week (350.98 €), and finally those who attend the Holy Week celebrations (333.33 €) and other activities (330.95 €).

If the relationship between total budget and approximate monthly income is analysed (Table 5), it is observed that for the people attending with income below 1000 €, the most frequent total budget is that of 50-150 € (50%), followed by up to 50 € (25%). However, in general, it is seen that as the monthly income increases, the proportions for the two lowest spending bands is reduced and the 150-300 € band increases, this being one of the two most frequent bands for the all the people attending the festival. This relationship is significant to 95%.

Table 4. Average overall budget according to the developed activity

	Overal budget (€)					Total
	Religious Music Week	Easter week	Visit to family/friends	Other	DK/NA	
Average	350.98	333.33	433.33	330.95	163.33	301.15

Table 5. Relationship between overall budget and monthly income

Percentage (%)		Monthly income (€)			
		< 1000 €	1000 - 1500	1501 - 2000	> 2000
Overall budget	Until 50 €	25.00	18.18	2.33	4.44
	50 - 150 €	50.00	18.18	18.60	16.67
	150 - 300 €	0.00	22.73	23.26	26.67
	300 - 500 €	12.50	9.09	20.93	24.44
	More than 500 €	0.00	31.82	32.56	25.56
	DK/NA	12.50	0.00	2.33	2.22
Total		100.00	100.00	100.00	100.00

Table 6. Relationship between overall budget and previous experience

Percentage (%)		Previous experience	
		I attend for the first time	I have previously attended
Overall budget	Until 50 €	6.35	5.98
	50 - 150 €	19.05	17.95
	150 - 300 €	22.22	27.35
	300 - 500 €	28.57	17.09
	More than 500 €	22.22	27.35
	DK/NA	1.59	4.27
Total		100.00	100.00

Analysing the evolution of the total budget according to prior experience of the Religious Music Week (Table 6), in general it is seen that the lowest budget bands (under 150 €) are more frequent among those attending for the first time, while for those who have previously attended the festival the most frequent budget bands are the highest one, over 500 € and 150-300 €, except for the 300-500 € spending band, which is more frequent for those who had not previously attended the Religious Music Week. However, this relationship is not significant. And also the average spending for those with no experience and those with is not very different 302.42 € and 300.45 €, respectively, though slightly higher for those attending for the first time.

Looking at whether the visit is carried out in company or not (Table 7), it is observed that those attending in company are in the higher total budget bands, 150-500 €; while those attending on their own are either in the top band (over 500 €) or in the lowest ones (under 300 €). This relationship is significant to 99%. The mean budget for the stay for those attending in company is 302.14 €, compared with 294.32 € for those attending alone.

The same thing happens when the condition of subscriber is examined: the total budget (359.48 €) estimated by people attending the festival as subscribers is greater than that of those who are not (291.91 €). This relationship is significant to 90%.

Table 7. Relationship between overall budget and attendance together

Percentage (%)		¿Accompanied?		
		Yes	No	DK/NA
Overall budget	Until 50 €	7.01	0.00	0.00
	50 - 150 €	17.20	27.27	0.00
	150 - 300 €	24.84	31.82	0.00
	300 - 500 €	22.93	9.09	0.00
	More than 500 €	24.84	31.82	0.00
	DK/NA	3.18	0.00	100.00
Total		100.00	100.00	100.00

Table 8. Relationship between overall budget and celebration date

Percentage (%)		Celebration date			
		Attending regardless of the date	Attending only if it coincides with the date of Easter	Attending only if performed on holiday	Not return, occasional visit
Overall budget	Until 50 €	5.77	10.26	2.78	0.00
	50 - 150 €	24.04	10.26	11.11	0.00
	150 - 300 €	30.77	17.95	19.44	0.00
	300 - 500 €	17.31	23.08	30.56	0.00
	More than 500 €	16.35	38.46	36.11	100.00
	DK/NA	5.77	0.00	0.00	0.00
Total		100.00	100.00	100.00	100.00

The total budget during the stay is also related with the date when the festival is held (Table 8). Those who declare that they would attend regardless of the date are found mainly in the middle spending bands, compared with those who would only attend if it coincided with either Holy Week or public holidays, who are found in the highest spending bands. This relationship is significant to 95 %.

The estimated total budget for the stay increases with the number of concerts the visitor plans to attend (relationship significant to 95%), and a positive relationship is also observed with the length of the stay (significant to 99%) and the chosen type of accommodation – the higher the chosen accommodation classification, the greater the expenditure level (significant to 90%).

3.2. Daily Spending per Service

In general, analysing the distribution of daily spending per visitor in different services, a preponderance of the lower spending bands in the established scale is observed, and often even the "Nothing" option is seen in the highest frequency rates (Table 9).

The highest spending of visitors to the Religious Music Week is on catering and entrance to the concerts. And the least important expenditure items are the acquisition of gifts and souvenirs/shopping, other spending, leisure and accommodation, which is a logical distribution of spending bearing in mind that 34.44% of the festival's public are from the province or city of Cuenca.

If the distribution of daily spending is analysed according to nationality (Table 10), it is observed that Spanish visitors are found in lower spending bands than those from abroad in accommodation, catering, gifts and other expenses, while in leisure and tickets for the Religious Music Week, foreigners are in lower spending bands than Spaniards.

Analysing the distribution of spending by expense items (Table 11), and differentiating the attending public living in Cuenca from the total, it is seen that the public which lives in Cuenca has the highest percentages in the option of not spending anything and the lowest for any spending segment in the items of accommodation, catering and souvenirs. This relationship is significant.

Table 9. Daily expenditure per person (%)

Accommodation	Until 30 €	Until 60 €	Until 100 €	More than 100 €	Nothing
	3.89	20.56	12.78	8.89	52.78
Catering	Until 20 €	Until 40 €	Until 60 €	More than 60 €	Nothing
	13.89	32.22	11.67	12.22	28.89
Entertainment	Until 10 €	Until 20 €	Until 30 €	More than 30 €	Nothing
	21.11	9.44	4.44	6.67	57.78
Purchase of gifts	Until 30 €	Until 60 €	Until 100 €	More than 100 €	Nothing
	15.00	2.78	1.67	1.11	78.89
Attendance at concerts	Until 20 €	Until 40 €	Until 60 €	More than 60 €	Nothing
	10.56	38.89	14.44	22.22	11.67
Other expenses	Until 20 €	Until 40 €	Until 60 €	More than 60 €	Nothing
	19.44	3.89	0.00	0.56	73.89

In the case of spending on leisure, those attending the festival who live in Cuenca have higher percentages than the rest of the public for spending levels of up to 20 €, more than 30 € and nothing. This relationship is significant to 95 %.

Table 10. Daily expenditure per person according to nationality (%)

Percentage (%)		Until 30 €	Until 60 €	Until 100 €	More than 100 €	Nothing	DK/NA	Sig.
Accommodation	Spanish	4.12	19.41	12.94	7.65	54.71	1.18	90 %
	Foreign	0	40	10	30	20	0	
		Until 20 €	Until 40 €	Until 60 €	More than 60 €	Nothing	DK/NA	
Catering	Spanish	13.53	32.94	10.59	11.18	30.59	1.18	90 %
	Foreign	20	20	30	30	0	0	
		Until 10 €	Until 20 €	Until 30 €	More than 30 €	Nothing	DK/NA	
Entertainment	Spanish	19.41	9.41	4.71	7.06	58.82	0.59	No sig.
	Foreign	50	10	0	0	40	0	
		Until 30 €	Until 60 €	Until 100 €	More than 100 €	Nothing	DK/NA	
Purchase of gifts	Spanish	13.53	2.35	1.76	1.18	80.59	0.59	No sig.
	Foreign	40	10	0	0	50	0	
		Until 20 €	Until 40 €	Until 60 €	More than 60 €	Nothing	DK/NA	
Attendance at concerts	Spanish	11.18	38.82	14.12	22.94	11.18	1.76	No sig.
	Foreign	0	40	20	10	20	10	
		Until 20 €	Until 40 €	Until 60 €	More than 60 €	Nothing	DK/NA	
Other expenses	Spanish	17.65	3.53	0	0.59	75.88	2.35	90 %
	Foreign	50	10	0	0	40	0	

Table 11. Daily expenditure per person according to place of residence (%)

Percentage (%)		Until 30 €	Until 60 €	Until 100 €	More than 100 €	Nothing	DK/NA	Total	Sig.
Accommo-dation	Total	3.89	20.56	12.78	8.89	52.78	1.11	100.00	99 %
	Cuenca	1.61	0.00	1.61	0.00	96.77	0.00	100.00	
	Total except Cuenca	5.08	31.36	18.64	13.56	29.66	1.69	100.00	
		Until 20 €	Until 40 €	Until 60 €	More than 60 €	Nothing	DK/NA		
Catering	Total	13.89	32.22	11.67	12.22	28.89	1.11	100.00	99 %
	Cuenca	11.29	12.90	1.61	3.23	70.97	0.00	100.00	
	Total except Cuenca	15.25	42.37	16.95	16.95	6.78	1.69	100.00	
		Until 10 €	Until 20 €	Until 30 €	More than 30 €	Nothing	DK/NA		
Enter-tainment	Total	21.11	9.44	4.44	6.67	57.78	0.56	100.00	95 %
	Cuenca	8.06	11.29	0.00	9.68	69.35	1.61	100.00	
	Total except Cuenca	27.97	8.47	6.78	5.08	51.69	0.00	100.00	
		Until 30 €	Until 60 €	Until 100 €	More than 100 €	Nothing	DK/NA		
Purchase of gifts	Total	15.00	2.78	1.67	1.11	78.89	0.56	100.00	95 %
	Cuenca	4.84	1.61	0.00	0.00	93.55	0.00	100.00	
	Total except Cuenca	20.34	3.39	2.54	1.69	71.19	0.85	100.00	
		Until 20 €	Until 40 €	Until 60 €	More than 60 €	Nothing	DK/NA		
Attendance at concerts	Total	10.56	38.89	14.44	22.22	11.67	2.22	100.00	No sig.
	Cuenca	19.35	45.16	14.52	11.29	8.06	1.61	100.00	
	Total except Cuenca	5.93	35.59	14.41	27.97	13.56	2.54	100.00	
		Until 20 €	Until 40 €	Until 60 €	More than 60 €	Nothing	DK/NA		
Other expenses	Total	19.44	3.89	0.00	0.56	73.89	2.22	100.00	No sig.
	Cuenca	6.45	1.61	0.00	1.61	88.71	1.61	100.00	
	Total except Cuenca	26.27	5.08	0.00	0.00	66.10	2.54	100.00	

For spending on tickets to concerts in the Religious Music Week, people living in Cuenca are more frequent in the expenditure items up to 60 €, while they are below the rest of the public in expenditure over 60 € and no spending. However, this relationship is not significant.

In other expenditures, the festival visitors from Cuenca have higher frequencies than other visitors for the option of expenditure over 60 € and nothing. This relationship is not significant.

Table 12. Daily expenditure per person according to experience (%)

Percentage (%)		Until 30 €	Until 60 €	Until 100 €	More than 100 €	Nothing	DK/NA	Sig.
Accommodation	First time	1.59	25.40	22.22	11.11	39.68	0	95 %
	Previously	5.13	17.95	7.69	7.69	59.83	1.71	
		Until 20 €	Until 40 €	Until 60 €	More than 60 €	Nothing	DK/NA	
Catering	First time	12.70	44.44	7.94	19.05	15.87	0	99 %
	Previously	14.53	25.64	13.68	8.55	35.90	1.71	
		Until 10 €	Until 20 €	Until 30 €	More than 30 €	Nothing	DK/NA	
Entertainment	First time	28.57	11.11	7.94	9.52	42.86	0	90 %
	Previously	17.09	8.55	2.56	5.13	65.81	0.85	
		Until 30 €	Until 60 €	Until 100 €	More than 100 €	Nothing	DK/NA	
Purchase of gifts	First time	26.98	4.76	0	1.59	66.67	0	95 %
	Previously	8.55	1.71	2.56	0.85	85.47	0.85	
		Until 20 €	Until 40 €	Until 60 €	More than 60 €	Nothing	DK/NA	
Attendance at concerts	First time	15.87	44.44	11.11	17.46	11.11	0	No sig.
	Previously	7.69	35.90	16.24	24.79	11.97	3.42	
		Until 20 €	Until 40 €	Until 60 €	More than 60 €	Nothing	DK/NA	
Other expenses	First time	23.81	6.35	0	0	68.25	1.59	No sig.
	Previously	17.09	2.56	0	0.85	76.92	2.56	

Looking at the distribution of daily spending according to experience (Table 12), it is observed that those attending for the first time are in higher expenditure bands than those who have attended before in accommodation, catering, leisure, gifts and other spending, while those who are repeating the experience are in higher spending bands for tickets to the Religious Music Week.

If the distribution of daily expenditure is analysed according to the length of stay (Table 13), spending on accommodation is greatest for three-day stays, followed by two days, and finally stays of four or more days. In catering, spending increases the longer the stay. In leisure, the highest spending is in three-day stays, followed by four-day stays. In gifts, spending is greater for longer stays and falls as the stay shortens. Spending on concerts is greatest for three-day stays, followed by four, one and two-day stays. And in other spending, the amount is higher the shorter the length of stay.

The amount the people attending the festival allocate to each spending item varies according to their monthly income (Table 14). Visitors with monthly income below 1,000 € are those who sped least on all items, and it should be remembered that these individuals were Cuenca residents or those who were only in the city for a day and did not spend the night there. Those with income between 1,500 and 2.000 € are those who spend most on accommodation, leisure, gifts and other spending. And those with income over 2,000 € are those who spend most on catering and Religious Music Week concerts.

Table 13. Daily expenditure per person according to length of stay (%)

Percentage (%)		Until 30 €	Until 60 €	Until 100 €	More than 100 €	Nothing	DK/NA	Sig.
Accommodation	One day	0	0	0	0	100	0	99 %
	Two days	0	44	24	16	12	4	
	Three days	0	50	25	16.67	8.33	0	
	Four or more days	12.73	25.45	20	14.55	25.45	1.82	
		Until 20 €	Until 40 €	Until 60 €	More than 60 €	Nothing	DK/NA	
Catering	One day	7.14	42.86	28.57	7.14	14.29	0	99 %
	Two days	28	36	12	16	8	0	
	Three days	8.33	66.67	8.33	16.67	0	0	
	Four or more days	14.55	34.55	21.82	20	5.45	3.64	
		Until 10 €	Until 20 €	Until 30 €	More than 30 €	Nothing	DK/NA	
Entertainment	One day	42.86	7.14	0	7.14	42.86	0	95 %
	Two days	24	8	8	0	60	0	
	Three days	29.17	12.50	12.50	8.33	37.50	0	
	Four or more days	27.27	9.09	5.45	5.45	52.73	0	
		Until 30 €	Until 60 €	Until 100 €	More than 100 €	Nothing	DK/NA	
Purchase of gifts	One day	35.71	0	0	0	64.29	0	95 %
	Two days	16	0	4	0	80	0	
	Three days	20.83	4.17	0	0	70.83	4.17	
	Four or more days	18.18	7.27	3.64	3.64	67.27	0	
		Until 20 €	Until 40 €	Until 60 €	More than 60 €	Nothing	DK/NA	
Attendance at concerts	One day	21.43	50	0	21.43	7.14	0	95 %
	Two days	8	48	16	8	20	0	
	Three days	8.33	25	16.67	37.50	12.50	0	
	Four or more days	1.82	29.09	16.36	34.55	12.73	5.45	
		Until 20 €	Until 40 €	Until 60 €	More than 60 €	Nothing	DK/NA	
Other expenses	One day	21.43	14.29	0	0	64.29	0	95 %
	Two days	16	8	0	0	72	4	
	Three days	29.17	4.17	0	0	58.33	8.33	
	Four or more days	30.91	3.64	0	0	65.45	0	

CONCLUSION

As the main conclusions and recommendations obtained from the research carried out, we can highlight the following:

The total budget each person attending Cuenca's Religious Music Week festival left in the city during his stay was on average 301.15€, quite a large amount, considering that the average stay is 3 days, which means an average daily expenditure in the destination of 100€ per person, including all costs: transport, accommodation, catering, tickets, leisure, gifts and other spending.

Looking at the breakdown of spending according to type of service, the most important expenditure items for those attending Cuenca's Religious Music Week are catering and tickets for the concerts and the least important are acquisition of gifts, leisure, accommodation and other expenses. This behaviour changes with the experience of those attending the festival: the greater the experience, the higher the spending on attending concerts and the lower it is on other items.

Distinguishing between the nationalities of those attending the festival highlights the larger spending made by foreign visitors, 480€ compared with the 290.24€ of Spanish visitors; this demonstrates the need to make a bigger effort to promote the festival abroad, to attract international visitors (who were only 5.56% in the edition analysed), because they spend more and the length of their stay is longer. And their biggest expenditure items are accommodation, catering and gifts, less on concert tickets and leisure.

The total budget is also different between those attending the festival who live in the province of Cuenca and those who live elsewhere, 171.43€ and 362.71€, respectively, because of their different needs during their stay, essentially in terms of accommodation. This result demonstrates the importance of extending the promotion of the festival in Spain.

Paradoxically, the highest total budget is that of those whose main activity is visiting family and friends, and the second that of those attending because of the Religious Music Week. This is because their spending on accommodation is low or zero, and they spend more on the other expenditure items analysed, such as concert tickets and catering.

The income level of the people attending the festival is a variable which contributes towards determining the available budget. In general, it is observable that the greater the income level, the higher the total budget for the stay.

On the other hand, the total budget of those who have attended a previous edition of the festival is slightly lower than those who are there for the first time, 302.42 € and 300.45 €, respectively. This is because they spend more on accommodation, catering, leisure and gifts, while those who have attended before choose more modest so cheaper accommodation, and get fewer gifts, their highest expenditure item being concert tickets.

The total budget for the whole stay is slightly higher for those who visit in company compared with those who do so on their own, 302.14 € and 294.32 €, respectively. Because visiting in company encourages a greater number of leisure activities.

Also, festival subscribers have a higher total budget than non-subscribers, so the festival organizers should offer attractive subscriptions to increase the demand for them, and even affect decisions such as whether to stay in the city of Cuenca.

Table 14. Daily expenditure per person according to monthly income (%)

Percentage (%)		Until 30 €	Until 60 €	Until 100 €	More than 100 €	Nothing	DK/NA	Sig.
Accommodation	< 1.000 €	0.00	0.00	0.00	0.00	100.00	0.00	No sig.
	1.000-1.500 €	9.09	18.18	4.55	13.64	54.55	9.09	
	1.501-2.000 €	9.30	25.58	13.95	6.98	44.19	9.30	
	More than 2.000 €	1.11	22.22	15.56	8.89	51.11	1.11	
		Until 20 €	Until 40 €	Until 60 €	More than 60 €	Nothing	DK/NA	
Catering	< 1.000 €	0.00	12.50	0.00	0.00	87.50	0.00	No sig.
	1.000-1.500 €	22.73	31.82	18.18	0.00	27.27	0.00	
	1.501-2.000 €	18.60	34.88	11.63	9.30	25.58	0.00	
	More than 2.000 €	11.11	32.22	10.00	18.89	25.56	2.22	
		Until 10 €	Until 20 €	Until 30 €	More than 30 €	Nothing	DK/NA	
Entertainment	< 1.000 €	12.50	0.00	0.00	0.00	87.50	0.00	No sig.
	1.000-1.500 €	18.18	4.55	4.55	4.55	68.18	0.00	
	1.501-2.000 €	25.58	16.28	0.00	6.98	48.84	2.33	
	More than 2.000 €	20.00	8.89	5.56	8.89	56.67	0.00	
		Until 30 €	Until 60 €	Until 100 €	More than 100 €	Nothing	DK/NA	
Purchase of gifts	< 1.000 €	25.00	0.00	0.00	0.00	75.00	0.00	No sig.
	1.000-1.500 €	31.82	0.00	0.00	0.00	68.18	0.00	
	1.501-2.000 €	2.33	4.65	0.00	2.33	90.70	0.00	
	More than 2.000 €	14.44	3.33	3.33	0.00	77.78	1.11	
		Until 20 €	Until 40 €	Until 60 €	More than 60 €	Nothing	DK/NA	
Attendance at concerts	< 1.000 €	37.50	37.50	25.00	0.00	0.00	0.00	No sig.
	1.000-1.500 €	13.64	45.45	4.55	22.73	9.09	4.55	
	1.501-2.000 €	6.98	44.19	16.28	18.60	13.95	0.00	
	More than 2.000 €	10.00	33.33	14.44	26.67	12.22	3.33	
		Until 20 €	Until 40 €	Until 60 €	More than 60 €	Nothing	DK/NA	
Other expenses	< 1.000 €	25.00	0.00	0.00	0.00	75.00	0.00	No sig.
	1.000-1.500 €	31.82	9.09	0.00	0.00	59.09	0.00	
	1.501-2.000 €	25.58	6.98	0.00	2.33	62.79	2.33	
	More than 2.000 €	15.56	2.22	0.00	0.00	81.11	1.11	

The total budget available also varies with the preferred date for holding the festival, being higher if the festival coincides with Holy Week or public holidays, compared with holding it on any other dates, because of the availability of more non-working days.

In the same way, the more concerts there are to attend, the longer the stays, the higher the category of accommodation and the greater the total budget, because the people attending the festival go to more concerts and stay more nights.

To sum up, all these figures show that Cuenca's *Semana de Música Religiosa* and the tourist activity associated with it generate important economic flows in the local and regional economy.

So the *Semana de Música Religiosa de Cuenca* has become one of Cuenca's important tourist attractions, which has provided important economic effects and made a substantial modification to the pattern of tourist behaviour of visitors, both in spending and in the services required.

REFERENCES

Cuadrado García, M. (2010). *Estudios culturales: Doce estudios de marketing*. Barcelona, Spain: Editorial UOC.

Cuadrado García, M. & Berenguer Contrí, G. (2002): *El consumo de servicios culturales*. Madrid, Spain: ESIC.

Falassi, A. (1997): Festival. In T. A. Green (Ed.), *Folklore, an encyclopedia of beliefs, customs, tales, music, and art*. Santa Barbara, California, USA: ABC-CLIO.

González Neira, A. & Ramírez Picón, J. (2008): *Impacto económico del XI Festival de Flamenco de Jerez*. Col. Cuadernos de Investigación Vigía, num. 5, Cádiz, Spain: Ed. Diputación Provincial de Cádiz.

Herrero Prieto, L. C. (2004). *Turismo Cultural e Impacto Económico de Salamanca 2002, Ciudad Europea de la Cultura*, Madrid (Spain): Ed. Civitas.

Mondéjar Jiménez, J. A., Cordente Rodríguez, M., Gázquez Abad, J. C., Pérez Calderón, E. & Milanés Montero, P. (2010). Visitor Profile of Cuenca Religious Music Week. *Journal of Business Case Studies*, 6 (7), 15-22.

National Statistics Institute (2010). *HBS. Household Budget Survey. Base 2006*. Madrid (Spain): National Statistics Institute.

Sassatelli, M. (2008). European Public Culture and Aesthetic Cosmopolitanism. http://www.euro-festival.org/docs/Euro-Festival_D1_MainReport.pdf

Semana de Música Religiosa (2009): *Claves de la 48 SMR de Cuenca*. http://www.smrcuenca.es/portal/lang__es-ES/tabid__11137/default.aspx

Chapter 4

LOOKING FOR A NORTHERN REGION: ATTRIBUTES OF CHOICE

Normand Bourgault[*1], *Patrice Leblanc*[2], *Judy-Ann Connelly*[2] *and Ann Gervais*[2]

[1]Université du Québec en Outaouais, Canada
[2]Université du Québec en Abitibi-Témiscamingue, Canada

ABSTRACT

It is not an easy task to attract tourists and new residents north of the 48th parallel in a country. This research was conducted in two phases. In the first phase, a qualitative approach was used to identify broader themes of concern to people when choosing not only a holiday destination, but also a place to live. Four focus groups generated some overall themes that form the consideration set of attracting variables. With this as a substrate, a quantitative approach was then applied in the second phase.

Fourteen items were developed. Importance and presence were measured in a statistically representative sample of 306 respondents from different regions of Québec Province, in Canada. The analysis showed that nearness to wildlife in safe conditions is the most strongly present factor of attraction, but people who are approachable and very hospitable toward travelers is quite a close second. Eleven other attributes of choice were tested. An advertising campaign theme, as revealed by the research, is presented.

1. INTRODUCTION

It is not an easy task to attract tourists and new residents north of the 48th parallel in a country. Prejudices against the northern regions of Canada lead to the presumption that tourists are not likely to find enjoyment, culture or leisure. Rather, they are led to suppose that good food is a rarity, and that a lack of things to do and to see is standard fare for the

---

[*] Contact author: Normand.Bourgault@uqo.ca

residents. Prejudices are often held by residents of a given country, but are not always in the minds of people from elsewhere.

In the summer of 2007, we carried out two research activities to identify the image of a northern region. We were interested in qualifying the Abitibi-Témiscamingue Region (AT), an area of the Canadian province of Québec. Rather than measure tourists' intentions to visit, we decided to determine if people from other regions of Québec province would be ready to live in Abitibi-Témiscamingue. We aimed to achieve three objectives:

1. Identify the attributes of choice for a region in which to live, work or study such as AT, for Quebecers who do not live in AT.
2. Measure the importance and the presence of these attributes of choice for AT and its two neighbours, Saguenay-Lac-Saint-Jean to the east, and Outaouais to the south.
3. Compute an attribute positioning mapping of the three regions.

To achieve these objectives, we carried out a two-phased research initiative. In the first phase, we proceeded using a qualitative method to produce a rich list of attributes of choice. In the second, we conducted a survey, streamlined the scales with quantitative techniques and determined a brand position for the area on the Québec market.

2. THE QUALITATIVE PHASE: IDENTIFYING ATTRIBUTES OF CHOICE

Because we were seeking to identify and measure attributes of choice, one issue in this research was to determine the concepts involved in our overall theme: choosing a region to live in or to visit. For this purpose, we had to define what constitutes a region's image to the consumer. In this research, we used a region of Québec province, Abitibi-Témiscamingue (AT), as an object contributing to the crystallization of what constitutes the image of a region. For the benefit of the reader, we briefly describe the region, even though marketing is not the intended goal, but only an incidental aspect. For the curious reader, at the end of this chapter, we will point out the marketing strategy and the axis of communications generated by this research.

2.1. Background Information on the Region

Abitibi-Témiscamingue is one of Québec's administrative regions, and occupies a land area of 65,000 km^2, roughly twice the size of Belgium. Figure 1 shows the location.

Abitibi-Témiscamingue has a population of 150,000 inhabitants spread out over a rather sparsely populated area, with 2 inhabitants/km^2. The language spoken in the home is French in 95% of households, English in 4% and Algonquin, an Amerindian language, in nearly 2% of cases. Its economy depends mainly on three major natural resources: 1) the forestry industry; 2) the mining industry, which has been exploiting rich deposits of gold and copper since early in the 20th century; and 3) agricultural production. Below, we shall see which attributes the focus groups used to describe this region.

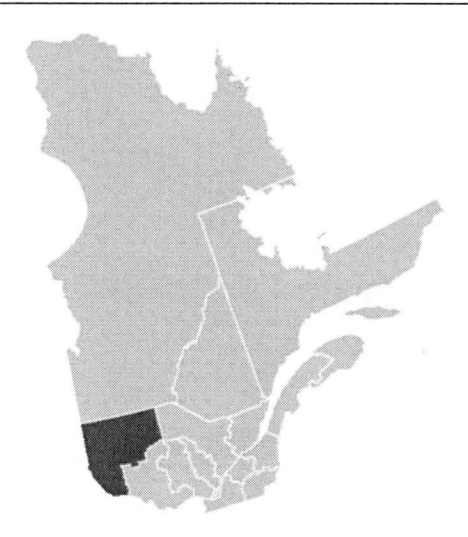

Figure 1.Geographic location of Abitibi-Témiscamingue in Québec, Canada.

2.2. The Image of an Area

Proshansky et al. (1983) stated that through their interactions with their physical environment, individuals develop cognitive structures, memorized representations, ideas, feelings, attitudes, values, preferences, meanings and behaviours connected with that environment. This conceptualisation postulates the image of a country as a multidimensional construct. Parameswaran and Pishandori (1994) follow a similar line of thinking. They define the image as a multidimensional construct whose different dimensions explain the political, economic, technological and social factors of the country. In the same vein, Martin and Eroglu (1993) assert that this image of a country corresponds to the total of all the descriptive, inferential and informational beliefs that an individual possesses on a particular country. This internalized conception is used by them to structure their system of reference to understand their experiences within their environment (Bonnes et Secchiaroli, 1995). It is reasonable to suppose that this same system of reference comes into play in evaluating other environments, such as other regions of a province as they do with products of other regions (Bourgault, 2007).

Rather than measure all the components bring into play to evaluate a region, we used a methodology that enables individuals belonging to groups targeted by the research to testify themselves to the attributes composing the image of the regions for which we wish to define the attributes of choice.

2.3. The Qualitative Instrumentation

We chose to use focus groups as a primary tool to identify the factors of attraction and the attributes that compose the perception of a region. Focus groups are particularly suitable for analysing complex situations. Sin, Cheung and Lee (1999) noted that this type of research protocol is occasionally used in cross-cultural research.

According to Baribeau (2009), the number of participants to a focus group should be 8 to 10 on average, which we set as a standard. We also have had to take care of representativity by inviting individuals from different socio-economic strata relevant to the situation. We thus formed focus groups of 6 to 12 adults, from different backgrounds and having different viewpoints on the image of Abitibi-Témiscamingue.

We then prepared a discussion leader's guide covering the image of AT as a living environment, as a working environment and as a sector of economic activity. Under each of these themes, a list of sub-themes, drawn from a review of the three compilations of questions from Bruner (1994, 1998, 2001), was examined.

2.4. Procedure Followed in the Focus Group Discussions

Three focus group discussions were held in August 2007 in different regions of the province of Québec: the first in the largest city, Montréal, the second and third in outlying areas, namely Mauricie and Bas-Saint-Laurent. In total, 31 participants took part in the three focus groups, including 18 men and 13 women. They were aged 22 to 80, with an average age of 49 years. More than 60% of the participants had a level of education equal to or higher than a University-level bachelor's degree, and the participants held various occupations. Sixteen of the participants had a real knowledge of the region, having spent time there. It is reasonable to suppose that these persons represent different socioeconomic spheres within Québec, different opinions on the object of the research even if groups are not statistically representative of the Québec population. That is one of the inherent limitations of focus groups.

2.5. Attributes Forming Part of the Region's Image

The descriptive attributes of the region were gathered using the following textual extraction method: once the entire discussions had been captured in textual form (verbatim), two researchers read through the text and systematically extracted the descriptive attributes. An initial version of this list was then drafted, with each entry on the list containing the concept, accompanied by sentences drawn from the verbatim, justifying the concept. A third researcher went through the list and deleted any repetitions and synonyms.

From this list, the first version of a questionnaire was prepared, using concepts drawn from the focus group discussions. For each concept, questions were identified in consulting thematic research handbooks (Bearden, Netemeyer et Mobley, 1993; Bruner and Hensel, 1994, 1998, 2001). This generated an instrument comprising nearly 200 questions, which we submitted to the Comité de valorisation de l'Abitibi-Témiscamingue (CVAT) for screening. This Committee, formed of regional development officials, was asked to retain only about ten questions on the relevant attributes of the region. Finally, as shown in figure 2, fourteen attributes were retained.

3. THE QUANTITATIVE PHASE: MEASURING ATTRIBUTES OF CHOICE

With the fourteen attributes retained, we then set in motion the quantitative phase and prepared the final questionnaire. Initially, we measured the importance to the respondents of the attributes listed in figure 2, by means of the following question:

First question. If you had to move to a region other than the one where you live now, for example for employment reasons, to pursue studies, or to find a new living environment, how much importance would you place on the following factors, on a scale from 1 to 10, where 1 = not at all important, and 10 = very important?

Secondly, we measured the degree of presence of these attributes for three regions of Québec: Abitibi-Témiscamingue, Outaouais and Saguenay-Lac-Saint-Jean.

Second question. According to what you know about the following three regions of Québec, Abitibi-Témiscamingue, Outaouais and Saguenay-Lac-Saint-Jean, at what level do you rate the presence of the following factors on a scale from 1 to 10, where 1 = not at all present in that region, and 10 = strongly present in that region?

We chose a decimal scale (with 10 increments) which we considered easy to use for the respondents. The CVAT committee was called upon to pre-test the measurement instrument. The conduct of the investigation will be presented briefly, followed by an assessment of the representativity of the sample, and then a presentation of the results.

We chose to survey a representative sample of the Québec population, excluding Abitibi-Témiscamingue. For this purpose, we used a stratified random sample of 306 respondents. We set quotas for each of the strata, in this case the populations of the other regions of Québec. We then randomly selected individuals assigned to each stratum, until the desired number was obtained.

1.	Cultural activities in the region
2.	Living close to nature
3.	Hospitable people
4.	Social and recreational activities
5.	Favourable weather conditions
6.	Presence of educational institutions
7.	Presence of health services
8.	Ease of finding employment
9.	Ease of joining groups
10.	Possibility of having an urban lifestyle
11.	Work environment
12.	Social environment
13.	Living environment
14.	Proximity to the major cities of Québec

Figure 2. Attributes retained.

The survey took the form of telephone interview from October 8 to 24, 2007, from 7:00 to 9:00 p.m. on weekdays and from 10:00 a.m. to 5:00 p.m. on Saturdays. They were conducted by an opinion survey firm. The response time varied from 10 to 14 minutes, and approximately 65% of those contacted refused to answer. The firm did not report any particular problems.

3.1. Representativity and Normality of Distribution

The population sample surveyed was proportional to the regions of Québec. With 306 respondents, the statistical error between the average for the population and for our sample should not exceed 5.6%, 19 times out of 20. The sample group comprised 60% women and 40% men. This proportion is in the confidence interval (54.5% - 65.5%, $p < 0.05$) for this sex variable. Considering the age, 18-24 year age range is less strongly represented in the sample than in the population, and the 45-65 year age group is over-represented. These anomalies are taken into account in the analysis. The respondents in the sample are more highly educated than the population at large. The proportion of individuals in the sample in the primary-secondary education category is lower than for the overall population; for those who attended university, the proportion is higher. The average annual family income in the sample within Québec stands at $56,826, slightly higher than the average for the population at large ($54,139), but within the confidence interval for the sample ($51,731 – $ 61,821, $p < 0.05$). Overall, it is reasonable to consider the sample consulted for the survey as representative of the population over most of the variables examined.

We verified the normality of the data from the questions on attributes of choice. Since none of the variables shows skewness or kurtosis exceeding set guidelines (between +3 and -3), the data are considered to be normally distributed. Since the sample is representative and the data are in a normal distribution, we shall examine the attributes of choice.

3.2. The Importance of Attributes of Choice for a Region

Which attributes would be given priority by Québec respondents, if they had to go and live in a region other than where they live now? The data presented in figure 3 show the importance placed on the attributes of choice. Above all, the respondents would assure themselves of the *presence of health services (9.01)*, and next in rank, the most general attribute, *living environment (8.81)*.

Hospitable people (8.55), *living close to nature* (8.10), *ease of finding employment* (8.10) and an attractive *social environment* (8.08) also form part of their concerns. Although they show averages below 8 out of 10, some of the other attributes nevertheless seem to be sought after by Québec respondents, should they have to go and live in a different region. These attributes include the *work environment* (7.87), the *ease of joining groups* (7.50), *social and recreational activities* (7.44) and *favourable weather conditions* (7.44).

Rank	Importance of attribute	Average	Standard deviation
1	Presence of health services	9.01	1.50
2	Living environment	8.81	1.32
3	Hospitable people	8.55	1.61
4	Living close to nature	8.10	2.08
5	Ease of finding employment	8.10	2.77
6	Social environment	8.08	1.80
7	Work environment	7.87	2.44
8	Ease of joining groups	7.50	2.17
9	Social and recreational activities	7.44	2.06
10	Favourable weather conditions	7.44	2.25
11	Presence of educational institutions	7.38	2.79
12	Proximity to the major cities of Québec	6.75	2.39
13	Possibility of having an urban lifestyle	6.55	2.32
14	Cultural activities in the region	6.32	2.43

Figure 3. Importance of attributes of choice.

3.3. The Presence of Attributes of Choice

In this research, it was possible to measure the presence of attributes of choice for the three regions referred to above. From this viewpoint, Québec respondents were asked to evaluate the presence of 14 attributes, for the three selected regions, on a scale ranging from *not at all present in that region (1) to strong presence in that region (10)*.

Attribute	Abitibi-Témiscamingue Average	Outaouais Average	Saguenay-Lac-Saint-Jean Average
Living close to nature	8.49	7.42	8.30
Hospitable people	8.06	7.48	8.50
Ease of joining groups	7.06	6.98	7.43
Social and recreational activities	6.52	7.36	7.39
Attractive social environment	6.41	6.93	7.18
Educational institutions	6.28	7.64	7.01
Presence of health services	6.27	7.38	7.01
Attractive social environment	6.09	6.97	7.16
Favourable weather conditions	5.53	7.23	6.48
Attractive work environment	5.49	6.94	6.57
Cultural activities in the region	5.41	7.04	6.77
Ease of finding employment	5.37	7.11	6.09
Possibility of having an urban lifestyle	5.30	7.04	6.46
Proximity to the major cities of Québec	4.49	7.15	6.09
Average	6.20	7.19	7.03

Figure 4. Degree of presence of attributes in three regions of Québec.

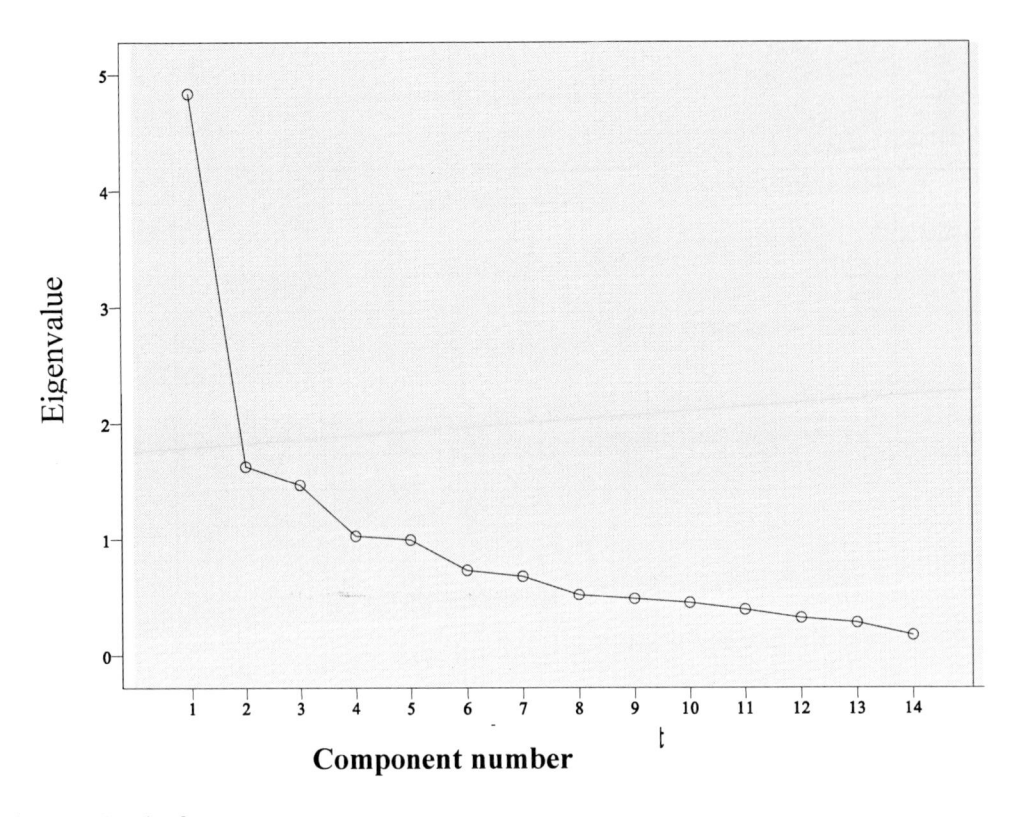

Figure 5. Graph of scree test.

The ratings of the presence of the attributes are presented in figure 4. The fact of *living close to nature* (8.49) and the presence of *hospitable people* (8.06) constitute the two main attributes of Abitibi-Témiscamingue. Next come *ease of joining groups* (7.06) and *social and recreational activities* (6.52).

Figure 4 also shows that for Abitibi-Témiscamingue, the average presence of attributes (6.20) is lower than that of Saguenay-Lac-Saint-Jean (7.03), a comparable region, and the difference is even greater when it is compared with Outaouais (7.19)

3.4. The Hidden Dimensions of Attributes of Choice

We then proceeded with a principle component analysis in order to identify the hidden dimensions of the attributes of choice. This calculation is of course based on the importance placed on the attributes, rather than on their degree of presence. According to Figure 5, the graph of eigenvalues (scree test) indicates that four components must be retained.

Figure 6 presents results that bring out four factors that take accounts of 64% of the total variance explained. The figure 6 shows that the *living environment* component explains 18.2% of the variance, followed by the components *career* (17.8%), *urban living* (14.9%) and *social life* (13.26%). This reveals that in the choice of a region in which to live, the *living environment* constitutes the main driver. It is composed of the importance placed on the possibility of meeting *hospitable people (0.75)*, of *living close to nature (0.67)*, of having a good *living environment (0.67)*, and the *presence of health services* (0.65).

Attribute	Factorial weighting	Component (factor)	% of the total variance explained
Hospitable people	.75	1. Living environment	18.2%
Living close to nature	.67		
Living environment	.67		
Presence of health services	.65		
Ease of finding employment	.92	2. Career	17.8%
Work environment	.90		
Presence of educational institutions	.63		
Proximity to the major cities of Québec	.81	3. Urban living	14.9%
Possibility of having an urban lifestyle	.69		
Favourable weather conditions	.53		
Cultural activities in the region	.73	4. Social life	13.2%
Social and recreational activities	.69		
Ease of joining groups	.51		
Cumulative %			64.1%

Extraction method: Principal component analysis. Varimax rotation with Kaiser normalization. The rotation converged in 14 iterations.

Figure 6. Clustering of attributes of choice under four factors.

3.5. Evaluating and Positioning the Regions

It is possible to obtain the citizens' evaluation of a region by multiplying the importance placed on a given attribute (figure 3) by its degree of presence (figure 4). When the product is divided by 10 (since the scale is from 1 to 10), the final result is consistent with the initial scale.

Calculating the positions involves calculating the average of the responses for each dimension, from the list of values for presence weighted according to the importance of attributes, as presented in figure 7. The coordinate of a region corresponds to the average of the presence values weighted according to the importance of the attributes that compose each dimension. For example, the dimension *living environment* for Abitibi-Témiscamingue is equal to:

(Hospitable people + living close to nature + living environment + presence of health services / 4 = (6.9 + 6.9 + 5.4 + 5.6)/4 = 6.2.

In addition to the regions' coordinates, there is an ideal point. This is defined using the average importance value of the attributes that compose the dimension, combined with a theoretical maximum value for their presence. In this case, there is no *a priori* reason for any attribute to have a desirable presence less than the maximum possible value, and therefore we have set this value at 10.

	Ideal point (theoretical)	Abitibi-Témiscamingue	Outaouais	Saguenay-Lac-St-Jean	Average
Living environment	8.6	6.2	6.3	6.7	6.4
Social life	7.1	4.5	5.1	5.1	4.9
Career	7.8	4.4	5.6	5.1	5.0
Urban living	6.9	3.5	4.9	4.4	4.3
Average		4.9	5.5	5.4	

Figure 7. Calculation of the positioning coordinates.

Figure 7 reveals that the region studied (AT) does not dominate over any one dimension. However, it is the *lifestyle* factor for which it obtains its best result (6.2). It is on this dimension and the next one that we base our positioning.

Graphic visualization of the placement of the regions on an artesian graph is shown in Figure 8. In this case, each region holds a specific position for each pair of components appearing in figure 7. For example, coordinates for the AT region are, for the pair *living environment* and *social life,* (6.2, 4.5). On the Figure 8, the dimension *living environment* extends on the X axis and the dimension *social life* on the Y one.

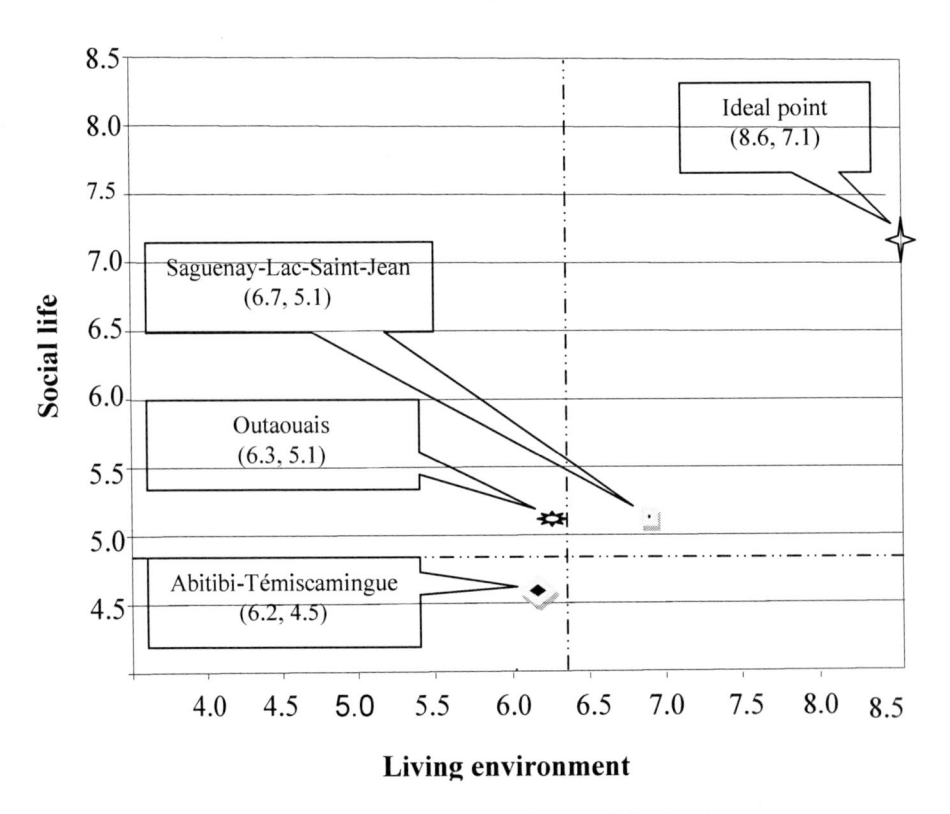

Figure 8. Position of the regions for the factors Social Life and Living Environment.

3.6. Strategic Positioning Options

The ideal point is of course the place where all regions would wish to be on a positioning table. There are four generic strategies that may be applied to improve the situation of any given point in relation to the ideal point:

1. Moving closer to the ideal point. To move closer to this point, it would be advantageous for AT to increase its score both on the *living environment* dimension, and on that of *social life*. It may do so by considering the constituent attributes of each dimension. It may either work to improve the importance of any of these attributes for which it is dominant, or work to improve the perception of the presence of that attribute.
2. Moving the ideal point closer to oneself. This strategy is similar to the above, since it is implemented by altering the importance of the attributes of choice of individuals to make them correspond more closely to one's own strengths.
3. Moving the other regions away from the ideal point. This strategy may be implemented based on two tactical approaches. The first involves minimizing the importance of the attributes for which the other regions are dominant. The second involves reducing the perception of the presence of these attributes for the individuals whom one aims to convince. This latter tactic, unfortunately, all too often leads to smear campaigns.
4. Launching a new product or opening a new perceptual category that corresponds in all ways to the ideal position. Since one obviously cannot launch a new region, the strategy would involve defining the region in a new way. This "new viewpoint" should be sufficiently different to press the targeted individuals to redefine the attributes of choice, their importance and how they rate their degree of presence. The result is that the individual then "opens" a new category of regions. The region involved is then the only one to be contained in that category.

The quantitative research finally showed that there is not evident strength dimension on which AT should base its future touristic development. Nonetheless, several arguments in favour of the first strategy (moving closer to the ideal point) were given in the qualitative research phase. The *living environment* dimension appears to be more promising than the others to define a niche area of excellence liable to attract people and retain them in the region.

CONCLUSION

During the preparatory discussions for the research, most of the stakeholders in the tourism sector of the region under study considered that the force of attraction of a northern region lay in the concept of a "pure" natural environment that is cold and wild; however our research has shown otherwise.

The qualitative research led to the realization that the northern wilderness presents something of a dichotomy. On the one hand, the northern lakes, mountains and forests do not have more elements of attraction than their counterparts further south. On the other hand, in order for the northern natural environment to reveal its true essence and beauty to visitors, they must expose themselves to risks connected with remoteness, the natural fierceness of the wildlife, or extreme weather conditions.

The quantitative phase brought out the attractive attributes of northern regions: the friendliness of people who give the region spirit, owing to their hospitality. What casual visitors as well as those who consider living in a northern environment seek is to live close to nature, with people who make it more attainable, and who give it a quality of tameness.

The two phases led us to recommend positioning strategy number one: moving closer to the ideal point. In order to achieve this, the region recently launched a communications campaign strengthening its image in terms of living environment and social life. The slogan "*à bras ouverts*" (with open arms) and the semiological symbols used show people who are active and hospitable, celebrating nature. A visit to the Tourisme Abitibi website (http://www.tourisme-abitibi-temiscamingue.org/activites-attraits/) illustrates this concept perfectly.

REFERENCES

Baribeau, Colette. 2009, "Analyse des données des entretiens de groupe", *Recherches qualitatives,* vol. 28(1), pp. 133-148.

Bearden, William O., Richard G. Netemeyer and Mary F. Mobley. 1993, *Handbook of Marketing Scales,* Publ. Sage Publications, London, United Kingdom, 353 pages.

Bonnes, Mirilia and Gianfranco Secchiaroli. 1995, *Environmental Psychology. A Psychosocial Introduction*, Thousands Oaks, London, 230 pages.

Bourgault, Normand. 2007, *Le lieu de production d'un produit agroalimentaire : un critère de choix*. Doctoral thesis. Université de Sherbrooke, 251 pages.

Bruner II, Gordon C., and Paul J. Hensel. 1994, *Marketing Scales Handbook*, American Marketing Association, Chicago, Illinois, 1315 pages.

Bruner II, Gordon C., Paul J. Hensel. 1998. *Marketing Scales Handbook*, Vol. 2, American Marketing Association, Chicago, Ill., 1045 pages.

Bruner II, Gordon C., Paul J. Hensel. 2001. *Marketing Scales Handbook*, Vol. 3, American Marketing Association, Chicago, Ill., 1045 pages.

Gagnon, André. 1993, *Les effets du pouvoir et de l'identité sociale sur la polarisation collective de la discrimination intergroupe.* Doctoral thesis, UQAM.

Hair, Joseph F. Jr., Rolph E. Anderson, Ronald L. Tathan and William C. Black. 1995, *Multivariate Data Analysis,* 4th ed., Prentice Hall, 745 pages.

Kline, Rex B., 1998, *Principles and practice of structural equation modeling*, Publ. Guilford Press, New York, 354 pages

Martin, Ingrid M. and Sevgin Eroglu. 1993, "Measuring a Multi-Dimensional Construct: Country Image", *Journal of Business Research*, Vol. 28, No. 3, pp. 191-210.

Parameswaran, Ravi and R. Mohan Pisharodi. 1994, "Facets of Country of Origin Image: An Empirical Assessment", *Journal of Advertising*, Vol. 23, No. 1, pp. 43-56.

Pettigrew, Denis, Saïd Zouiten and William Menvielle. 2002, *Le consommateur : acteur clé en marketing*, Coll. Marketing, Publ. SMG, Trois-Rivières, Québec, 469 pages.

Proshansky, Harold M., A. K. Fabian and R. Kaminoff. 1983, "Place-identity: Physical World Socialization of the Self", *Journal of Environmental Psychology*, Vol. 3, pp. 57-83.

Rabbie, J. M. and M. Horwitz, 1969, "Arousal of Ingroup-outgroup Bias by a Chance Win or Loss". *Journal of Personality and Social Psychology*, Vol. 13, pp. 269-277.

Sherif, Muzafer and Carolyn W. Sherif. 1964, *Reference groups: exploration into conformity and deviation of adolescents*, Harper and Row, New York, 370 pages. (Republished, 1972, H. Regnery.)

Sin, Leo Y. M., Gordon W. H. Cheung and Ruby Lee. 1999, "Methodology in Cross-Cultural Consumer Research: A Review and Critical Assessment", *Journal of International Consumer Marketing*, Vol. 11, No. 4, pp. 75-96.

Tajfel, Henri, C. Flament, M. G. Billig and R. P. Bundy. 1971, "Social Categorization and Intergroup Behaviour", *European Journal of Social Psychology*, Vol. 1, pp. 149-178.

Tajfel, Henri. 1981, *Human Groups and Social Categories*, Cambridge University Press, Cambridge, 369 pages.

In: Research Studies on Tourism and Environment
Editors: J. Mondejar-Jimenez et al.
ISBN: 978-1-61209-946-0
© 2012 Nova Science Publishers, Inc.

Chapter 5

VISITORS PROFILE OF A MEDITERRANEAN GREEN CORRIDOR: CASE STUDY OF THE PARC FLUVIAL DEL TÚRIA (VALENCIA, SPAIN)

María José Viñals[*], Zeina Halasa and Mireia Alonso-Monasterio
Universidad Politécnica de Valencia, Spain

ABSTRACT

The *Parc Fluvial del Túria*, Valencia (Spain) is a green corridor with a riverside shared-use 30km path along the towns of Vilamarxant, Benaguasil, Riba-roja de Túria, L'Eliana, Paterna, Manises, Quart de Poblet, Mislata and the city of Valencia. A multi-disciplinary team of researchers from the Universidad Politécnica de Valencia has carried out a survey between March and April 2009, before the official inauguration of the Park. This campaign has been done for monitoring the visitors' behaviors and attitudes, determining the visitor's profile of this corridor and unfolding people's thoughts and perceptions, in order to design the signage of the Park as well as the etiquette and the code of ethics. Basic descriptive statistics has been applied over the 1,824 visitor respondent forms. Additionally, attention has been given to the interpretation of the richness of the data obtained. The findings of this survey show that the prevailing visitors' profile is dominated by a slight majority of male visitors'(63.3%), age ranges between 36-50 years old, working in the service sector, and coming from the residential adjacent areas. Pedestrian activities with 82.7% (including walking, walking the dog and jogging) are the preferable activities developed in the site. Cycling activities (64.2%) are also very popular. Other activities such as observing nature, going for a picnic and horse riding were also practiced in a regular manner. The results show a high level of satisfaction related with the greenway setting and the most significant problem detected was the conflict between different users.

---

[*] Contact author: mvinals@cgf.upv.es

1. INTRODUCTION

The *Parc Fluvial del Túria* (PFT) is a green corridor[1] that connects the city of Valencia (eastern Spain) with 8 country upstream towns (Mislata, Quart de Poblet, Manises, Paterna, L'Eliana, Riba-roja de Túria, Benaguasil, Vilamarxant) along 30 km encompassing the Turia River Valley covering 250 ha of riparian forest and flood plains (Figure 1).

This Park has been promoted by the Jucar River Basin Agency (Confederación Hidrografica del Jucar, Ministerio de Medio Ambiente, Medio Rural y Marino) which has carried out a restoration project and furthermore, the valley has been adapted to suite a green corridor that offers optimum wild life conditions and facilitates the access to recreational uses for more than 1Milion people living near the river. Thus, several facilities have been constructed; among them a multi-shared use path has been built in order to provide access to the visitors. Two visitor centers (one near the entrance of the Park in Quart de Poblet, and the other at the end, in Vilamarxant) and signage facilities (directional and informational post signs) and two amenity areas help visitors with their stay in the Park.

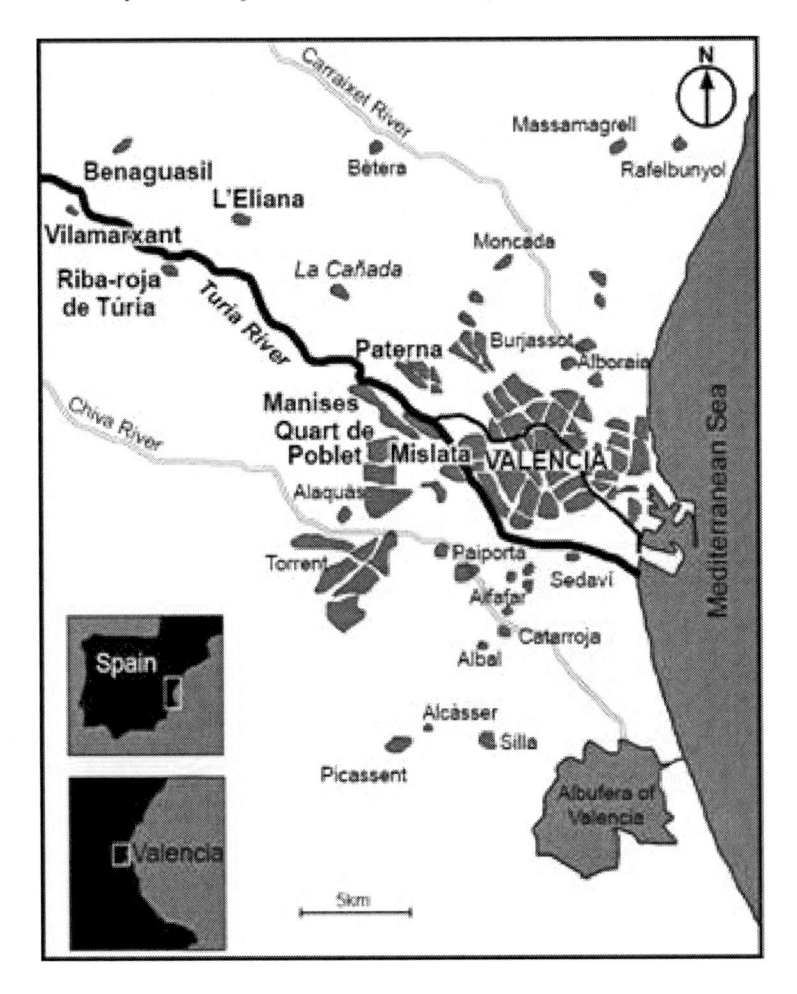

Figure 1. Location map of the *Parc Fluvial del Túria*. Source: Own elaboration.

---

[1] Following Fabos (1995), this is a recreational greenway.

A survey campaign has been conducted in order to know the profile of the current visitors of the Park and the Use Level in some crucial points of the Park. This kind of information is especially relevant in green corridors located in the urban-rural fringe as these areas tend to receive a wide spectrum of visitors each with different recreational styles. Thus, this sometimes implies the presence of conflicts among users as some authors have already outlined (Coles & Bussey, 2000; Galloway, 2002; Payne et al., 2002; Tinsley et al., 2002; Sasidharan et al., 2005; Sanesi & Chiarello, 2006; Yilmaz et al., 2007).

Obtaining data about the visitors in green areas is becoming an ongoing practice worldwide, as it provides the essential instruments for developing policies focused on improving the compatibility of the social and ecological uses.

This chapter outlines the results of this survey campaign conducted over the spring of 2009 as to give evidence, as commented above, about the visitors' profile, their characteristics and motivations to assess whether the provisions made in the Park do meet people's needs and expectations or not. To know the main characteristics of the visitors and the activities they usually practice in the Park has proved to be a very useful management tool especially for managers and urban planners as some studies might show for e.g. Parks' Visitor Activity Management Process (VAMP) developed by Graham et al. (1988) in Canada, which incorporated data about users, their characteristics and satisfactions with data about the natural environment.

Other authors as Ishiuchi et al. (2007) analyzes park management focused on user activity and seasonal changes. In addition, park use management has been obtaining researches attention this has been materialized in the first international conference on Monitoring and Management of Visitors Flows in Recreational and Protected Areas (2002) held in Vienna, Austria.

Many are the studies that address issues related with the characteristics of tourists, based usually on socio-economic, demographic, and geographic variables. However, less attention has been given to examine the differences in visitors behaviors, flow patterns in the site, psychological comfort of visitors, activities developed, etc. A good example of multi-variable studies is the one carried out by Schipperijn et al. (2009); these authors have defined the factors influencing the use of green space in Denmark.

As green corridors are complex open spaces that normally receive a wide-range of visitors each with different demands and specific needs, it is necessary to study them including these last variables. In this way, Furuseth (1991) carried out a survey about the socioeconomic, demographic and residential characteristics of greenway users on four trails in the Capital Area Greenway System in Raleigh, North Carolina (USA), and Briffett et al. (2004) conducted a study about green corridors and the quality of life in Singapore using on-site questionnaires and household surveys.

2. METHODOLOGY

Surveys, beyond presenting mere socio-demographic data, are a good method to study people's attitudes and values (García, 1998); this is the reason behind developing the surveys campaign: from one hand to obtain useful information about the visitors and from the other to consider the results in further management programs.

The sample size in the PFT includes a total of 1,824 interviews. Though this figure represents a higher number of visitors that we estimated in 5,500 as most of the surveyed people were part of a group and they filled in one survey per group. The surveys campaign was conducted during March, April and May of 2009, divided into 21 days and it was carried out before the official inauguration of the Park.

The aim of this campaign was to monitor the visitors' attitudes and behaviors with the purpose of determining the visitors' profile and to unfold people's thoughts and perceptions, in order to design the signage system of the Park, as well as, the etiquette and code of ethics. The main tool for managing visitors in the PFT is nowadays the signage system as this green corridor does not have a specific plan for managing visitors (Public Use Plan, Management Action Plan, etc.).

Some of the technical aspects that have been taken into consideration when undertaking the surveys were related with the selection of different types of days. Therefore, weekends, mid-week days, bank holidays and the Easter holidays have been surveyed. Other important aspects have been the time fringe (morning, midday and evening) and the weather conditions (sunny and cloudy days).

The surveys were face-to-face interviews held with park visitors and done by two pollsters. The survey included a total of 26 questions and did not exceed the maximum of 30 questions proposed for a survey that addresses the general public as Cadoche et al., (1998) suggests.

The survey was divided into three main sections: (1) Personal information (gender, age, place of residence, level of studies and working sector). These questions included closed-answers and were part of the introduction. (2) The second section involved the recreational activities developed in the Park. This section was very important as it held the bundle of important information related to the activities in the Park, thus it was placed in the middle of the survey (Cadoche, et al. 1998). The answers were closed-answers too. The activities registered in the Park, have been included in 3 categories: entertainment and leisure activities, educational-interpretative activities and adventure-sportive activities as Viñals & Bernabé (1999) defined. (3) The third section of the survey examined the feelings and opinions of the visitors about the Park. This section included both open and close answers and at the end of the survey people could make observations or comment what they wanted.

In addition to the survey, a Use Level study was done to evaluate how many people usually go to the Park in general terms.

3. RESULTS AND DISCUSSION

3.1. Visitors Profile

From the data gathered at the personal information section of the survey, we had been able to highlight a basic visitor's profile. The majority (63.3%) of the visitors are male, between 36-50 years old (36.9%) working in the service sector (46.0%) with a high school studies level (36.1%). Probably, and as we will see below, these characteristics determine what are the main activities developed in the Park. In fact, authors as Roovers et al. (2002)

wrote about how visitor characteristics are very important variables explaining recreational activities (Table 1).

It must be also emphasized that when pollsters where asked about issues related with security the majority answered that they didn't feel uneasy during day-hours. However, results indicate that only 5.3% of women who visit the site do it alone in comparison to 23.8% of men. These results show that significant interaction terms can be found between gender and the potential predictor variable. In fact, some surveys carried out in other parks have reported residents' feelings of insecurity associated with vandalism, and fear of crime in deserted places (Melbourne Parks, 1983; Grahn, 1985; Bixler & Floyd, 1997). From this we conclude that insecurity may play an important role in visitation pattern.

Most of visitors (84.17%) arrive from the nearby riverside towns. This percentage appears to be rather high because, as we can see in other studies, it is presumed that proximity to the site is a crucial factor influencing the visit frequency of urban forests (Lindhagen, 1996; Schipperijn et al., 2010). Thus, the smaller distance the higher frequency of use. Usually critical distances, if shown consideration for travel time, are between 0 and 3 km for pedestrian, between 0 and 10 km for bikers and less than 25 km for car transportation (Roggeman, 1982) although if a green space is well-known and well-liked, respondents are likely to underestimate the distance, if it is less-known and disliked, distance is likely to be overestimated (Scott et al., 2007).

3.2. Activities Done in the Park

In this part of the study we wanted to know what activities the users of the PFT indulge in. The pollsters were asked to mark the activity or activities which they practiced; usually they marked more than one answer.

Table 1. Visitors profile in the PFT

Gender	Frequency	%	Level of studies	Frequency	%
Male	1149	63.3	None	42	2.3
Female	665	36.7	Elementary	542	29.9
Age	Frequency	%	High School	655	36.1
<14	171	9.4	B.A	544	30.0
14-18	70	3.9	Work sector	Frequency	%
19-25	145	8.0	Student	264	14.6
26-35	456	25.1	Primary sector	19	1.0
36-50	669	36.9	Service sector	834	46.0
51-65	236	13.0	Industrial sector	355	19.6
>65	67	3.7	Retired	159	8.8
			Others	158	8.7

Table 2. Activities developed in the *Parc Fluvial del Túria*

Activities	Percentage (%)	Total (%)
Walking	57.5	
Walking the dog	13.1	
Jogging	12.1	82.7
Mountain biking	37.2	
Bicycling	27.0	64.2
Picnic	11.7	
Fruit gathering	6.7	
Fishing	6.4	
Horse riding	3.3	
Observing nature	2.9	
Photography	2.1	
Swimming	1.5	
Sunbath	0.7	
Reading	0.6	
Drawing	0.1	
Others	0.2	

Pedestrian activities are the most popular ones (82.7%), including in this group walking, walking the dog and jogging. In second place, are cycling activities with a (64.2%) of practitioners, including in this group mountain bikers and bicyclists (Table 2). This is not rare: as many authors conclude, walking is internationally the most important activity in forest recreation (Lindhagen, 1996; Guyer & Polland, 1997; Roovers et al., 2002) and within the PFT there is a great area where visitors can enjoy the landscape offered by the riparian forest. Therefore, walkers who want to spend a nice day out totalize 64.5% (walking 57.4% and walking dog 13.1%); this kind of users can be considered as leisure and entertainment practitioners. Another reason for this is based on the proximity of the Park, many of the residents have direct access to the Park.

Mountain bikers and bicyclers have stated that the reason behind their high affluence resides on the suitable characteristics of the trail: the length and adequacy of the greenway, vehicular traffic free and paving materials.

A specific consideration must be given to the high use level of mountain bikers (37.2%) and joggers (12.1%). These special sport practitioners, are considered to belong to the 'more-more' profile described by (Corbett, 2006), as people who work for long hours during the day and who also tend to be more active in their leisure time. They might find their work boring and dull, and long to escape from everyday routine, believing that they can find excitement in their outdoor activities. This concept is very close to vigorexy, defined as the addiction to exercise (Pope et al., 2001). We can find interaction terms between this variable and gender, because this profile is also endorsed by the majority of male visitors (63.3%) who normally are much more related to vigorexy than women.

It is also worth to mention that the PFT receives horse riders. Though they only represent 3.3% of the visitors, they are quite visible in the greenway. Thus, they represent an important group to be taken into account when elaborating the etiquette code as we will see below.

Many interaction terms were detected in the visitors' answers in relation with the practice of more than one activity. Statistically it is very difficult to provide accurate conclusions about this because many combinations of variables have been detected. For example, going for a picnic (11.7%) and observing nature (2.9%) are suspected to be connected to one another, as most of the users included both answers.

The frequency of visits is almost equal between those who first come to the site and those who come once per month with (17.9% and 17.5%) respectively. While visiting the site 2-4 times per week is highest with 52.5%. Once more this last result could be related with those 'more-more' visitors because they come to the Park several times per week. On the other hand this latter data could also indicate that the majority of the visitors are from nearby towns.

Results in relation with the time spent in the site vary depending on the activity performed. Pedestrians required less time in the park (1-2 hours), while cyclers remained in the greenway between 2 to 3 hours. As for the more specialized visitors who do horse riding, fishing, or a picnic stay between 3 to 4 hours.

The peak hours in which the site receives most of the visitors is between 10:00-12:00 (40.8%) and consecutively from 12:00-14:00 (26.1%). The results indicate that during weekends and bank holidays, visitors tend to spend the morning there. Peak hours in mid-week days have not been found as the site receives the majority of visitors in weekends, holidays and bank holidays and because the greenway has not been used by commuters until the moment. Coming to the Park with friends indicates to be the most popular trend (34.1%) in relation with the visitor group variable. Visiting the site with the company of family (27.8%) seems to be equally important especially in weekends and holidays. Visiting the site alone was mostly selected by those who use the site to exercise.

The accumulation of visitors in a short interval (2 hours per day) could promote a sense of stress or overcrowding but, as we will see below and due to the typology of the site and the visitors' profile, this does not happen.

4. PERCEPTIONS OF THE VISITORS

In this part of the survey the pollsters were asked about their feelings and opinions regarding the Park. They were asked to answer four open questions about what they like most in the Park, what they disliked and whether they perceived a sense of overcrowding or not.

The analyses of the results have showed that visitors choose the 'Natural Setting' (76.9%) as the predominant answer. This is not a surprise given that the Park bestows feelings of relaxation and calmness. As a matter of fact, several studies all over the world highlighted the importance for people to access green areas, especially for those living in urban areas. Contemporary research on the use of urban parks and forests, for example, verifies beliefs about stress-reduction benefits and mental health (Hartig et al., 1991; Conway, 2000).

The second most popular option is 'adequacy and length of the trail'. As we commented above, this is probably due to the high level of cyclists in the Park, where they find a vehicular traffic free greenway immersed in a natural setting.

We can state that 87% of all respondents felt that their overall enjoyment of the visit and their experience was very good and they manifested the intention to come back again.

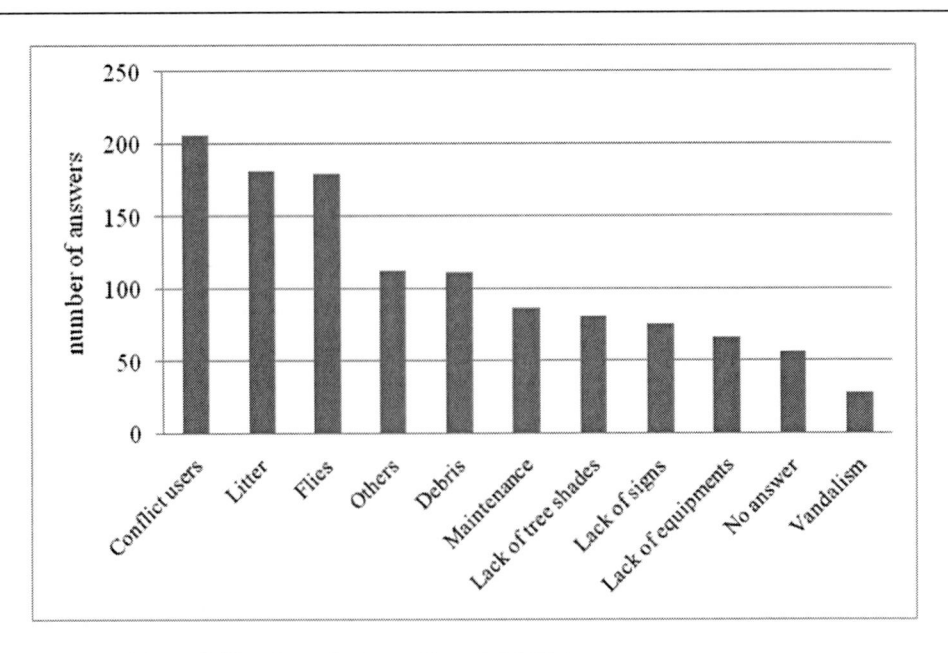

Figure 2. What visitors dislike about the *Parc Fluvial del Túria*.

Contrary to this, visitors perceive some conflicts between users due to the existence of different groups concentrated in the same spatial and temporal areas in the greenway. These kinds of conflict have been interpreted by some authors as Ryan (1993) as an indicator of the greenway's popularity. In order to reduce this conflict, etiquette and code of ethics were necessary assets, a that have been especially designed for the Park to minimize the problems by showing the expected attitude and behaviour from the different park users.

The presence of litter was another negative aspect outlined by respondents (Figure 2). Visitors between 36-50 years old were the most worried about litter (37.6%), followed by visitors between 26-35 years old (23.8%). Environmental education and awareness campaigns involving local people pose as necessary actions to be undertaken in this matter.

Flies are another question that worries visitors but we must take into account that the PFT is located in a Protected Area so fumigation with chemical products is not always a good solution.

In relation to other matters (debris, luck of facilities, etc.), we must recall that surveys were done before the official inauguration, when the greenway and other facilities were under construction and these issues, for obvious reasons, are not taken into account in this assessment.

Regarding the sense of overcrowding, (51.2%) of visitors think that the Park is somehow crowded but they do not mind. This feeling is also an important element in the final evaluation of satisfaction since the Psychological Comfort carries a lot of weight in determining the quality of the visitors experience and how they perceive the site (Viñals et al., 2007).

Probably, this linear green space implies that people meet other people through the trail but they do not feel stress or overcrowding. We must also consider that most of the visitors who meet during peak hours (they normally visit the Park accompanied by family or friends and they spend the weekend in the Park) do not mind the overcrowding of the site; otherwise they usually like sharing the place with other families and in a familiar environment (Morant & Viñals, 2008).

CONCLUSION

We could conclude that the main visitors' profile of the PFT is: man, between 36-50 years old with a high school level of studies and working in the service sector. This kind of visitors probably fit in the 'more-more' profile seeing that most of them use the Park as a sport ground.

The rest of visitors tend to visit the Park on foot, on weekends and bank holidays, usually accompanied by families and practice activities like going on picnic and enjoying the natural setting. These visitors are identified as 'standard profile visitors' and they are accustomed to share the area with other families so they do not feel overcrowding.

The demand on natural areas, especially in urban environments is an increasing phenomenon, given that they provide the adequate setting for relaxation and practicing sports. The typology of this open green space seems to be a determinant factor that would inform about the activities developed in the site and thus the prevailing visitors' profile.

Visitors' management seems to be necessary at those parks where several users' groups happen to meet. In this sense, the code of ethics and the etiquette code represent the minimal requirement to achieve harmony among users.

REFERENCES

Bixler, R.D. & Floyd, M. (1997). Nature is Scary, Disgusting and Uncomfortable. *Environment and Behavior*, v.29, No. 4, pp.443-467.

Briffett, C., Sodhi, N., Yuen, B. & Kong, L. (2004). Green corridors and the quality of urban life in Singapore. Singapor: In Shaw et al. eds. Proceedings 4th International urban wildlife symposium. pp.56-63.

Cadoche, L., Stegmayer, G., Burioni, J.P. & Bernardez, M. (1998, March 12). Material del Seminario de Encuestas en Educación. Universidad Nacional del Litoral y Universidad Tecnológica Nacional. Retrieved March 20, 2010, from http://www.unl.edu.ar/fave/sei/encuestas/index.html

Chiesura, A. (2004). The role of urban parks for the sustainable city. *Landscape and Urban Planning*. No. 68, pp.129-138.

Coles, R.W., Bussey, S.C. (2000). Urban forest landscapes in the UK – progressing the social agenda. *Landscape and Urban Planning* No.52, pp.181–188.

Conway, H. (2000). Parks and people: the social functions. In: Woudstra, J., Fieldhouse, K. (Eds.), *The Regeneration of Public Parks* (pp. 9-20). London: E& FN Spon.

Corbett, J. B. (2006). *Communicating Nature: How we create and understand environmental messages*. Island Press.

Fabos, J. Gy. 1995. Introduction and Overview: the greenway movement, uses and potentials of greenways. *Landscape and Urbanplanning*, No.33, pp: 1-13.

Furuseth, O.J. & Altman, R.E. (1991). Who's on the greenway: socioeconomic, demographic, and locational characteristics of greenway users. *Environmental Management*, v.15, No.3, pp. 329–336.

Galloway, G. (2002). Psychographic segmentation of park visitor markets: evidence for the utility of sensation seeking. *Tourism Management*, No. 23, pp.581–596.

García, P.A. (1998). *Elementos del Método Estadístico*. (México, Universidad Nacional Autónoma de México).

Graham, R., Nilsen, P. & Payne, R. J. (1988). Visitor Management in Canadian National Parks. *Tourism Management*, v.9, No.1, pp. 44-61.

Grahn, P. (1985). *Man's Needs for Urban Parks, Greenery and Recreation*. Institute for Landscape Planning. (Alnarp Swedish Agricultural University).

Guyer, G. & Polland, J. (1997). Cruise visitor impression of the environment of the Shannon-Erne Waterways system. *Environmental Management*, v.51, pp.199-215.

Hartig, T., Mang, M. & Evans, G.(1991). Restorative effects of natural environments experiences. *Environment and Behavior*, No.23, pp. 3–26.

Ishiuchi, T., Kuwahara, Y. & Koyanagi, T. (2007). A New Proposal for Park Management focused on User Activity and Seasonal Changes. *Papers on Environment*. Information Science, No.21, pp.153-158.

Lindhagen, A. (1996). *Forests Recreation in Sweden. Four Case Studies using Quantitative and Qualitative Methods*. Dissertation, Swedish University of Agricultural Science.

Melbourne Parks. (1983). *A Survey of the Use of Selected Sites*. Melbourne and Metropolitan Board of Works, Ministry of Planning and Environment, Melbourne.

Morant, M. & Viñals, M.J. (2008). La Capacidad de Carga Recreativa en la gestión de los visitantes. El Caso del Parque Natural del Carrascal de la Font Roja (Alicante, España). *Revista de Análisis Turístico*, No.5, pp.66-74.

Payne, L.L., Mowen, A.J. & Orsega-Smith, E. (2002). An examination of park preferences and behaviours among urban residents: the role of residential location, race, and age. *Leisure Sciences*, No.24, pp.181–198.

Pope, H.G., Phillips, K.A. & Olivardia, R. (2001). *The Adonis Complex: the secret crisis of male body obsession*. New York: Bargain Books..

Roggeman, G., (1982). *Bepalingen van de recreatieve van het bos te Lembeke-Waarschoot*. Thesis RUGent.

Ryan, K. L. (ed.) (1993). *Trails for the Twenty-First Century planning, design, and management manual for multi-use trails. Rails to Trails Conservancy*. Washington: Island Press.

Roovers, P., Hermy, M. & Gulink, H. (2002). Visitor profile, perceptions and expectations in Forests from a gradient of increasing urbanization in central Belgium. *Landscape and Urban Planning*, No. 59, pp. 129-145.

Sanesi, G. & Chiarello, F. (2006). Residents and urban green spaces: The case of Bari. *Urban Forestry & Urban Greening*, No.4, pp.125-134.

Schipperijn, J., Ekholm, O., Stigsdotter, U.K., Toftager, M., Bentsen, P., Kamper-Jǿrgensen, F. & Randrup, T.B. (2010). Factors influencing the use of green space: Results from a Danish national representative survey. *Landscape and Urban Planning*, No. 95, pp. 130-137.

Scott, M.M., Evenson, K.R., Cohen, D.A., & Cox, C.E. (2007). Comparing perceived and objectively measured access to recreational facilities as predictors of physical activity in adolescent girls. *Journal of Urban Health*, No. 84, pp.346–359.

Tinsley, H. E. A., Tinsley, D. J., & Croskeys, C. E. (2002). Park usage, social milieu, and psychological benefits of park use reported by older urban park users from four ethnic groups. *Leisure Sciences*, No.24, pp.199-218.

Yilmaz, S., Zengin, M., Demircioglu & Yildiz, N. (2007). Determination of user profile at city parks: a sample from Turkey. *Building and Environment*, No. 42, pp.2325–2332.

Viñals, M.J. & Bernabé, A. (1999). Los espacios naturales y rurales. Nuevos escenarios del turismo sostenible. In: M.J. Viñals & A.Bernabé (Eds.), *Turismo en espacios naturales y rurales* (pp. 13-34). Valencia: Universitat Politècnica de València.

Viñals, M.J., Morant, M., Hernández, C., Ferrer, C., Quintana, R. D., Maravall, N., Cabrelles, G. Ramis, J. & Bachiller, C. (2004). Albufera de València (Spain): Measuring carrying capacity in a fragile ecosystem. In: *World Tourism Organization, Indicators of sustainable development for tourism destinations: A Guidebook* (pp.330-337). Madrid: WTO.

In: Research Studies on Tourism and Environment
Editors: J. Mondejar-Jimenez et al.
ISBN: 978-1-61209-946-0
© 2012 Nova Science Publishers, Inc.

Chapter 6

SEGMENTATION OF TOURISTS AND DAY-TRIPPERS IN CASTILLA-LA MANCHA (SPAIN): MAIN DIFFERENCES AND SIMILARITIES[*]

Águeda Esteban-Talaya, Carlota Lorenzo-Romero[1] and María-del-Carmen Alarcón-del-Amo

University of Castilla-La Mancha, Spain

ABSTRACT

According to tourist official data, the region of Castilla-La Mancha in Spain ranks in the fifth position as favourite destination for Spanish nationals tourists. Due to the importance of tourist sector in Spain, and specifically in this autonomous region, this chapter carries out a comparative analysis of the different groups of tourists who visited Castilla-La Mancha during 2008, taking into consideration the period of travel (Easter Holidays vs. weekends). In order to achieve this objective, this research has used a cluster analysis which has allowed us to classify tourists and day-trippers in different groups according to two variables: kind of tourism (cultural, rural, nature, etc.), and activities carried out by the tourists (rest, gastronomy, nightlife, etc.). The association between the group to which a tourist belongs and these two variables have also been taken into consideration in this analysis, as well as other characteristics like age, marital status, and autonomous region of origin, which have been analyzed separately for each of the two types of visitors, tourists and day-trippers, in the wider context of the separation between Easter holiday travel and weekend journey. The results of this research show the differences and similarities between the visitors analyzed, which should be of interest for the tourist industry, especially when planning marketing strategies adapted to each of the tourist profiles described here.

---

[*] This chapter is part of a longer research project on "Construction and Implementation of a System for Tourist Research in Castilla-La Mancha (SITdCLM)", funded by the Consejería de Cultura y Turismo de la Junta de Comunidades de Castilla-La Mancha, Spain, 2008-2010.
[1] Contact author: Carlota.Lorenzo@uclm.es

Keywords: Tourist and day-tripper, segmentation, cluster analysis, differences and similarities

1. INTRODUCTION

The tourist sector is one of the most dynamic in the Spanish economic system. Improving quality in the tourist industry has several positive consequences on the experience perception, travel habits, and services demand of tourists and day-trippers (visitors who do not stay overnight). The region of Castilla-La Mancha in Spain ranks in the fifth position as favourite destination for Spanish nationals tourists. Specifically, Castilla-La Mancha has extraordinary resources. This Region has two cities declared World Heritage Sites by UNESCO: Toledo and Cuenca; a high amount of historical sites and monuments; and two National Parks, in addition to natural parks and nature reserves. Therefore, in Castilla-La Mancha the importance of nature is vital. Moreover, in the urban environment are improving the built and natural ecological systems, which act as main supporting features to achieve and ensure dynamism and sustainability in the cities (Aminzadeh and Khansefid, 2010). On the other hand, environmental protection by companies is increasing. Castilla-La Mancha is among the six first regions from Spain whose expenditure on environmental protection double in recent years (Vargas-Vargas *et al.*, 2010). Since rural and nature tourism has experienced high growth over the past 20 years (Ferrari *et al.*, 2010), Castilla-La Mancha is making a great effort to preserve its environment.

The frequency and characteristics of tourists' and day-trippers' travel in Castilla-La Mancha are subject to modification depending on socioeconomic, cultural and lifestyle factors which determine the preference for some types of travel over others and delineate the behaviour of individuals travelling to one specific tourist destination in relation to the specific profile of visitors.

Handling updated data on tourists' and day-trippers' profiles and having information about the main differences and similarities existent between both tourist segments is therefore essential for the Spanish tourist sector in general, as well as for the regional tourist sector in Castilla-La Mancha. This chapter thus carries out a comparative analysis of the different groups of tourists who visited Castilla-La Mancha during 2008, taking into consideration the period of travel (Easter Holidays vs. weekends). In order to achieve this objective, this research has used a cluster analysis which has allowed us to classify tourists and day-trippers in different groups according to two variables: kind of tourism (cultural, rural, nature, etc.), and activities carried out by the tourists (rest, gastronomy, nightlife, etc.). The association between the group to which a tourist belongs and these two variables have also been taken into consideration in this analysis, as well as other characteristics like age, marital status, and autonomous region of origin, which have been analyzed separately for each of the two types of travellers, tourists and day-trippers, in the wider context of the separation between Easter holiday travel and weekend journey. The results of this research show the differences and similarities between the travellers analyzed, which should be of interest for the tourist industry, specially when planning marketing strategies adapted to each of the tourist profiles described here.

2. CURRENT SITUATION IN THE TOURIST SECTOR. TOURIST MARKET SEGMENTATION

Improved quality in the tourist services is devised to meet the growing national and international demand, as well as to adapt to the changes in tourists' and day-trippers' habits.

Market segmentation allows identifying existent variables and to divide the market in different groups in relation to those variables, in order to different implement marketing strategies for each group (Kotler *et al.*, 2006). The main objectives of the segmentation of the tourist market are (1) to improve efficiency in resource management, (2) to personalize services more adequately and (3) to attract and retain the most profitable tourist segments. Due to the fact that market segmentation precedes marketing strategies, several studies have used different variables to develop a market segmentation focusing on the tourist sector (Jang *et al.*, 2004; Kim and Agrusa, 2208). To be more precise, there have been two main sets of variables used in tourist market segmentation (Blázquez *et al.*, 2010): Socio-demographic, geographic or psychographic variables (e.g. Cha *et al.*, 1995; Juaneda and Sastre, 1999; Anderson *et al.*, 2000; González and Esteban, 2000; Jang Morrison and O'Leary, 2002; Sirakaya *et al.*, 2003; Sung, 2004) and variables related to the characteristics of travel, such as travel experience, activities performed in the destination, travel duration, information sources used, etc. (e.g. Hsieh *et al.*, 1992; Lang *et al.*, 1993; Dellaert *et al.*, 1995; Stemerding *et al.*, 1996; Formica and Uysal, 1998; Choi and Tsang, 1999; Moscardo *et al.*, 2000; Neal, 2003; Jang *et al.*, 2004; Albadalejo and Díaz, 2005; Tisotsou and Vasioti, 2006; Huang and Sariöllü, 2007). Other studies (e.g. Frochot and Morrison, 2000; Jang *et al.*, 2002; Frochot, 2005; Tisotsou, 2006; Andriotis *et al.*, 2008) have used cluster analysis to identify homogeneous segments or profiles of tourists which would help tourist managers finding new business opportunities.

Apart from these variables, several other factors also have a direct influence on the frequency and characteristics of travel: social and economic, political and cultural, technological and environmental factors and vary every year and determine the main characteristics of the tourists' and day-trippers' visits to regional, national or European destinations.

2.1. Tourist Profiles

In the European context, according to the Report España en Europa: Comportamiento Turístico de los Residentes en la Unión Europea [Spain in Europe: Tourist Behaviour of European Union residents] (2005) carried out by the Instituto de Estudios Turísticos (IET) [Institute of Tourist Studies, ITS] using data coming from the EUROSTAT and the Movimientos Turísticos de los Españoles [Spanish Tourist Movements] (FAMILITUR) statistics for Spain, one third of Spanish residents have travelled abroad, and more than half of them have chosen a EU country as foreign destination. However, Spain is the country with more domestic travel in the EU-27. These domestic journeys are normally made by car, and a fourth of the long journeys made in Spain take place in August, as it happens in Italy. Nearly 75 percent of Spanish tourists use individual accommodation, while other countries like Holland prefer collective accommodation. Spain is also the fifth country in Europe by its offer of accommodation establishments, and the fourth in hotel beds. In 2005, Spain was the country with the highest tourist income. German residents rate first in tourist expenses abroad.

In the Spanish context, the most significant reports on tourist habits in Spain are *Hábitos Turísticos de los Residentes en España [Habits of Tourists Residing in Spain], Movimientos Turísticos de los Españoles [Spanish Tourist Movements] 2007 (FAMILITUR), Movimientos Turísticos en Fronteras [Tourist Movements in Borders] 2007 (FRONTUR),* all three carried out by the ITS, and *Encuesta de Ocupación Hotelera [Survey on Hotel Occupation] 2007 (EOH),* obtained by the Instituto Nacional de Estadística (INE) [National Institute of Statistics, NIS]. According to these surveys, most Spanish have been tourists inside the country and half of them have travelled abroad. People between 25 and 29, with high education, living in a couple and with children, and residing in a tourist destination travel most often. Spaniards travel mainly during summertime and less during Christmas, and they choose hotels to stay. One third of journeys are made during the weekend, and this is characteristic of people living in Madrid. One tenth of tourism is a consequence of business travel, half of which happens abroad. More than one third of households have some kind of affective relationship with another region in Spain, most of them because their families come originally from these regions. The countries with which the Spaniards have more affective connections are France, Italy and the United Kingdom.

In the regional context, within the specific context of Castilla-La Mancha, there is a significant lack of reports with information about regional tourist behaviour, especially if we consider tourist destination. There are however other regional studies carried out in Galicia, the Basque Country and Asturias, Granada and in the town of Toledo, among others. The report *Hábitos Turísticos de los Residentes en España [Tourist Habits of Spanish Residents]* carried out by the ITS, provides information about regional destinations, and concludes that Castilla-La Mancha is the fifth Spanish region as destination for domestic tourist travel. Contrarily, Castilla-La Mancha stands among the four regions with lowest travel frequencies, because 50% of its population never travels. Andalusia and the Valencia autonomous regions are the region's favourite tourist destinations, while tourists in Castilla-La Mancha normally come from Madrid and Catalonia. If we consider rural tourism, Castilla-La Mancha is the autonomous regions with the highest growth in 2007.

2.2. Day-tripper Profile

Day-tripping[2] is determined by the lack of overnight stay in destination, and is particularly common in Castilla-La Mancha.

In the national context, the most significant surveys on the habits of Spanish day-trippers are *Hábitos Turísticos de los Residentes en España [Tourist Habits of Spanish Residents], Movimientos Turísticos de los Españoles [Tourist Movements in Spain] 2007 (FAMILITUR) and Movimientos Turísticos en Fronteras [Tourist Movements in Borders] 2007 (FRONTUR),* all carried out by the ITS. They show that during 2007 Spanish residents made a total of 212.2 million recreational day trips, and also that approximately half of Spanish

---

[2] According to the survey *Movimientos Turísticos de los Españoles [Tourist Movements in Spain], FAMILITUR 2007,* day-tripping is: "One-day travel without overnight stay."

residents make some kind of day-trip, normally in a family group and in all the periods of the year. Day-trippers are usually between 30 and 39 or under 14, they have high education and employment (they tend to be employers themselves), and their households consist of couples with children. They also possess a second home and have some relatives in a town different form where they live. The town where they reside has commonly between 100,000 and 499,999 inhabitants. The industrial sector is the occupation that provides more day-trippers. If we consider day-tripping coming from abroad, Catalonia, Castilla y León, Galicia and the Basque Country received the highest number of foreign day-trippers. More than half of tourists visiting these regions were day-trippers. These regions are close to the border, which facilitates day-tripping.

In the regional context, as it happens with the analysis of tourist profiles, there is an important lack of surveys and reports analyzing the behaviour of day-trippers in regional contexts. Using the information about regional differences available in the reports mentioned above, it is noticeable that autonomous regions which highest levels of day-tripping are Andalusia, Catalonia, Valencia and Madrid. Most day-trips done in Spain in 2007 had the same autonomous region as destination (intra-regional day-tripping).

3. METHODOLOGY

A personal survey has been carried out to question Castilla-La Mancha tourists in two different periods: Easter and weekends. The selection of the sample population has used a number of statistical sources that have allowed us to obtain a statistically significant sample in order to increase the representation of the whole sample population. To be more precise: (1) information obtained by the Instituto de Promoción Turística de Castilla-La Mancha (Castilla-La Mancha Institute for Tourist Promotion, ITP) from the specialists in charge of tourist promotion in each province, from Tourist Information Offices, and the Internet; (2) Infotur 2008 report carried out by the ITS; (3) information provided by the Tourist Information Offices of each of Castilla-La Mancha five provinces; (4) the Journal *Castilla-La Mancha Tierra de Don Quijote 2007*; (5) other statistics coming from the Statistical National Institute with information about number of hotel nights, rural accommodation, hotel beds, restaurants, etc., including the Encuesta de Ocupación Hotelera [*Survey on Hotel Occupation*], the Encuesta de Ocupación de Turismo Rural [*Survey on Rural Tourism Occupation*], the Encuesta de Ocupación de Apartamentos [*Survey on Appartment Occupation*], the Registro de establecimientos turísticos [Registrar of Tourist Establishments] provided by the General Tourism and Handcrafts Administration of the Castilla-La Mancha regional government, and the Registro de la Junta de Comunidades de Castilla-La Mancha [General Registrar of the Castilla-La Mancha Government]. The survey was carried out by survey takers and through questionnaires distributed by Tourist Information Offices in the five provinces of Castilla-La Mancha, using as sample size 49 villages and cities distributed around this region. Table 1 displays the technical information of the survey.

Table 1. Technical Specifications of Survey

Periods	1st Period (Easter)	2nd Period (Weekends)
Population	Castilla-La Mancha Residents and non-residents, 2007 (*FAMILITUR*) 11,367,767 tourists 7,726,428 day-trippers Residing abroad, 2007 (*FRONTUR*) 219,000 tourists 17 day-trippers **Total Sample of Castilla-La Mancha destination** **11,586,767 tourists** **7,726,445 day-trippers**	
Sample Size	**Total Sample Population:** **2,222 tourists** **801 day-trippers**	
	1,974 (Survey takers: 1,750 enc.) (Tourist Info Offices: 224 enc.) 602 (Survey takers: 522 enc.) (Tourist Info Offices: 80 enc.)	248 tourists 199 day-trippers
Reliability Level	95.5% (K=2)	
Sample Error C-LM	Castilla-La Mancha (2,222 **tourists**): ±2,12% Castilla-La Mancha (801 **day-trippers**): ±3,53%	

Table 1. (Continued)

Tecnique	Personal Survey carried out by survey takers Voluntary Survey in Tourist Info Offices	Personal Survey
Field Work	Easter 2008 (20-23 March, 2008)	Weekends in May and first half of June, 2008
Software	SPSS 15.0®	

For this purpose, this research has used a conglomerate or cluster analysis which has allowed us to classify tourists in groups that are internally homogeneous but different from each other according to the value of the variables considered.

Once the main groups have been identified, the profile of tourists and day-trippers has been described using variables such as age, marital status, or autonomous region of residence, because other variables like sex, level of education, employment status and monthly income have not provided relevant differences between the two groups. In order to determine the relationship between the group and each of these variables, a *chi-squared independence test* has been carried out.

4. SEGMENTATION OF TOURISTS AND DAY-TRIPPERS IN CASTILLA-LA MANCHA

In this section there is a detailed analysis of the types of tourists and day-trippers in Castilla-La Mancha during 2008.

4.1. Tourist Types in Castilla-La Mancha

In order to analyze the types of tourist in Castilla-La Mancha, this chapter has segmented tourists in two groups, according to their differences: Easter and weekends.

4.1.1. Easter Tourists in Castilla-La Mancha

Easter tourists have been classified further in four clusters of groups according to the kind of tourism carried out and the activities they performed (see Table 1). These four groups have been called as follows: Group 1. Cultural and Passive Tourists; Group 2. Cultural and Gastronomic Tourists; Group 3. Natural-Cultural and Gastronomic-day-tripping Tourists; and Group 4. Multi-product and Hyperactive Tourists. Besides, as Table 2 shows, there is a connection[3] among the group a tourist belongs to and the variables kind of tourism, activities, age, marital status and autonomous region of origin, so that it is possible to affirm that the distribution of those variables is different in each group.

---

[3] In order to determine whether there is a relationship between the group a tourist belongs to and each of the variables considered, a *chi-squared independence test* has been carried out. If the significance level is under 0.05, for a reliability level of 95%, the null hypothesis can be rejected, and it is therefore possible to conclude that there is a connection among the contrasted variables.

Taking into account the information in Table 2, it is possible to describe the following characteristics for each of the tour groups identified:

Table 2. Segmentation of Easter Tourists

Variable	Scale	Group 1 (36.0%)	Group 2 (19.1%)	Group 3 (32.9%)	Group 4 (12.0%)	X^2 Value	Sig.
Kind of tourism	Cultural	71.4%	91.0%	71.1%	87.7%	87.527	0.000
	Rural	21.1%	26.5%	55.9%	69.4%	288.458	0.000
	Nature	9.6%	4.0%	96.9%	88.1%	1,489.916	0.000
	Health	3.7%	8.2%	4.9%	28.5%	162.454	0.000
	Education	1.5%	1.6%	1.2%	6.4%	25.677	0.000
	Business	2.4%	3.2%	1.1%	11.5%	63.133	0.000
	Oenology	1.3%	10.6%	4.8%	21.3%	128.263	0.000
Type of activity	Relax/rest	67.9%	80.7%	79.3%	88.5%	60.937	0.000
	Gastronomy	0.0%	99.7%	67.1%	91.1%	1,325.435	0.000
	Nightlife	14.1%	14.8%	18.2%	69.4%	348.231	0.000
	Sports	5.3%	4.8%	10.6%	56.6%	455.113	0.000
	Shows and Entertainment	10.3%	21.2%	8.5%	49.4%	241.220	0.000
	Shopping	5.1%	15.9%	4.2%	71.5%	712.552	0.000
	Trips	21.9%	29.9%	68.5%	82.1%	462.256	0.000
	Visit friends and relatives	30.5%	24.1%	19.9%	52.3%	94.277	0.000
Age	<25	19.3%	9.3%	10.9%	31.6%	94.764	0.000
	25-34	22.3%	25.9%	27.3%	27.8%		
	35-44	24.9%	26.7%	29.9%	17.9%		
	45-54	19.5%	21.4%	18.3%	11.1%		
	55-64	8.8%	11.4%	7.4%	4.7%		
	>65	5.3%	5.3%	6.2%	6.8%		
Marital Status	Single	29.1%	22.0%	27,.8%	42.7%	53.936	0.000
	Living with a partner	13.9%	14.8%	14.1%	16.7%		
	Married	52.3%	59.3%	53.1%	32.1%		
	Divorced	2.7%	2.1%	3.1%	3.8%		
	Widow(er)	2.0%	1.9%	1.9%	4.7%		
Autonomous Region of origin	Andalusia	7.8%	5.9%	5.3%	7.7%	72.382	0.026
	Aragón	3.0%	6.1%	4.2%	3.8%		
	Asturias	3.0%	1.6%	2.0%	1.3%		
	Valencia Region	16.0%	17.3%	17.6%	12.4%		
	Cantabria	0.4%	0.8%	0.9%	0.0%		
	Castilla y León	6.3%	6.1%	5.5%	7.3%		
	Castilla-La Mancha	17.1%	14.9%	13.2%	24.4%		
	Catalonia	7.1%	8.8%	11.1%	8.5%		
	Extremadura	2.0%	3.2%	3.1%	2.6%		
	Galicia	1.0%	1.6%	1.7%	1.7%		
	Madrid	23.3%	20.2%	20.4%	18.4%		
	Murcia	3.0%	2.7%	4.7%	6.4%		
	Navarra	0.9%	0.8%	0.9%	0.4%		
	Basque Country	3.9%	2.4%	4.4%	0.9%		
	Other regions	2.2%	3.7%	2.8%	1.3%		
	Abroad	3.0%	4.0%	2.2%	3.0%		

- *Group 1: "Cultural and Passive Tourists."* This is the largest group, including 36% of Easter tourists who visit Castilla-La Mancha. They are tourists who practice cultural tourism mainly, because 71.4% of them have a religious and/or historic interest in the region. Besides, there are not active, for their main activity is to relax (67.9%), while their second most common activity is visiting friends and relatives (30.5%). 47.2% of them are in the age range from 25 to 44, and more than half of them are married. They reside in the following autonomous regions: Madrid (23.3%), Castilla-La Mancha (17.1%) and Valencia (16%). Furthermore, the comparative analysis shows that the percentage of tourists residing in Andalusia (7.8%) and Asturias (3.0%) is bigger that in the other groups.
- *Group 2: "Cultural and Gastronomic Tourists."* This group includes 19.1% of all Easter tourists. They do mainly cultural tourism (91%) and prefer activities related to gastronomy (99.7%) and rest (80.7%). 55.6% of them are between 25 and 44 and the percentage of individuals in the 45-64 age range (32.8%) is higher than in the other groups. It is also significant that nearly 60% of them are married. They come mainly from Madrid (20.2%), Valencia (17.3%) and Castilla-La Mancha itself (14.9%). Finally, it is worth mentioning that tourist from the Aragon region (6.1%), from abroad (4%) and from Extremadura (3.2%) is higher than in the other three groups.
- *Group 3: "Natural-Cultural and Gastronomic-day-tripping Tourists."* This group contains 32.9% of Easter tourists. They prefer nature tourism (96.9%), although they are also interested in other kinds of tourism like cultural (71.1%) or rural (55.9%) tourism. The analysis of their favorite activities allows us to deduce that they opt for activities related to relax (79.3%), day trips (68.5%) and gastronomy (67.1%). 57.2% of them are between 25 and 44 and (53.1%) of them are married. Their autonomous regions of origin vary from Madrid (20.4%), Valencia (17.6%), Castilla-La Mancha (13.2%) and Catalonia (11.1%). In this sense, apart from the high number of Catalonian tourists, it is significant to mention the percentage of tourists from the Basque Country (4.4%), which is higher than in other groups.
- *Group 4: "Multiproduct and Hyperactive Tourists."* This is the smallest group, consisting only of 12% of Easter tourists. This groups show a special inclination for natural (88.1%) and cultural (87.7%) tourism. Moreover, the comparison with the other groups demonstrates that they tend to do rural (69.4%), health (28.5%), oenology (21.3%) and business (11.5%) tourism. This chapter shows that they demand a large offer of complementary tourist activities, because they are prone to go on day trips (82.1%), do shopping, (71.5%), go out at night (69.4%), practice sports (56.6%), visit friends and relatives (52.3%) or attending at shows (49.4%). This group consists mainly of tourists under 35 (59.4%) who are single (42.7%) or live with their partner (16.7%). 25% of them come from Castilla-La Mancha, while the percentage residing in the two major autonomous regions sending tourists to Castilla-La Mancha (Madrid and Valencia) is lower than in the other groups.

Table 3. Segmentation of Weekend Tourists

Variable	Scale	Group 1 (29.8%)	Group 2 (24.9%)	Group 3 (33.1%)	Group 4 (12.2%)	X^2 Value	Sig.
Kind of tourism	Cultural	94.5%	18.0%	81.5%	100%	121.335	0.000
	Rural	13.7%	50,.8%	28.4%	13.3%	26.505	0.000
	Nature	4.1%	83.6%	98.8%	86.7%	175.743	0.000
	Health	1.4%	1.6%	3.7%	23.3%	25.478	0.000
	Educational	1.4%	3.3%	0.0%	0.0%	3.517	0.319
	Business	5.5%	0.0%	1.2%	0.0%	6.477	0.091
	Oenology	0.0%	0.0%	1.2%	6.7%	9.007	0.029
Type of activity	Relax/Rest	76.7%	70.5%	86.4%	86.7%	6.732	0.081
	Gastronomy	45.2%	16.4%	88.9%	90.0%	92.566	0.000
	Nightlife	11.0%	13.1%	9.9%	26.7%	5.898	0.117
	Sports	1.4%	24.6%	2.5%	13.3%	28.256	0.000
	Shows and Entertainment	9.6%	3.3%	9.9%	20.0%	6.674	0.083
	Shopping	9.6%	1.6%	0.0%	100.0%	189.200	0.000
	Trips	26.0%	68.9%	66.7%	90.0%	48.678	0.000
	Visit friends and relatives	4.1%	14.8%	28.4%	16.7%	16.725	0.001
Age	<25	4.1%	13.1%	0.0%	0.0%	60.361	0,000
	25-34	21.9%	45.9%	21.0%	33.3%		
	35-44	26.0%	21.3%	25.9%	20.0%		
	45-54	28.8%	14.8%	34.6%	6.7%		
	55-64	8.2%	4.9%	16.0%	16.7%		
	>65	11.0%	0.0%	2.5%	23.3%		
Marital Status	Single	17.8%	32.8%	8.8%	10.0%	32.573	0.001
	Living with a partner	5.5%	21.3%	16.3%	16.7%		
	Married	67.1%	44.3%	68.8%	60.0%		
	Divorced	5.5%	1.6%	5.0%	3.3%		
	Widow(er)	4.1%	0.0%	1.3%	10.0%		
Autonomous region of origin.	Andalusia	6.8%	3.3%	6.2%	24.1%	71.743	0.007
	Aragón	11.0%	4.9%	0.0%	13.8%		
	Asturias	1.4%	0.0%	1,.2%	0.0%		
	Valencia region	17.8%	19.7%	11.1%	10.3%		
	Cantabria	1.4%	1.6%	0.0%	0.0%		
	Castilla y León	12.3%	4.9%	11.1%	6.9%		
	Castilla-La Mancha	9.6%	16.4%	7.4%	0.0%		
	Catalonia	5.5%	3.3%	9.9%	10.3%		
	Extremadura	1.4%	0.0%	4.9%	0.0%		
	Galicia	0.0%	1.6%	1.2%	0.0%		
	Madrid	19.2%	36.1%	28.4%	24.1%		
	Murcia	8.2%	6.6%	9.9%	0.0%		
	Navarra	1.4%	0.0%	0.0%	0.0%		
	Basque Country	1.4%	1.6%	0.0%	3.4%		
	Other regions	0.0%	0.0%	0.0%	3.7%		
	Abroad	2.7%	0.0%	8.6%	3.4%		

4.1.2. Weekend Tourists in Castilla-La Mancha

Weekend Tourists in Castilla-La Mancha can also be divided into tour major clusters or groups (see Table 3). These groups have been called: Group 1. Cultural and Passive Tourists; Group 2. Nature-loving and Day-tripping Tourists; Group 3. Natural-cultural and Gastronomic-day-tripping Tourists; and Group 4. Cultural-natural and Shopping-day-tripping Tourists. As Table 3 shows, there is a connection between the group they belong to and most of the variables considered. It is necessary to say, however, that there is no significant variation in the distribution of the variables referring to educational and business tourism, and to the rest, nightlife and shows activities in any of these groups.

After the analysis of the major characteristics of each of these groups (see Table 3), it is possible to describe them as follows:

- *Group 1: "Cultural and Passive Tourists."* This groups is similar to the one described in Easter tourists, but in this case it consists of 29.8% of weekend tourists visiting Castilla-La Mancha, which means they are the second larger group. They are mainly tourists interested in culture (94.5%), and they are not active, because most of them prefer to spend their time resting (76.7%). 54.8% of them are between 35 and 54, and two-thirds are married. They live in the autonomous regions of Madrid (19.2%), Valencia (17.8%), Castilla y León (12.3%), and Aragón (11%).

- *Group 2: "Nature-loving and day-tripping Tourists."* This groups amounts up to 24.9% of weekend tourists. They opt for tourism in nature (83.6%) and in the rural context (50.8%). Apart from wanting to rest (70.5%), going on day trips is also one of their most significant preferences (68.9%). If we consider age, 67.2% of them are between 25 and 44 and the percentage of under-25 (13.1%) is higher than in the other groups. 44.3% of them are married, although the number of singles (32.8%) and of tourists living with their partners (21.3%) are also important. They come mostly from Madrid (36.1%), Valencia (19.7%) and Castilla-La Mancha itself (16.4%).

- *Group 3: "Natural-cultural and Gastronomic-day-tripping Tourists."* This is the largest group, consisting of 33.1%the sample of weekend tourists. They are a type of tourists who, apart from showing interest on nature (98.8%), also engage in cultural tourism (81.5%). The analysis of the main activities they perform demonstrates that they focus on gastronomy (88.9%), rest (86.4%) and day trips (66.7%). They are primarily in the 35-54 age range (60.5%) and married (68.8%), and they come from Madrid (28.4%) in the highest quantity, Castilla y León (11.1%) and Valencia (11.1%). Finally, it is relevant to mention that the percentage of tourists from Murcia (9.9%), from abroad, (8.6%) and from Extremadura (4.9%) is significantly higher than in the other three groups.

- *Group 4: "Cultural-natural and Shopping-day-tripping Tourists."* The smallest group, it contains only 12.2% of weekend tourists. These tourists definitely prefer cultural (100%) and natural (86.7%) tourism, and they are noteworthy for their enthusiasm for health (23.3%) and oenology (6.7%) tourism. Apart from that, they are eager for rest (86.7%), shopping (100%), day-trips (90%) and for the gastronomy of the destination (90%). Age range varies greatly: 33.3% of them are between 25 and 34, but the greater percentage (40%) corresponds to tourists over 55 and is higher than in the rest of groups. If we consider their marital status, 60% are married.

Finally, 24.1% comes from Andalusia, the same percentage of tourists residing in Madrid. Besides, it is worth mentioning that the number of them coming from Aragón (13.8%), Catalonia (10.3%), and the Basque Country (3.4%) is higher than in the other groups.

4.2. Types of Day-trippers in Castilla-La Mancha

No significant differences in behavior have been found between Easter and Weekend Day-trippers in Castilla-La Mancha, so the segmentation considers a group that includes both.

The same *chi-squared independence test* has been applied for this group of day-trippers, resulting in five clusters or day-tripper groups in Castilla-La Mancha, according to the kinds of tourism and the activities performed (see Table 4). These five groups have been called as follows: Group 1. Cultural and Passive Day-trippers; Group 2. Cultural and Gastronomic-shopping Day-trippers; Group 3. Natural-cultural and Gastronomic Day-trippers; Group 4. Naturales-cultural and gastronomic-day-tripping Day-trippers; and finally Group 5. Multiproduct and Active Day-trippers. Besides, as it is shown on Table 4, there is a connection between the group the tourist belongs to and the variables kind of tourism, activities, age, marital status, and autonomous region of origin[4], consequently, it is possible to state that the distribution of these variables is different in each of the five groups.

Looking at the information displayed in Table 4 it is possible to describe the main characteristics of the five different groups described as follows:

- *Group 1: "Cultural and passive Day-trippers."* This one is the largest group, for it contains 30% of day-trippers visiting Castilla-La Mancha. These day-trippers are practice cultural tourism mostly, because 90.4% have a religious or historic interest for the region. They are hardly active, as their main intention in to relax (56.7%), while day trips (25.6%) is the activities that ranks second among their preferences. 31.2% of them are between 25 and 34, and 47.1% are married. They come mainly from Madrid (41.8%) and Castilla-La Mancha (36.6%). Besides, comparison with other groups will show that the percentage of visitors coming from Extremadura (2.6%) and Andalusia (1.6%) is higher.
- *Group 2: "Cultural and Gastronomic-shopping Day-trippers."* This group describes 6.8% of day-trippers. They carry out cultural tourism only (96.3%) and prefer activities related to shopping (100%), relax (64.8%) and gastronomy (59.3%). Their age range is 35 to 44, to which 43.4% of them belong; besides 63% of them are married. If we consider their regions of origin, Madrid is first (44.4%) and Castilla-La Mancha second (38.9%). Finally, it is worth mentioning that the percentage of day-trippers from Aragón (5.6%) and the Basque Country (1.9%) is higher than in the other four groups.

---

[4] There is an association between the Spanish autonomous region of origin and the group, but at a 0.1 significance level, that is at a reliability level of 90%.

Table 4. Segmentation of Day-trippers

Variable	Scale	Group 1 (39.0%)	Group 2 (6.8%)	Group 3 (30.2%)	Group 4 (19.7%)	Group 5 (4.3%)	X^2 Value	Sig.
Kind of tourism	Cultural	90.4%	96.3%	67.8%	68.4%	91.2%	67.651	0.000
	Rural	4.8%	5.6%	17.4%	16.5%	79.4%	147.862	0.000
	Nature	0.0%	33.3%	69.8%	60.1%	58.8%	335.756	0.000
	Health	0.6%	1.9%	0.8%	1.9%	2.0%	85.060	0.000
	Educational	1.3%	1.9%	1.2%	0.0%	8.8%	16.256	0.003
	Business	3.2%	5.6%	0.8%	1.9%	8.8%	10.724	0.030
	Oenology	0.3%	1.9%	1.2%	40.0%	20.6%	57.647	0.000
Type of activity	Relax/Rest	56.7%	64.8%	100.0%	0.0%	85.3%	410.568	0.000
	Gastronomy	0.0%	59.3%	59.5%	60.8%	91.2%	325.040	0.000
	Nightlife	5.1%	7.4%	0.8%	5.1%	55.9%	159.218	0.000
	Sports	1.3%	1.9%	5.0%	6.3%	44.1%	114.796	0.000
	Shows and Entertainment	3.2%	11.1%	5.0%	15.8%	79.4%	210.879	0.000
	Shopping	0.0%	100.0%	0.0%	1.3%	41.2%	672.126	0.000
	Trips	25.6%	48.1%	35.5%	56.3%	64.7%	56.100	0.000
	Visit friends and relatives	16.7%	7.4%	7.9%	8.9%	47.1%	47.739	0.000
Age	<25	14.0%	13.2%	9.9%	12.2%	30.3%	40.025	0.005
	25-34	31.2%	11.3%	23.1%	28.8%	39.4%		
	35-44	26.0%	43.4%	28.1%	23.7%	12.1%		
	45-54	18.5%	17.0%	24.8%	23.1%	9.1%		
	55-64	7.8%	9.4%	10.3%	10.3%	3.0%		
	>65	2.6%	5.7%	3.7%	1.9%	6.1%		
Marital Status	Single	34.7%	24.1%	29.0%	28.0%	58.8%	37.563	0.002
	Living with a partner	12.3%	7.4%	9.5%	15.9%	5.9%		
	Married	47.1%	63.0%	56.4%	51.6%	26.5%		
	Divorced	4.2%	3.7%	4.1%	3.2%	0.0%		
	Vidow(er)	1.6%	1.9%	0.8%	1.3%	8.8%		
Autonomous regions of origin	Andalusia	1.6%	0.0%	3.7%	2.6%	0.0%	73.369	0.060
	Aragón	0.7%	5.6%	0.4%	1.3%	0.0%		
	Valencia region	4.9%	1.9%	9.5%	7.7%	8.8%		
	Castilla y León	4.2%	1.9%	2.9%	1.9%	8.8%		
	Castilla-La Mancha	36.6%	38.9%	38.0%	41.7%	38.2%		
	Catalonia	2.3%	0.0%	1.7%	0.6%	2.9%		
	Extremadura	2.6%	1.9%	0.8%	1.9%	0.0%		
	Madrid	41.8%	44.4%	26.4%	35.3%	32.4%		
	Murcia	0.7%	0.0%	2.9%	3.8%	2.9%		
	Basque Country	0.0%	1.9%	1.2%	0.6%	0.0%		
	Other regions	1.4%	0.0%	0.4%	2.5%	0.0%		
	Abroad	3.3%	3.7%	2.1%	0.0%	5.9%		

- *Group 3: "Natural-cultural and Gastronomic Day-trippers."* 30.2% of day-trippers belong to this group. They are mainly interested in natural tourism (69.8%), although other kinds like cultural and rural tourism also score high (67.8%) and (17.4%) respectively. The activities they perform are mainly relax (100%) and gastronomy (59.5%). 28.1% of them is 35 and 44 and the percentage for the 45-54 group is higher than in the other groups. 56.4% of these day-trippers are married. The highest percentage of them comes from Castilla-La Mancha itself (38%) and from Madrid (26.4%). The percentage of residents from Valencia (9.5%) and Andalusia (3.7%) is also noteworthy, because it is higher than in other groups.
- *Group 4: "Natural-Cultural and Gastronomic-day-tripping Day-trippers".* This segment contains 19.7% of day-trippers. They care mostly about cultural (68.4%) y de natural (60.1%) tourism only, and they stand out as the group with the highest inclination to practice oenology tourism (40%). Their favorite activities are gastronomic (60.8%) and day-tripping (56.3%). Besides, 28.8% of them belong in the age interval 25-34, and 51.6% are married; the percentage of those living with a partner (15.9%) is higher than in the rest of groups. Their autonomous regions of origin are Castilla-La Mancha (41.7%) and Madrid (35.3%), although the comparison with other groups reveals a higher proportion of day-trippers from Murcia (3.8%).
- *Group 5: "Multiproduct and Active Day-trippers."* This is the smallest group, as it includes only 4.3% of day-trippers. They tend to practice cultural (91.2%), rural (79.4%) and natural (58.8%) tourism, and they are likelier to carry out educational (8.8%), business (8.8%) and health (2%) tourism than day-trippers in the other groups. The analysis of their preferences in activities shows that these are mainly concerned with gastronomy (91.2%), relax (85.3%), shows and entertainment (79.4%), day trips (64.7%) and nightlife (55.9%). The percentage of them who visit their friends or relatives (47.1%) or practice sports is also significantly higher than in the other groups. Day-trippers belonging to this group are mostly between 25 and 34 (39.4%), and it is also relevant to note that the percentage of under 25 and over 65 among these day-trippers scores higher than in the rest of groups. Besides, 58.8% of them are single and 8.8% are widow(er)s. Also, 38.2% comes from Castilla-La Mancha and 32.4% from Madrid, even though the percentages of day-trippers residing in Castilla y León (8.8%), abroad (5.9%), and in Catalonia (2.9%) are higher than in the other four groups.

CONCLUSION

Tourism is one of most important sectors of Spanish economy, and continues to be (it has even grown slightly) in the difficult current economic situation. The environmental characteristics and the habits of tourists and day-trippers determine the consumption of tourist products, which means that knowing the profiles of European, national and regional tourists, as well as their differences and similarities, is critical in order to provide tourist managers with an updated and well-informed vision of the habits and profiles of tourists which allows them to adapt their marketing strategies to the demands of their market targets.

The analysis of the habits of tourists and day-trippers in Castilla-La Mancha reveals that there are four major Easter tourist groups whose differences can be attributed mainly to the kind of tourism they practice and the activities they perform: "cultural and passive tourists," "cultural and gastronomic tourists," "natural-cultural and gastronomic-day-tripping tourists," and "multiproduct and hyperactive tourists."

In relation to weekend tourists who visit Castilla-La Mancha, it is possible to differentiate four main groups, three of which are different from the groups mentioned above for Easter tourists: "cultural and passive tourists", "nature-loving and day-tripping tourists," "natural-cultural and gastronomic-day-tripping tourists," and "multiproduct and hyperactive tourists."

More specifically, this research has shown that there are two more groups if we consider day-trippers: "cultural and passive day-trippers," "cultural and gastronomic-shopping day-trippers," "natural-cultural and gastronomic day-trippers," "natural-cultural and gastronomic-day-tripping day-trippers" and "multiproduct and hyperactive day-trippers".

In general, there are several differences and similarities among the different segments analyzed here, both inside the tourist and the day-trippers groups.

To be more precise, out of the thirteen segments of travelers obtained in this research, three of them are common for each type of traveler analyzed (Easter tourists, weekend tourists and day-trippers) due to the similarities found in the variables used in their analysis (type of tourism and activity, age, marital status and autonomous region of origin): "cultural and passive," "natural-cultural and gastronomic-day-tripping," and "multiproduct and hyperactive".

On the other hand, the following main differences among the profiles, according to the three types of travelers analyzed here, are noteworthy:

- "Cultural and passive": Weekend tourists are older and come from northern Spain. Resides this group ranks second in importance among the other four, while among Easter tourists and day-trippers the "cultural and passive" is the largest. Another significant difference is that Easter tourists come from both northern and southern Spain.
- "Natural-cultural and gastronomic-day-tripping": This group is the largest among the weekend tourists, and it is clearly preferred by older age tourists in comparison with Easter tourists and day-trippers. It is also relevant that the most common marital status among day-trippers is living with a partner, whereas the weekend and the Easter tourists are mainly married.
- "Multiproduct and hyperactive": The most relevant differences here can be found in the age range of weekend and day-trippers when compared with Easter tourists. The former tend to both earlier and later age ranges than the latter. Besides, day-trippers come from Madrid and Castilla-La Mancha mainly, while weekend tourists have traveled mainly from Madrid and Andalusia and Easter tourists from Castilla-La Mancha itself.

A great number of differences can also be found among the following profiles obtained here:

- "Cultural and gastronomic": Day-trippers are also buyers, not only cultural and gastronomic like Easter tourists. Contrarily, there is no specific profile that describes the major characteristics of weekend tourists, because they are mixed in the tour profiles described in the results.
- "Nature-loving and day-tripping": This profile belongs characteristically to the weekend tourist.
- "Natural-cultural and gastronomic": This profile described the day-tripper travelers.

In order to be able to adapt properly to the activities and kind of tourism demanded, it would interesting to know what type of tourist we are dealing with, so as to adapt the offer to their needs with, for instance, holiday promotions. It would also be convenient for destinations to be able to offer all kinds of tourism and all the activities required in order to respond to the needs of all tourist profiles.

As a future line of research we propose a more thorough analysis of the different types of tourists and day-trippers in the different provinces of Castilla-La Mancha. This chapter could carry out a latent analysis to complement the work accomplished here, in order to identify any existent gaps between the offer and demand of tourist types and therefore improve the services locally in each destination.

REFERENCES

Albadalejo, I.P. & Díaz, M.Y. (2005). Rural Tourism Demand by Type of Accommodation. *Tourism Management*, v.26, No.6, pp. 951–959.

Aminzadeh, B. & Khansefid, M. (2010). Improving the Natural and Built Ecological Systems in an Urban Environment. *Int. J. of Environ. Res.*, v.4, No.2, pp. 361-372.

Anderson, V., Prentice, R. & Watanabe, K. (2000). Journeys for Experiences: Japanese Independent Travelers in Scotland. *J. of Travel & Tourism Marketing*, v.9, No.1/2, pp. 129–151.

Andriotis, K., Agiomirgianakis, G. & Mihiotis, A. (2008). Measuring Tourist Satisfaction: A Factor–cluster Segmentation Approach. *J. of Vacation Marketing*, v.14, No.3, pp. 221–235.

Blázquez, J.J., Esteban, A. & Molina, A. (2010). Determinación y análisis de un modelo de lealtad hacia el destino turístico: Aplicación de la nueva lógica dominante del servicio [Identification and analysis of loyalty model to the destination: Implementation of the service dominant new logic]. *Dissertation*, University of Castilla-La Mancha, Spain.

Cha, S., McCleary, K.W. & Usal, M. (1995). Travel Motivations of Japanese Overseas Travellers: A Factor–Cluster Segmentation Approach. *J. of Travel Res.*, v.34, No.1, pp. 33–39.

Choi, W.M. & Tsang, C.K.L. (1999). Activity based Segmentation on Pleasure Travel Market of Hong Kong Private Housing Residents. *J. of Travel & Tourism Marketing*, v.8, No.2, pp. 75–97.

Confederación Española de Hoteles y Alojamientos Turísticos (2008): Observatorio de la Industria Hotelera Española. Perspectivas para el Segundo Cuatrimestre de 2008

[Observatory of the Spanish hotel industry. *Prospects for the Second Quarter of 2008*], Price Water House Coopers, Madrid.

Dellaert, B., Borgers, A. & Timmermans, H. (1995). A Day in the City: Using Conjoint Choice Experiments to Model Urban Tourists' Choice of Activity Packages. *Tourism Management,* v.16, No.5, pp. 347–353.

Esteban, A., Martín-Consuegra, D., Molina, A. & Díaz, E. (2005). *Turismo y Consumo: el Caso de Toledo [Tourism and Consum: The case of Toledo]*, Centro de Estudios de Consumo de la Universidad de Castilla-La Mancha y de la Junta de Comunidades de Castilla-La Mancha, Toledo.

Ferrari, G., Mondéjar-Jiménez, J. & Vargas-Bargas, M. (2010). Environmental Sustainable Management of Small Rural Tourist Enterprises. *Int. J. of Environ. Res.,* v.4, No.3, pp.407-414.

Formica, S. & Uysal, M. (1998). Market Segmentation of an International Cultural–Historical Event in Italy. *J. of Travel Res.,* v.36, No.4, pp. 16–24.

Frochot, I. (2005). A Benefit Segmentation of Tourists in Rural Areas: A Scottish Perspective, *Tourism Management,* v.26, No.3, pp. 335–346.

Frochot, I. & MORRISON, A.M. (2000). Benefit Segmentation: A Review of its Applications to Travel and Tourism Research. *J. of Travel & Tourism Marketing,* v.9, No.4, pp. 21–45.

González, A.M. & Esteban, A. (2000). *Valores y Estilos de Vida en el Análisis de la Demanda Turística* [Values and Life Styles in Tourism Demand Analysis], Ed. FITUR (International Tourist Fair), Madrid.

Hsieh, S., O'Leary, J. & Morrison, A.M. (1992). Segmenting the international travel market by activity. *Tourism Management,* v.13, No.2, pp. 209–223.

Huang, R. & Sarigöllü, E. (2007). Benefit Segmentation of Tourists to the Caribbean. *J. of Int. Consumer Marketing,* v.20, No.2, pp. 67–83.

Instituto de Estudios Turísticos [Tourist Studies Institute] (2005). Comportamientos Turísticos de los Residentes en la Unión Europea 2005 [*Tourist Behaviour of the European Union Residents 2005*], Secretaría de Estado de Turismo y Comercio, Ministerio de Industria, Turismo y Comercio, Madrid.

Instituto de Estudios Turísticos (2008a). Hábitos Turísticos de los Residentes en España 2007 [Resident Tourist habits in Spain 2007*], Secretaría de Estado de Turismo y Comercio, Ministerio de Industria,* Turismo y Comercio, Madrid.

Instituto de Estudios Turísticos (2008b). Movimientos Turísticos de los Españoles 2007 [Spanish Tourist Movements 2007], Sec*retaría de Estado de Turismo y Comercio, Ministerio de Industria,* Turismo y Comercio, Madrid.

Instituto de Estudios Turísticos (2008c). Movimientos Turísticos en Fronteras 2007 [Frontier Tourist Movement 2007], *Secretaría de Estado de Turismo y Comercio, Ministerio de Industria,* Turismo y Comercio, Madrid.

Instituto Nacional de Estadística [Statistical National Institute] (2008a). *Encuesta de Ocupación en Alojamientos de Turismo Rural 2007* [Accommodation Occupancy Survey for Rural Tourism 2007], INE, Madrid.

Instituto Nacional de Estadística (2008b): Encuesta de Ocupación Hotelera (varios años) [Hotel Occupancy Survey (various years)], INE, Madrid.

Jang, S., Morrison, A. & O'Leary, J.T. (2004). A Procedure for Target Market Selection in Tourism. *J. of Travel & Tourism Marketing,* v.16, No.1, pp. 19–33.

Jang, S.C., Morrison, A. M. & O'Leary, J.T. (2002). Benefit Segmentation of Japanese Pleasure Travelers to the USA and Canada: Selecting Target Market Based on the Profitability and Risk of Individual Market Segments. *Tourism Management*, v.23, No.4, pp. 367–378.

Juaneda, C. & Sastre, F. (1999). Balearic Islands Tourism: a Case Study in Demographic Segmentation. *Tourism Management*, v.20, No.4, pp. 549–552.

Junta de Comunidades de Castilla-La Mancha [Castilla-La Mancha Gouverment] (2006). Consultas en Oficinas de Información Turística de Castilla-La Mancha [Enquiries Tourist Information Offices in Castilla-La Mancha], *Revista Castilla-La Mancha Tierra de Don Quijote*, JCCM. From www.jccm.es/revista/206/clmvistapor.htm

Kim, S.S. & Agrusa, J. (2008). Segmenting Japanese Tourists to Hawaii According to Tour Purpose. *J. of Travel & Tourism Marketing*, v.24, No.1, pp. 63–80.

Lang, C.T., O'Leary, J. & Morrison, A. (1993). Activity Segmentation of Japanese Female Overseas Travelers. *J. of Travel & Tourism Marketing*, v.2, No.4, pp. 1–21.

Moscardo, G., Pearce, P., Morrison, A., Green, D. & O'Leary, J.T. (2000). Developing a Typology for Understanding Visiting Friends and Relatives Markets. *Journal of Travel Research*, v.38, No.3, pp. 251–259.

Neal, J.D. (2003). The Effect of Length of Stay on Travelers' Perceived Satisfaction with Service Quality. *J. of Quality Assurance in Hospitality and Tourism*, v.4, No.3/4, pp. 167–176.

Sirakaya, E., Uysal, M. & Yoshioka, C.F. (2003). Segmenting the Japanese Tour Market to Turkey. *J. of Travel Res.*, v.41, No.3, pp. 293–304.

Sistema de Información Turística de Asturias [Tourist Information System from Asturias] (2002). *El Turismo Activo en Asturias en 2002* [The Active Tourism in Asturias in 2002], SITA. From www.sita.org

Sistema de Investigación Turística de Castilla-La Mancha [Tourist Research System from Castilla-La Mancha] (2008). *Boletín Anual de Turismo* (ANUALd2007), No. 2, SITdCLM, Instituto de Promoción Turística de Castilla-La Mancha and University of Castilla-La Mancha.

Sistema estadístico del turismo Vasco [Basque tourist statistic system] (2006). Estudio sobre los Viajeros, Excursionistas y Turistas en la Comunidad Autónoma de Euskadi [*Survey on Travelers, day-trippers, and tourists in the Autonomous Community of Euskadi*], Departamento de Industria, Comercio y Turismo.

Stemerding, M.P., Oppewal, H., Beckers, T.A.M. & Timmermans, H.J.P. (1996). Leisure Market Segmentation: An Integrated Preferences/constraints–based Approach. *J. of Travel & Tourism Marketing*, v.5, No.3, pp. 161–185.

Sung, H.H. (2004). Classification of Adventure Travelers: Behavior, Decision Making, and Target Markets. *J. of Travel Res.*, v.42, No.4, pp. 343–356.

Tsiotsou, R. (2006). Using Visit Frequency to Segment Ski Resorts Customers. *J. of Vacation Marketing*, v.12, No.1, pp. 15–26.

Tsiotsou, R. & VASIOTI, E. (2006). Satisfaction: A Segmentation Criterion for 'Short Term' Visitors of Mountainous Destinations. *J. of Travel & Tourism Marketing*, v.20, No.1, pp. 61–73.

Turgalicia (2007). Análise Estatística sobre o Excursionismo en Galicia en 2006, Procedente de Asturias, León, Zamora e Rexión Norte de Portugal [*Statistic analysis on day-trip in*

Galicia in 2006 come from Asturias, León, Zamora and Portugal North region], Universidad de Santiago de Compostela.

Unidad de Estudios y Estadística de Murcia [*Studies and Statistic Deparment from Murcia*] *(2006). Estudio sobre los Hábitos de Consumo de los Turistas en la Región de Murcia* [*Study on the consumption habits of tourists in the Region of Murcia*], Consejería de Turismo, Comercio y Consumo.

Vargas-Vargas, M., Meseguer-Santamaría, M.L., Mondéjar-Jiménez, J. & Mondéjar-Jiménez, J.A. (2010). Environmental Protection Expenditure for Companies: A Spanish Regional Analysis. *Int. J. of Environ. Res.*, v.4, No.3, pp. 373-378.

In: Research Studies on Tourism and Environment
Editors: J. Mondejar-Jimenez et al.
ISBN: 978-1-61209-946-0
© 2012 Nova Science Publishers, Inc.

Chapter 7

TOURISTIC SECTOR COMPANIES AT THE WARSAW STOCK EXCHANGE

Krzysztof Kompa and Dorota Witkowska
Warsaw University of Life Sciences, Poland

ABSTRACT

The paper is divided into two main parts. The first one describes the selected aspects of touristic sector in Poland. The aim of the research, discussed in the second part, is to present and analyze the companies belonging to that sector and being listed at the Warsaw Stock Exchange. Research is conducted applying comparative analysis, trend function and taxonomic measures based on real data.

1. INTRODUCTION

The industries which collectively provide travel and tourism[1] services worldwide are some of the fastest growing anywhere. Therefore the role of touristic sector in national economies has been increasing. The main characteristic of the touristic "product" are resource immobility, capacity constraints, seasonality and consumers' inability to experience the product before purchase[2]. The measure of influence of the touristic sector on national economy is extremely difficult since the majority of impact is indirect. Direct effect can be measured by the touristic expenditures and earnings of employees or enterprises operating in that sector. Indirect effects occur through the tourism value chain since tourism is connected with food and beverages consumption, transport and hotel services that may cause the development of other sectors as construction, furniture, motor industry, etc.

---
[1] Tourism is interdisciplinary area of research therefore different methodology is used. The paper Dann, Nash and Pearce, 1988 attempts to highlight some areas of tourism research which are believed to lack sufficient methodological sophistication.
[2] Bull, 1995 that discusses the problems of touristic sector from both micro- and macro-economic point of view.

Many countries, attracted by the potential economic benefits, have embarked upon the development of tourism. This has often taken place without an adequate appreciation of the associated costs[3] such as:

- pushing up local prices and state's exchange rate,
- deprive local people of access to natural resources on which they rely, such as fishing area, forests, waters, etc.,
- environmental effects[4].

Recent literature review[5] for the World Bank by the Overseas Development Institute concludes that tourist industry has substantial impact on economy and people especially in developing countries, where tourism has been often developing in rural areas that is an alternative to unemployment and migration to urban areas[6]. Encouraging tourism in poor countries is, surprisingly often, an effective way of achieving poverty reduction through inclusive economic growth. But not always. Sometimes international tourism development does little for the local economy and the livelihoods of the poor.

According to the estimations made by United Nation World Tourism Organization (UNWTO) the country level broadly-defined tourism accounts for between 2% and 12% of GDP in advanced, diversified economies, and up to 40% of GDP in developing countries, and up to 70% of GDP in small island economies.

Analyzing touristic industry in Poland one must remember that free traveling all around the world became possible for the Polish citizens after the political changes that took place in 1989. It is also worth mentioning that the first years of the political system transformation were characterized by deep economic crisis. Therefore the development of touristic sector in Poland has been observed only in the last decade.

Describing the role of the touristic sector in national economy it is important to discuss:

1. the tourist attractions of the country or region,
2. changes in count and structure of visitors,
3. the share of the tourist industry in GDP and exports,
4. position of enterprises belonging to the touristic sector.

Therefore the paper contains 3 sections. The first two describe the selected aspects of the touristic sector in Poland. The third section contains the results of research that are provided for the companies belonging to that sector and being listed at the Warsaw Stock Exchange (WSE). We chose firms that are quoted at WSE because of importance of this institution for transition of Polish economy and transparency of the listed companies since these companies are required to publish verified quarterly reports about their activity. Analysis is conducted applying comparative analysis, trend function and taxonomic measures based on real data.

---

[3] The discussion on that subject can be found in Mathieson, Wall, 1982 among others.
[4] See Jenner, Smith, 1992.
[5] See Ashley at al. 2007, p. 8, Mitchell, Ashley, 2009.
[6] See Gannon A, 1994.

1.1. Poland as the Tourist Attraction

Poland is a state with over a thousand-year history and culture. It lies in the heart of Europe and the geometric centre of the continent is right here. The total surface area is 312,600 sq km of land. This makes Poland the ninth largest country in Europe, after Russia, Ukraine, France, Spain, Sweden, Germany, Finland and Norway, and the 63rd largest in the world. The population in Poland is over 38 millions that gave Poland in 2008 the 34-th place in the world.

Poland's territory extends across several geographical regions. One can find more or less everything in Poland that is alpine mountains, wide beaches, clean lakes, deep forests, world-class historic monuments. Polish history started in the 10-th century but there are also archeological findings dated 700BC.

Poland has a moderate climate with both maritime and continental elements. This is due to humid Atlantic air which collides over its territory with dry air from the Eurasian interior. As a result, the weather tends to be capricious and the seasons may look quite different in consecutive years. Climate, with often changing and being unpredictable in medium term weather, seems to be the main disadvantage for tourists both foreign and domestic who prefer to spend their holidays in places where the weather is nearly "guaranteed".

Poland's greatest attraction is nature especially areas of outstanding natural value, both Europeanwide and worldwide. There are still places hardly touched by the civilization. The most valuable gems of Poland's flora include the several hundred ancient oak trees. There are 33 thousands of animal species, animals that have since died out in other parts of Europe still survive in Poland. White storks occur almost anywhere in the country except for the higher parts of the mountains. Poland has become a region with the highest density of storks in Central Europe. There is also a great population of bisons.

1.2. Touristic Sector in Poland

When Poland belonged to the Soviet block, Polish citizens could not freely travel abroad from political and economic reasons, especially since Polish currency was inconvertible. In the same time, Poland as a country being East of the Iron Curtain (till 1989), seemed not to be a good place for visitors. It caused that tourist industry did not play an important role in the Polish economy. Although foreign visitors were treated as a source of convertible currency and there were special shops and tourist agencies (with goods and services unavailable at regular market) where payment was made in foreign (convertible) currencies.

Introduction the market economy in Poland and the collapse of the communist system removed formal constrains of the tourist sector development. However in nineties of the 20th century, Poles could not afford many goods and services therefore they resigned from tourist activity at the first place. In that period many resorts, especially the ones that belonged to the huge state enterprises, bankrupted or went to rack and ruin. Also state travel agencies started the privatization process although at the beginning of transition period all – also cooperative or even private – touristic enterprises suffered from lack of clients. One should also notice that during first years of political and economic transformation Poland was not attractive for foreigners, especially because:

Table 1. Role of touristic sector in Polish economy

Years	2001	2002	2003	2004	2005	2006	2007
Share of touristic sector in GDP [%]	6.7	7.4	6.4	6.2	5.7	6.1	6.0
Share of touristic sector in export [%]	×	×	×	5.5	5.6	5.2	5.0

Source: http://www.mg.gov.pl/TURYSTYKAWPKB3.doc

- in the past there were restrictions that could be difficult to obey by visitors,
- foreigners expected low standard of tourist service,
- level of public safety seemed not to be high enough.

However it is necessary to add that during first years of transition the so called "trade tourism", that based on disparities between prices of certain goods in different states (especially countries that belonged to the former Soviet Union), was quite popular, and informal export and import (being the "grey" economic sector) influence the national economy essentially.

At present, Poland is a part of the global tourism market with constantly increasing number of visitors, particularly after joining the European Union. Poland's main tourist offers are sightseeing within cities and out-of-town historical monuments, business trips, qualified tourism, agro-tourism, and mountain hiking, among others. Poland is the 17th most visited country by foreign tourists in 2008[7]. In late nineties, due to economic growth and improvement of the life standard, also Poles started to consume touristic services in Poland and travel abroad.

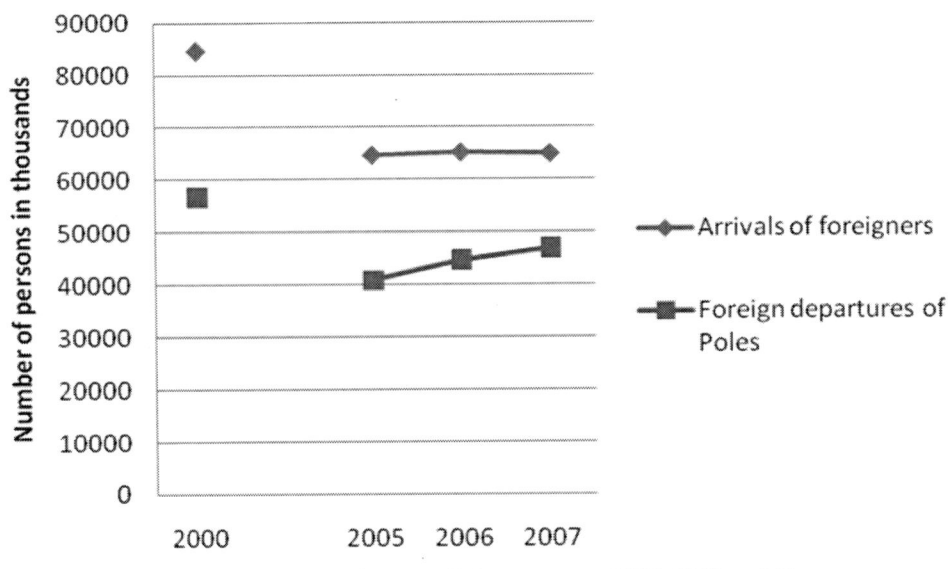

Source: Own elaboration on basis of Mały Rocznik Statystyczny, 2008, GUS, p. 269.

Figure 1. Arrivals of foreigners to Poland and foreign departures of Poles.

---

[7] "UNTWO World Tourism Barometer, Vol.5 No.2".

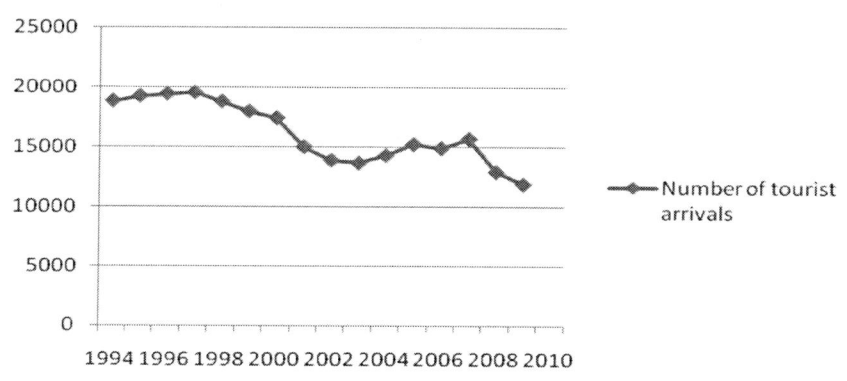

Number of tourist arrivals

Source: Estimation by Institute of Tourism.

Figure 2. Number of tourist arrivals [in thousands].

Worldwide, the touristic sector creates about 10% of global GDP and 8% of all employees work in that sector. The size of the tourism industry can be measured both in terms of the number of visitors (or trips) and the amount they spend. However it is difficult to define "touristic sector" and measure its influence because of indirect effects, there is no doubts that it plays an important role in worldwide economy. In further consideration we assume that touristic sector direct output contains only expenditures on hotels, restaurants and travel agencies services.

Analyzing the share of the tourist industry in Polish GDP and export, one may notice that the role of this sector in Poland is smaller than the world average (Table 1).

The impact of touristic sector can be measured by number of border crossings that was the biggest in the year 1999 – 89.1 mil. persons. At Figure 1 we present arrivals of foreigners to Poland and foreign departures of Poles[8]. It is visible that the number of border crossings decreased by over than 25% in the year 2005 in comparison to the year 2000. The lowest number of foreigners visited Poland is observed in the year 2002 while the lowest number of Poles went abroad in the year 2004[9]. Of course not all of travelers are tourists and changes of the border crossings numbers cannot be directly addressed to the situation of the touristic sector. According to the definition of the World Tourism Organization (UNWTO) tourism "includes activities of persons travelling for purposes other than that connected with earning income, i.e. mainly for recreation, work-related, religious and other purposes, and remaining outside their usual environment not longer than one year."

As one can notice on Figure 2, in 1998 there were almost 20 millions of foreign tourists in Poland and this number has been decreasing below 12 millions in 2009[10].

Taking into account the purpose of traveling to Poland one may distinguished: business trips, touristic journeys, transit, shopping, visiting relatives and friends. The structure of visitors who came to Poland in years 2001 – 2007 for different purposes is presented at Figure 3 and Table 2. While the dynamic of changes in that structure is presented at Figure 4.

---

[8] It is worth mentioning that just before Christmas of 2007, Poland have joined Schengen Zone. Data for 2007 do not comprise traffic on border crossings with Germany, Czech republic, Slovakia, Lithuania and sea border crossings between 21 and 31 December since there is no border control within the zone.
[9] Mały Rocznik Statystyczny, 2008, GUS, p. 270.
[10] Data concerning border traffic in years 2008 and 2009 are estimated.

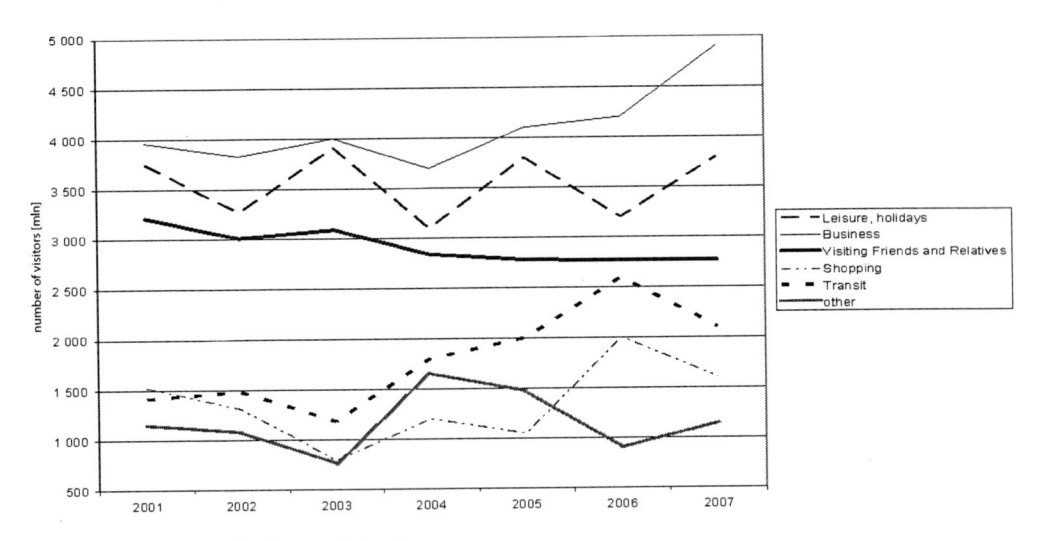

Source: estimation by Institute of Tourism.

Figure 3. Number of visitors in Poland in years 2001 -2007 by purpose of their visits.

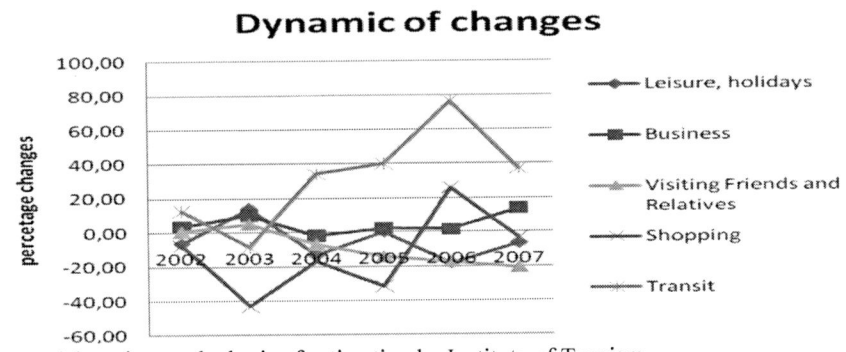

Source: own elaboration on the basis of estimation by Institute of Tourism.

Figure 4. Dynamic of changes in the count and the structure of visitors in years 2002 – 2007 in comparison to 2001 [in percentages].

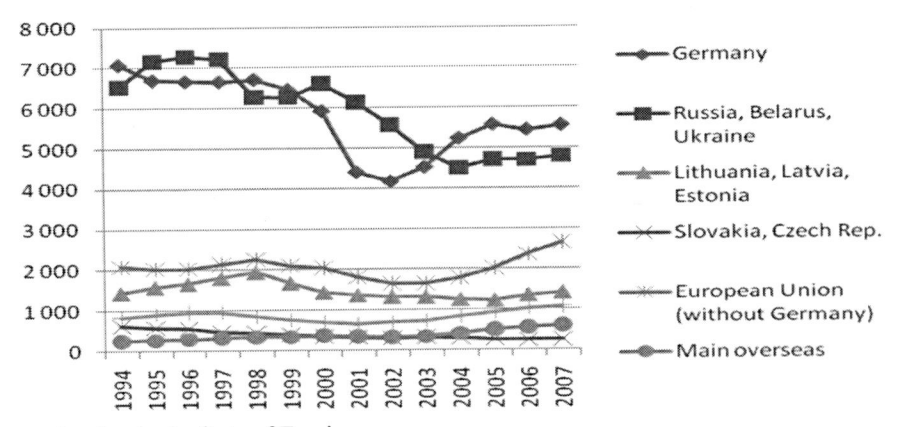

Source: estimation by Institute of Tourism.

Figure 5. Arrivals to Poland by country [in thousands].

Table 2. Structure of visitors in percentages

Years	2001	2002	2003	2004	2005	2006	2007
Leisure, holidays	25.00	23.39	28.48	21.69	25.00	20.42	23.28
Business	26.40	27.40	29.13	25.89	26.97	26.80	30.02
Visiting Friends and Relatives	21.40	21.53	22.51	19.87	18.29	17.68	16.97
Shopping	10.13	9.37	5.75	8.40	6.91	12.76	9.80
Transit	9.40	10.59	8.59	12.60	13.16	16.59	12.87
other	7.67	7.73	5.54	11.55	9.67	5.74	7.05

Source: Bartoszewicz W., Łopaciński K. (2007).

As one can see the most important purposes for visiting Poland are: business, leisure and holidays, and visiting friends and relatives. The dynamics of the structure of arrivals evaluated for the year 2001 as the base period is presented on Figure 4. It can be noticed that the biggest dynamic is observed for transits. While the percentage of trips on business, holidays and to visit friends and relatives did not change essentially in analyzed years.

Another important issue is to describe arrivals by the countries (see Fig.5). The majority of visitors come from Germany and Russia, Belarus and Ukraine that are neighbors of Poland together with Lithuania, Slovakia and Czech Republic.

Taking into account departures of the Polish residents, one can notice that the biggest number of trips abroad was observed in the year 2000 (Figure 6). The structure of departures in terms of purposes for traveling abroad is presented at Figure 7.

Rest, holidays, country tours, i.e. typical tourist purposes have been recently the basic incentives for Polish foreign departures. In recent (2008 and 2009) years leisure tourism was 52%, while in 2007 – 42% of all departures. Stays with families and friends keep the level around 25%, while the share of business trips varies from year to year: in 2007 it was 26%, and in the next years – barely 18% and 17%.

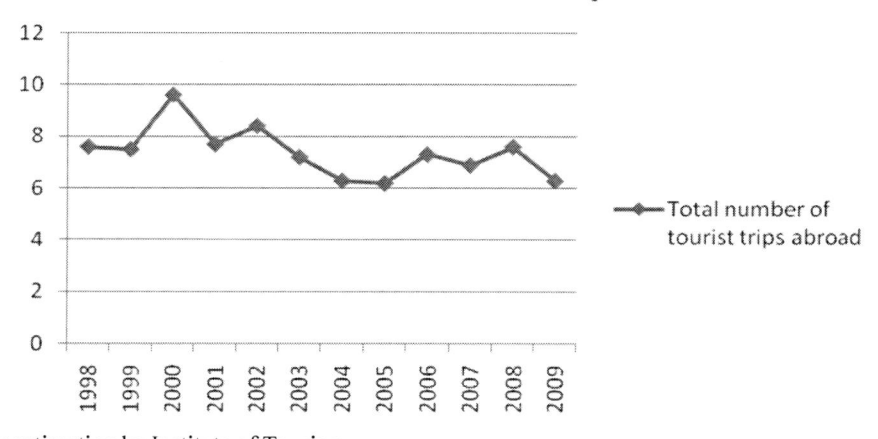

Source: estimation by Institute of Tourism.

Figure 6. Total number of tourist trips abroad made by the Polish residents in years 1998 – 2009 [in millions].

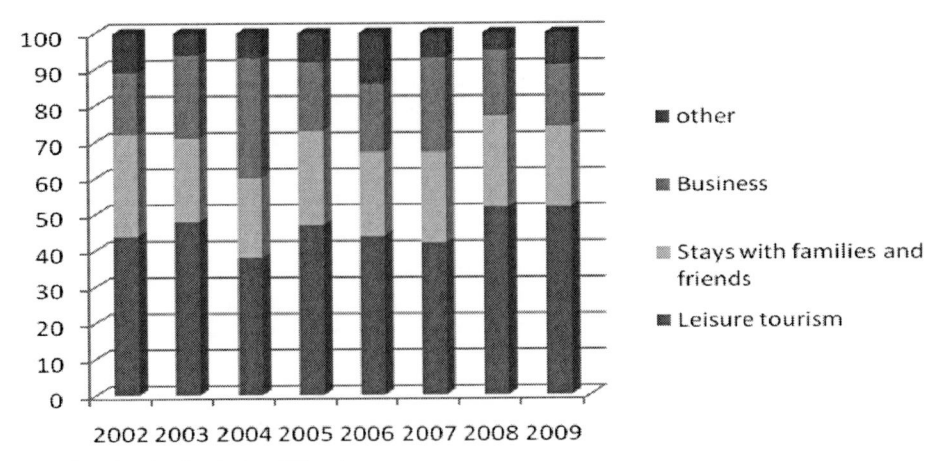

Figure 7. Structure of tourist trips abroad made by Polish residents by purposes in years 2002 – 2009 [in percentage].

Another measurement of tourist industry impact on the economy is the value of expenditures that are spent for travels and tourism as it is presented at Figure 8. As one can see in years 2001 – 2005, state expenditures for tourist increased from 2.0 to 2.7 bil. PLN, business traveling expenses from 8.0 to 10.7, and domestic expenditures of Polish residences for foreign travels from 3.6 to 4.8. Foreigners' expenditures in Poland also increased from 19.2 to 20.3 bil. PLN while expenditures of Polish residences for domestic travels decreased from 17.4 in the year 2001 to 15.7 bil. PLN in the year 2005.

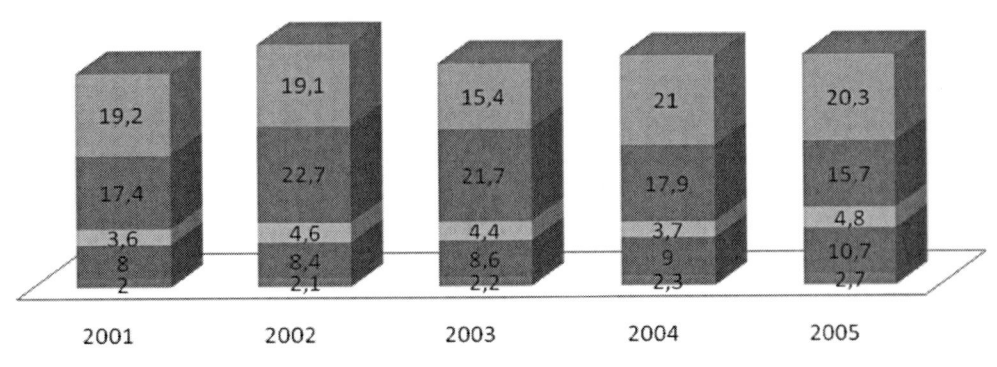

Source: Institute of Tourism.

Figure 8. Tourist sector in Poland in years 2001 – 2005 [in billions PLN].

2. REPRESENTATION OF TOURISTIC SECTOR BY COMPANIES LISTED AT THE WARSAW STOCK EXCHANGE

According to the data presented in Table 1, touristic sector in Poland is less important economic branch as the worldwide average since it creates only 6.4% of GDP (in average evaluated for years 2001-2007). That fact is also visible in position of the firms that belong to the tourist sector in comparison to the enterprises that represent other economic branches.

In our analysis we consider only companies quoted at The Warsaw Stock Exchange (WSE) that enjoys a good reputation among investors and financial institutions both in Poland and abroad. It is worth mentioning that the Polish capital market is characterized by the state-of-the-art infrastructure, EU standard compliant regulations, professional service, transparent rules and strong economic performance. Therefore majority of developing companies, that usually need to raise the funds necessary to pursue its strategic objectives, try to be listed at The Warsaw Stock Exchange, and we believe that these firms are the most transparent and reliable.

The Polish capital market is the medium size among European capital markets. The Warsaw Stock Exchange evaluates five indexes that describe the whole Polish capital market and its segments distinguished on the basis of companies' sizes or origin, and several sectoral sub-indexes that allow to evaluate the efficiency of investments into businesses in various sectors of economy. In the sub-indexes' portfolios there are the same weightings as in the index of all WSE companies - WIG portfolio, but selected on sectoral criterion basis. There are 10 economic branches that are represented by companies that are listed at The Warsaw Stock Exchange. These sectors are: banking, construction, chemical, developers, energy, IT, media, oil & gas, food and telecommunication.

Unfortunately the tourist sector is not represented by any sectoral WSE sub-indexes, therefore one may claim that the number of companies that belong to that sector and their capitalization is not essential. Due to the companies' declarations we can consider only firms that belong to the branch: hotels and restaurants since there is also no tourist sector as the branch selection possibility. Among 339 companies that are listed at the main market[11] we found only 7 i.e.: AmRest, Interferi, Olympic, Orbis, PolRest, Polskie Jadlo, Sfinks that are defined as hotels and restaurants, and one - Rainbow quoted at parallel market (among 40 companies). Two of these firms are foreign companies - Olympic from Estonia and AmRest from the Netherlands. All companies, except PolRest, are used for WIG calculation while AmRest and Orbis are also used for mWIG40 calculations. Capitalization of

- AmRest and Orbis exceeds 250 mil. euro,
- Olympic belongs to the interval (50; 250) mil. euro;
- Interferi, Polskie Jadlo, Sfinks and Rainbow belongs to the interval (5; 50) mil. euro, while
- PolRest is smaller than 5 mil. euro.

---

[11] Data from http://www.gpw.pl/zrodla/gpw/spws/pol/spolrypl.html May, 5, 2010.

Number of companies in touristic sector

■ >250 mil. Euro　■ <50; 250> mil. Euro　■ <5; 50> mil. Euro　■ <5 mil. Euro

Total number of companies

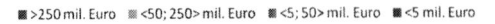

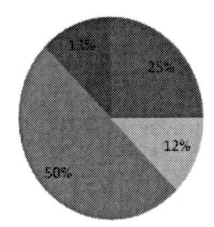

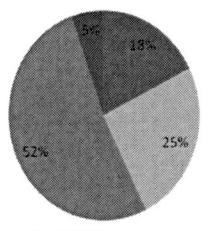

■ >250 mil. Euro　■ <50; 250> mil. Euro　■ <5; 50> mil. Euro　■ <5 mil. Euro

Source: own elaboration on the basis of http://www.gpw.pl/zrodla May, 5, 2010.

Figure 9. Comparison of the companies' capitalization.

Taking into account the value of the company capitalization (see Figure 9) one may say that the structure of companies from the tourist sector (that represents 2.1. % of all companies listed) is quite similar to the structure of all firms quoted at The Warsaw Stock Exchange.

2.1. Description of Companies

The Polish travel agency Orbis[12] was founded in 1920. In the year 1939, the company had 136 branches in Poland and 19 abroad, with 500 employees, and Orbis rendered services to over 5 million clients. During the Second World War the major part of its property was damaged and after the war Orbis started to rebuild its position. However one must remember that during communist time traveling abroad was strictly limited and there was not many foreigners who were willing to visit Poland – the country behind the Iron Curtain.

Now Orbis is the largest hotel group in Poland and Central Europe. Apart from operating hotels belonging to the Group and building new ones, the Group also runs travel agency, car rental and lease business as well as international and local couch communication. At the beginning of the third millennium, the Orbis Group is a modern, dynamic hotel company that maintain the leading position on the growing hotel market in Poland and Lithuania.

In the year 1991, the state-owned enterprise Orbis was transformed into a company wholly owned by the State Treasury. On June 27, 1997, the General Meeting of Shareholders decided to offer Orbis S.A. shares on the public market. Orbis S.A. shares have been quoted on the Warsaw Stock Exchange since November 20, 1997.

AmRest[13] was established in October 2000, as a joint venture of American Retail Concepts (ARC) and the Yum Brands company (earlier operating as Tricon Global Restaurants). AmRest is the franchisee of KFC, Pizza Hut and Burger King restaurants. In 2007 AmRest has signed joint-venture agreements with Starbucks Coffee Company on the opening and operation of Starbucks outlets. The group also operates its own restaurants: freshpoint and Rodeo Drive. AmRest is currently present in Poland, Czech Republic, Hungary, Bulgaria, Russia and Serbia. AmRest Holdings SE is the largest independent restaurant operator in Central and Eastern Europe. Today, AmRest operates over 430 category

---

[12] See www.orbis.pl, and www.bankier.pl.
[13] See www.amrest.euwww.bankier.pl.

leading Quick Service and Casual Dining restaurants. The company has been quoted at WSE since April, 27, 2005.

Sfinks Polska S.A.[14] was established in 1999 as a joint stock company which previously had been registered as a sole proprietor business. Now Sfinks Polska is a network of 97 restaurants operated under the name of "Sphinx" (casual dining segment) placed in various locations across Poland and abroad, of 9 restaurants under the name of "Chłopskie Jadło" (premium segment) and 6 "WOOK" restaurants (casual dining segment). With regard to the total sales and the number of outlets, Sphinx is Poland's largest restaurant network in the casual dining segment. At the same time, Sfinks Polska is the third largest foodservice company in Poland in terms of its total sales turnover. Sfinks Polska S.A. is a Polish company that manages a capital group of companies (subsidiaries). These capital group member companies operate in Poland, the Czech Republic, Hungary, and Germany. It has been quoted at The Warsaw Stock Exchange since June, 8, 2006.

Table 3. Shares by trading value in years 2007-2009

Company	Year	Turnover value [mil. PLN]	Share in turnover [%]	Average volume per session (shares)	Average number of transactions	Average number of orders
Orbis	2007	1838.3	0.40	48206	78	278
	2008	1170.1	0.37	45045	59	216
	2009	483.2	0.15	27021	47	189
AmRest	2007	1838.3	0.40	32008	73	205
	2008	1235.3	0.39	29591	49	177
	2009	649.5	0.20	21518	55	182
Sfinks	2007	289.1	0.06	11880	55	92
	2008	126.4	0.04	13305	30	167
	2009	27.2	0.01	5235	25	136
Interferie	2007	379.4	0.08	77775	165	525
	2008	24.9	0.01	9837	12	99
	2009	17.0	0.01	7527	8	70
PolRest	2007	23.0	0.01	3993	22	95
	2008	6.1	0.00	1127	8	53
	2009	178.9	0.05	107553	148	442
Olympic	2007	10.9	0.00	4385	9	44
	2008	4.8	0.00	1635	5	40
	2009	7.2	0.00	5586	11	74
Rainbow	2007	2.9	0.00	4776	9	33
	2008	14.8	0.00	4156	14	109
	2009	50.0	0.02	37359	43	167
Polskie Jadlo	2007	×	×	×	×	×
	2008	19.9	0.01	14708	32	111
	2009	39.3	0.01	14847	27	136

Source: Rocznik Giełdowy 2008, 2009, 2010.

---

[14] See www.sfinks.pl and www.bankier.pl.

Interferie S.A.[15] in Lubin has been present on the tourist market since 1992. It was initially a limited liability company, yet in December 2004 it became a joint-stock company. In the beginning, the company took over several leisure centers, hotels and health centers that belonged to KGHM Polska Miedź S.A. Interferie S.A. is the greatest tourist company in the region of Lower Silesia and one of the biggest in Poland. The company has got several tourist centers, from high-class hotels to holiday camps. They are located in three climatically and geographically diverse regions of Poland i.e. at the Baltic seaside, in the mountains and at the lakes. Interferie has also got a travel agency and it is a member of the Polish Chamber of Tourism. Interferie has been quoted on WSE since August, 8, 2006.

PolRest[16] was founded in 1998 under the name Korporacja Krakowska S.A. but is has been functioning as PolRest since 2006. It operates 19 restaurants in two segments casual dining and premium. The company runs the net of 8 casual dining restaurants under the name Rooster in selected cities in Poland and several premium restaurants localized in Zakopane and center of Cracow. PolRest shares has been quoted on the Warsaw Stock Exchange since June, 6, 2007.

The first Olympic Casino opened its doors in Estonia at Tallinn Olympic Yachting Centre in the year 1993. Olympic Entertainment Group17 that operates under the Olympic Casino brand name is the largest provider of casino entertainment in the region. The Group operates casinos in Estonia, Latvia, Lithuania, Belarus, Romania, Poland and Slovakia. Shares of Olympic Entertainment Group are traded in the main list of the Tallinn Stock Exchange, and the Warsaw Stock Exchange since September, 26, 2007.

Rainbow Tours[18] was founded in 1990, and in the year 2003 it became a joint stock company. It is a travel agency that sells own tourist services and operates as tour-operator. It has 22 branches in Poland. Since October, 9, 2007 Rainbow Tours has been quoted at WSE.

Grupa Kosciuszko Polskie Jadlo S.A.[19] is well known entity which operates in the sector of gastronomic services in Poland. Activity of the company which was established in 1996 concentrated on development of "Chlopskie Jadlo" restaurant chain, which became one of the most recognizable and popular restaurants in Poland. The first public stock issue of Grupa Kosciuszko Polskie Jadlo S.A took place on June, 10, 2008.

2.2. Position of Touristic Companies at Polish Capital Market

To evaluate the role of the enterprises from the touristic sector in Poland, we compare companies, that belong to the sector hotels &restaurants, to all companies listed at The Warsaw Stock Exchange taking into account selected features. In Table 3 we present analysis of the touristic companies taking into account the trading values of shares. It is visible the dramatic drop in turnover value in 2009 in comparison to 2007 and 2008. For two the most important companies i.e. Orbis and AmRest share in turnover (in the total WSE turnover) varies from 0.40% in the year 2007 to 0.15 and 0.20 % respectively, in the year 2009. While turnover observed for other firms is smaller than 0.10%, and for Olympic and Rainbow

---

[15] See www.interferie.pl and www.bankier.pl.
[16] See www.polrest.pl and www.bankier.pl.
[17] See www.olympic-casino.com and www.bankier.pl.
[18] See www.rainbowtours.pl and www.bankier.pl.
[19] See www.grupakosciuszko.pl and www.bankier.pl.

smaller than 0.01%. It is worth mentioning that the share in turnover of all touristic companies decreased in following years from 0.95 in the year 2007, 0.82 – in 2008, to 0.45 in 2009. The biggest average numbers of orders, transactions and volume per session are observed for Interferie in the year 2007 and for PolRest in the year 2009.

Taking into account share prices, one can notice (Table 4) that in the year 2008 all companies (except Polskie Jadlo that was not listed in 2008) are characterized by negative annual rate of return. That was caused by drops of share prices in years 2007 - 2009. The negative returns are observed for Sfinks and Interferie also in the year 2007. The situation was improved in 2009 when all firms, except PolRest and Polskie Jadlo, obtained positive annual rate of return. The highest annual returns are observed for AmRest in the years 2007, 2009, and for Rainbow in the year 2009. The highest losses at the end of the year 2008 obtained investors who bought shares of Olympic, AmRest, Rainbow and PolRest at beginning of the year.

Table 4. Shares by return in years 2007-2009

Company	Year	Market capitali-zation [mil. PLN]	Market share [%]	Rate of return since year beginning		Share	price	in PLN
Orbis	2007	3202.35	0.63		euro	low	high	end
	2008	1499.81	0.56	10.6	17.6	57.00	95.00	69.50
	2009	2043.98	0.49	-52.7	-59.1	30.90	70.00	32.55
AmRest	2007	1898.86	0.33	36.3	38.4	23.23	56.90	44.36
	2008	680.95	0.34	80.8	92.1	67.55	158.90	134.00
	2009	1163.28	0.39	-64.2	-69.0	41.11	136.50	48.00
Sfinks	2007	243.16	0.05	70.8	73.5	30.00	95.40	82.00
	2008	100.23	0.04	-47.6	-44.3	22.02	78.50	26.20
	2009	104.41	0.02	-58.8	-64.4	10.80	26.29	10.80
Interferie	2007	75.44	0.01	3.1	4.71	6.00	16.20	11.25
	2008	60.88	0.02	-45.5	-42.1	4.66	13.53	5.18
	2009	77.92	0.02	-19.3	-30.3	3.92	5.93	4.18
PolRest	2007	123.75	0.02	28.0	30.0	3.71	5.98	5.35
	2008	48.38	0.02	×	×	×	×	×
	2009	10.25	0.00	-60.9	-66.2	6.91	20.50	7.74
Olympic	2007	2005.28	0.35	-78.8	-78.5	1.50	7.74	1.64
	2008	347.30	0.18	×	×	×	×	×
	2009	422.80	0.14	-82.4	-84.8	2.00	13.60	2.30
Rainbow	2007	93.24	0.02	21.7	23.7	1.80	3.70	2.80
	2008	35.40	0.01	×	×	×	×	×
	2009	59.30	0.01	-62.0	-67.2	2.83	9.50	2.95
Polskie Jadlo	2007	×	×	66.8	69.4	1.01	5045	4.92
	2008	130.51	0.05	×	×	×	×	×
	2009	66.12	0.02	×	×	×	×	×

Source: Rocznik Giełdowy 2008, 2009, 2010.

As one can notice in Table 4, Orbis has been the company with the biggest capitalization among investigated firms. The share of the market capitalization of enterprises that are listed as hotels & restaurants in total capitalization of WSE was 1.11% in the year 2007, 1.22 in the year 2008 and 1.07 in 2009.

2.3. Trend Analysis

Further analysis is provided for the economic results gained by considered companies in years 2006 – 2007 due to quarterly reports published by the companies. Comparison of revenues obtained by the companies let us to divide them into 3 groups[20]:

1. Orbis, Rainbow and Interferies that are characterized by strong seasonality since they are operating as tourist agencies (see Figure 10);
2. Olympic, AmRest and Sfinks that are characterized by relatively high incomes and operate abroad (see Figure 11);
3. PolRest and Polskie Jadlo with realitively small number of restaurants and incomes (see Figure 12).

Applying linear trend models with additional binary variables to describe seasonality or additional factors that caused changes in the revenue path we try to describe the tendencies in the companies' development thru time.

To describe revenues of these companies we estimate the trend functions with seasonal coefficients in following form:

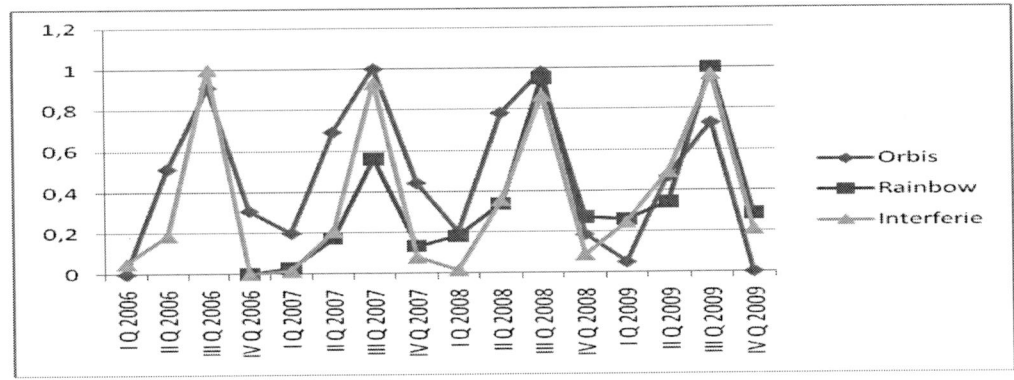

$$y_t = \beta_0 + \beta_1 t + \beta_2 z_1 + \beta_3 z_2 + \beta_4 z_3 + \beta_5 z_4 + \varepsilon_t \qquad (1)$$

Source: Own elaboration on basis of data www.bankier.pl 5.05.2010.

Figure 10. Quarterly normalized revenues of Orbis, Rainbow and Interferie in years 2006-2009.

---

[20] To make revenues (that are presented at Figures 10 – 12) comparable, they are normalized due to formula: $y_{it} = \frac{x_{it} - x_i^{min}}{x_i^{max} - x_i^{min}}$; where yit – is normalized value of revenue, xit – is original value of revenue, x_i^{min} ; x_i^{max} - are minimal and maximal values respectively.

Table 5. Trend function with seasonality parameter estimates

Company		z_4	z_3	z_2	z_1	trend	const.
Orbis	$\hat{\beta}_i$	-36033.90	73537.16	23624.97	-61128.23	-1662.94	285960.10
	t	$R^2 = 0.88$	7.32	2.35	-5.98	-1.28	22.99
Rainbow	$\hat{\beta}_i$	-24337.29	56865.37	-10018.46	-22509.62	3883.84	19313.38
	t	$R^2 = 0.95$	10.88	-1.94	-4.31	4.94	2.30
Interferie	$\hat{\beta}_i$	-2637.71	5388.72	-374.85	-2376.17	124.18	8567.15
	t	$R^2 = 0.96$	16.32	-1.14	-7.08	2.92	20.97

Source: Own calculations.

where: z_1, z_2, z_3, z_4 - binary variables that are dedicated to the seasonal effects in quartiles Q_i (i = 1, 2, 3, 4), $\beta_1, \beta_2, \beta_3, \beta_4, \beta_5$ – trend and seasonal coefficients, t – time variable, ε_t – random coefficient.

Trend functions are estimated using OLS method. Parameter estimates[21], values of *t*-Student statistics and determination coefficients R^2 are presented in Tables 5 - 7.

According to the results presented in Table 5, we may claim that operating of three companies has seasonal character since all (but one) seasonal coefficients essentially differ from zero at the significance level $\alpha = 0.05$. Seasonal effects are positive in the third quarter and negative in the first and fourth quarters. In the second quarters the seasonal effects are positive and significant for Orbis, negative and significant for Rainbow and insignificant for Interferie. It is worth mentioning that trend coefficient is insignificant for Orbis that means that this company has already achieved certain level of development and the situation of the enterprise is stabilized. While "young" companies Rainbow and Interferie have been still developing so the trend coefficients are positive and significant.

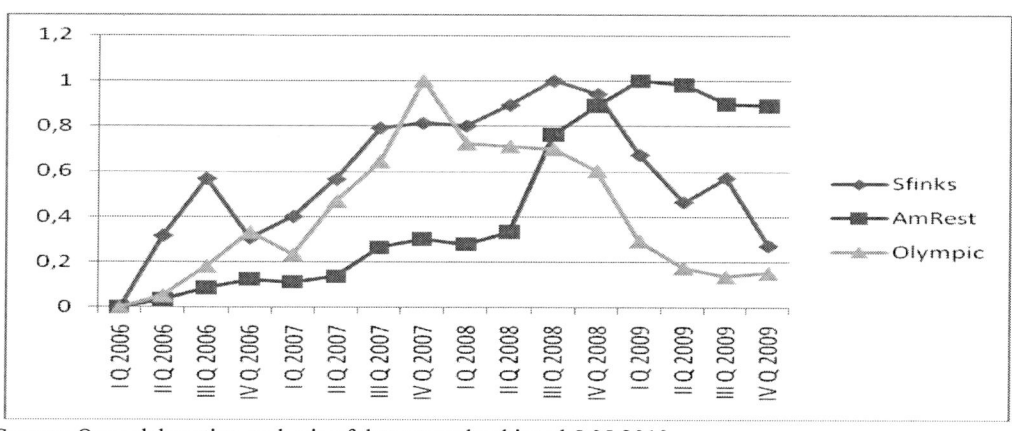

Source: Own elaboration on basis of data www.bankier.pl 5.05.2010.

Figure 11. Quarterly normalized revenues of Sfinks, AmRest and Olympic in years 2006-2009.

---

[21] Estimation of model in the form (1) is impossible because matrix XTX is singular. Therefore the model was transformed but quarterly effect for the last period of the year is not estimated but it is calculated since the sum of all seasonal effects is equaled zero.

Table 6. Trend function parameter estimates

Company		sample	trend	const.	sample	trend	const.
AmRest	$\hat{\beta}_i$	t = 1,..., 16	29467.24	56904.55			
	t	$R^2 = 0.87$	9.84	1.96			
Olympic	$\hat{\beta}_i$	t = 1,..., 8	60833.23	249524.86	t = 9,..., 16	-49717.20	781931.04
	t	$R^2 = 0.88$	6.76	5.45	$R^2 = 0.89$	-6.87	21.41
Sfinks	$\hat{\beta}_i$	t = 1,..., 11	2041.93	33520.35	t = 12,... 16	-3436.70	56054.70
	t	$R^2 = 0.86$	7.47	18.07	$R^2 = 0.84$	-3.94	19.37

Source: Own calculations.

Revenues gained by the chains of restaurants and casinos that operate in Eastern and Central Europe seem to develop with constant tendency however revenues have been increasing only for AmRest while for Sfinks and Olympic[22] two sub-periods with the opposite tendencies can be distinguished. Therefore we use linear trend function:

$$y_t = \beta_0 + \beta_1 t + \varepsilon_t \qquad (2)$$

estimated for the whole sample or for both sub-samples separately.

As one can notice in Table 6 trend coefficient for AmRest, estimated for the whole sample is positive and significant for the significance level α = 0.05 while the slope of the trend function did not significantly differ from zero and determination coefficients are extremely small $R^2 = 0.02$ for Olympic and $R^2 = 0.15$ for Sfinks. Therefore for both companies we estimate trend functions separately for distinguished subsamples. Olympic revenues started to decrease in the first quarter 2008 so both trend functions are estimated for 8 observations. The decreasing tendency for Sfinks was observed 3 quarters later i.e. in the last quarter 2008 so the first trend function is estimated using 11 elements, and the second one – for 5 observations. Parameter estimates prove that both companies offering more luxuries services have been suffering from crisis.

PolRest and Polskie Jadlo has been listed at The Warsaw Stock Exchange for relatively short period of time so the time series of data that characterize both companies contains only 15 and 11 quarterly observations. Therefore it is more difficult to anlyse these enterprises, especially that the plots of the revenues obtained by both firms are not linear. The simple trend function (2) explains the revenue changes in 1% ($R^2 = 0.008$) in case of PolRest and in 2% ($R^2 = 0.022$) in case of Polskie Jadlo. We assume that:

- there may be some disturbances that can be represent by binary variable d, and we estimate trend models as following:

$$y_t = \beta_1 t + \beta_2 d + \varepsilon_t \qquad (3)$$

- the seasonality appears then we estimate model (1).

---

[22] There was lack of data for the second quarter Q2 in the year 2008 and this missing observation was replaced by the average for neighboring quarters i.e. Q1 2008 and Q3 2008.

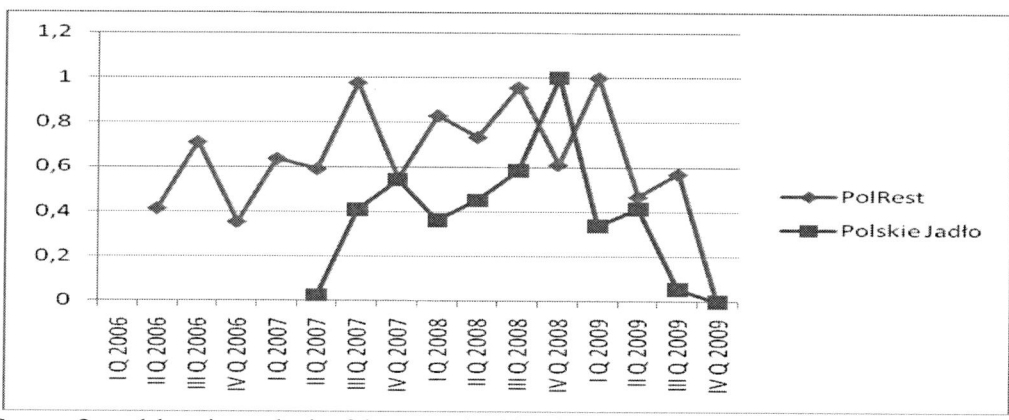

Source: Own elaboration on basis of data www.bankier.pl 5.05.2010.

Figure 12. Quarterly normalized revenues of PolRest and Polskie Jadlo in years 2006-2009.

For both companies (see Table 7) trend and additional effects represented by the binary variable d are significant at the significance level $\alpha = 0.05$. In case of PolRest in the estimated model (1) seasonal coefficients are significant at $\alpha = 0.1$ and $R^2 = 0.52$. Therefore we may claim that for this company slight seasonal effects appear.

Table 7. Trend function with seasonality parameter estimates

Company		$d=1$ for QIV 2008	trend	Company		$d=1$ for QIV 2009	trend
PolRest	$\hat{\beta}_i$	-13015.58	1013.04	Polskie Jadlo	$\hat{\beta}_i$	8469.57	565.92
$R^2 = 0.79$	t	-2.55	6.98	$R^2 = 0.69$	t	2.08	3.13

Source: Own calculations.

Table 8. Characteristic of the company activity

Company	Travels abroad	Domestic tourism	Hotels and resorts	Transportation	Tour-operator	Restaurants	Casino
AmRest						×	
Interferi		×	×				
Olympic							×
Orbis	×	×	×	×	×	×	
PolRest						×	
Polskie Jadlo						×	
Rainbow	×		×		×		

Source: Own elaboration on the basis of websites of the companies.

Table 9. List of variables

Symbol of variable	Variable	Description of variable
x_1	ROR	Return On Revenue
x_2	ROTA	Return On Total Assets
x_3	RONA	Return On Net Assets
x_4	EBITDA	Erning Before Interest, Taxes, Depreciation and Amortization
x_5	ROE	Return On Equity
x_6	ROA	Return On Assets
x_7	SER	Shareholders Equity Ratio
x_8	NRS	Net Revenue per Share
x_9	NOI	Net Operating Income
x_{10}	GIPS	Gross Income Per Share
x_{11}	EPS	Net Income per Share
x_{12}	BVS	Book Value per Share
x_{13}	CFS	Cash Flow per Share
x_{14}	EBITDAPS	EBITDA per Share
x_{15}	P/E	Price – Earnings Ratio (Market Value per Share/Earnings per Share)
x_{16}	P/BV	Price /Book Value
x_{17}	P/CF	Price to Cash Flow Ratio

Source: Own elaborations.

2.4. Multivariate Comparable Analysis

Investigated companies, that are recognized by the WSE as hotels and restaurants, operate on different markets: domestic and abroad, they offer various services, and they address they offer to different clients (see Table 8). One should also notice that each firm is characterized by different background and they have been running for varous length of time. Therefore it is imposible to compare them directly so we construct synthetic taxonomic measures to make such comparison possible.

The synthetic taxonomic measure SMR_{it} is defined as (see Hellwig, 1968):

$$SMR_{it} = 1 - \frac{q_{it}}{\bar{q}_t + 2 \cdot S_{qt}} \quad i = 1, 2, ..., n;\ t = 1, 2, ..., T \qquad (4)$$

where q_{it} is the distance of the i-th object from the benchmark:

$$q_{it} = \sqrt{\sum_{j=1}^{k}(z_{jt}^i - z_{jt}^0)^2} \quad z_{jt}^0 = \begin{cases} \min_{i=1,2,...n}\{z_{jt}^i\}\ for\ x_{jt}^i \in D \\ \max_{i=1,2,...n}\{z_{jt}^i\}\ for\ x_{jt}^i \in S \end{cases} \qquad (5)$$

evaluated for standardized variables z_{jt}^0, z_{jt}^i that describe the benchmark and the i-th investigated company, respectively for each period t and the j-th variable:

$$z_{jt}^i = \frac{x_{jt}^i - \bar{x}_{jt}}{S_{jt}^x} \qquad (6)$$

where for the i-th company and t-th period: z_{jt}^i - standardized variables, x_{jt}^i - observations of raw variables, $\bar{x}_{jt}$, S_{jt}^x - average and standard deviation, respectively. D and S are sets of destimulants and stimulants, $\bar{q}_t$, S_{qt} - the average and the standard deviation of distances q_{it}, respectively:

$$\bar{q}_t = \frac{1}{n}\sum_{i=1}^n q_{it} \qquad S_{qt} = \sqrt{\frac{1}{n}\sum_{i=1}^n (q_{it} - \bar{q}_t)^2} \qquad (7)$$

Table 10. Ranking of companies made due to values of the indicator SMR_{it}

No.	Set of	No. of class	2007	2008	2009
1	17 variables	I	Orbis	AmRest	AmRest
2			Olympic	Orbis	Orbis
3		II	AmRest	Polskie Jadlo	Interferie
4			PolRest	PolRest	Olympic
5		III	Interferie	Interferie	Rainbow
6			Sfinks	Olympic	Polskie Jadlo
7			Rainbow	Rainbow	Sfinks
8		IV	Polskie Jadlo	Sfinks	PolRest
1	6 variables	I	Olympic	PolRest	Orbis
2		II	Orbis	Polskie Jadlo	Olympic
3			PolRest	Olympic	Sfinks
4		III	AmRest	AmRest	Interferie
5			Interferie	Orbis	AmRest
6			Polskie Jadlo	Sfinks	Polskie Jadlo
7			Sfinks	Interferie	PolRest
8		IV	Rainbow	Rainbow	Rainbow

Source: Own calculations.

To construct the synthetic measures (4) we use annual observations of 17 variables[23] that are presented in Table 9, variables $x_1 - x_{14}$ are stimulants, while $x_{15} - x_{17}$ are de-stimulants. Taxonomic indicators (4) are calculated for 2 different sets of variables that contain: either all 17 variables: $x_1 - x_{17}$ or 6 variables: $x_4 - x_6$ and $x_{15} - x_{17}$.

On the basis of the synthetic measures SMR_{it}, it is possible to construct a ranking of all companies and to classify them into homogenous groups in terms of the financial and economic performance in years 2007 – 2009. We distinguish four classes of firms[24]:

I. the best firms in the sector for $SMR_{it} \geq MSR_t + S_{SMRt}$;
II.– good firms in the sector for $SMR_t + S_{SMRt} > SMR_{it} \geq SMR_t$;
III.–catching up firm in the sector for $SMR_t > SMR_{it} \geq SMR_t - S_{SMRt}$;
IV.–poor firms in the sector for $SMR_{it} < SMR_t - S_{SMRt}$;

$$SMR_t = \frac{1}{n}\sum_{i=1}^{n} SMR_{it} \quad S_{SMRt} = \sqrt{\frac{1}{n}\sum_{i=1}^{n}(SMR_{it} - SMR_t)^2} \tag{8}$$

The ranking of firms is presented in Table 10, and it differs due to the sets of variables that are taken into account. Orbis belongs to the first or the second class in all classifications except the one made for 2008 using 6 variables. AmRest is recognized as good or very good firm by the indicator constructed for 17 variables while as the catching up company if the index contains only 6 descriptors. Rainbow seems to be the worst enterprise when measure with reduced number of variables is employed while using SMR_{it} with 17 features, it is the third class firm.

CONCLUSION

Touristic industry plays an important role in the national economy, although in Poland the share of this sector in GDP and in export is smaller than the average for the world. The biggest incomes of this sector are created by the expenditures of Polish residences for domestic travels and the foreigners' expenditures in Poland.

Analysis of the companies belonging to this sector and being listed at the Warsaw Stock Exchange proved the hypothesis that their position in terms of: number of firms, market capitalization, turnover value and rate of returns is not essential at Polish capital market.

The lost in capitalization of these companies in the year 2009 is 50% in comparison to the year 2007 and this lost is higher than another sectors of the Polish economy. Restoring of the former level will be very difficult and it requires time and it depends on the company size.

Among eight companies that are classified by WSE as hotels and restaurants, Orbis is the firm with the longest tradition and it was listed at The Warsaw Stock Exchange as the first one. This firm keeps good position at the touristic market in terms of synthetic measures and

---

[23] Discussion about variables that should be included into synthetic taxonomic measures to evaluate the financial and economic condition of firms can be found in Łuniewska, Tarczyński, 2006, p. 46 -53, , p. 203 -331.
[24] See Nowak 1990.

trend analysis. Also enterprises that address their services to regular clients seem to be in good economic and financial conditions. While the companies that offer more luxuries services seem to face the economic crises.

REFERENCES

Ashley C., De Brine P., Lehr A., Wilde H., *The Role of the Tourism Sector in Expanding Economic Opportunity*, Harvard University, The fellows of Harvard College, Overseas Development Institute, International Business Leaders Forum, 2007 http://www.hks.harvard.edu/publications/report_23_EO%20Tourism%20Final.pdf

Bartoszewicz W., Łopaciński K., Prognozy wyjazdów z poszczególnych krajów i przyjazdów do Polski w latach 2006-2009 oraz prognoza globalna przyjazdów do Polski do 2013 roku, http://www.pot.gov.pl/dokumenty/dane-i-wiedza/badania-i-analizy/Ekspertyza.pdf/, Instytut Turystyki, Warszawa, 2007.

Bull, A., *The Economics of Travel and Tourism*, Longman Cheshire Pty Ltd., 1995.

Dann, G., Nash, D., Pearce, P., Methodology in Tourism Research, *Annals of Tourism Research*, 1988.

Gannon A., Rural Tourism as a Factor in Rural Community Economic Development for Economies in Transition, *Journal of Sustainable Tourism*, Volume 2, Issue 1 & 2, 1994, p. 51 – 60.

Hellwig Z., Zastosowanie metody taksonomicznej do typologicznego podziału krajów ze względu na poziom ich rozwoju oraz zasoby i strukturę kwalifikowanych kadr, *Przegląd Statystyczny*, 1968.

Jenner, P., Smith, C., Special Report - Economist Intelligence Unit, Economist Intelligence Unit, 1992.

Łuniewska M., Tarczyński W., *Metody wielowymiarowej analizy porównawczej na rynku kapitałowym*, PWN, Warszawa, 2006. *Mały Rocznik Statystyczny*, GUS, Warszawa, 2008.

Mathieson, A., Wall, G., *Tourism, Economic, Physical and Social Impacts*, Longman, 1982.

Mitchell J., Ashley C., *Tourism and Poverty Reduction. Pathways to Prosperity,* Earthscan, London, 2009.

Nowak E., *Metody taksonomiczne w klasyfikacji obiektów społeczno - gospodarczych*, PAN, Warszawa 1990.

Rocznik Giełdowy 2008. *Dane statystyczne za rok 2007*, Giełda Papierów Wartościowych w Warszawie, Warszawa, 2008.

Rocznik Giełdowy 2009. *Dane statystyczne za rok 2008*, Giełda Papierów Wartościowych w Warszawie, Warszawa, 2009.

Rocznik Giełdowy 2010. *Dane statystyczne za rok 2009*, Giełda Papierów Wartościowych w Warszawie, Warszawa, 2010.

Tarczyński W., *Rynki kapitałowe. Metody ilościowe*, Placet, Warszawa, 2001.

"UNTWO World Tourism Barometer, Vol.5 No.2". www.tourismroi.com. http://www.tourismroi.com/Content_Attachments/27670/File_633513750035785076.pdf. Retrieved 2009-10-12.

Witkowska D., *Podstawy ekonometrii i teorii prognozowania*, Oficyna Ekonomiczna, Kraków, 2005.

In: Research Studies on Tourism and Environment
Editors: J. Mondejar-Jimenez et al.
ISBN: 978-1-61209-946-0
© 2012 Nova Science Publishers, Inc.

Chapter 8

OVERBOOKING AS A DOUBLE-SELLING PRACTICE IN THE TOURIST INDUSTRY: A BRIEF VIEW FROM SPANISH JURISPRUDENCE

María Valmaña Ochaíta[*]
Universidad de Castilla-La Mancha, Spain

ABSTRACT

Overbooking is a commercial strategy that has became very relevant in the last years in the tourist industry. Although overbooking is well known in sectors like air transport overbooking is also used by other kind of companies like hotels in order to optimize resources and infrastructures.

Despite overbooking being regulated in the European Union by an specific regulation in order to protect consumers there are still many questions debatable from economic as well as legal point of view.

1. INTRODUCTION

This work tries to offer a general vision of the practice of overbooking and its most relevant aspects. The analysis of these aspects includes those decisions of the Spanish jurisprudence in the particular cases in that there are interesting uprisings on the raised questions. The perspective of the analysis is fundamentally legal but also takes into account others, of economic nature, as long as the practice of overbooking plays an important role in the commercial strategy of the company. Most literature on overbooking is of economic nature.

---

[*] Contact author: maria.valmana@uclm.es

2. DEFINITION OF OVERBOOKING

Overbooking is a term, which literally means excess reserves (*over* and *booking*). But nowadays the concept of overbooking is different and *overbooking* is understood as a situation of oversold (*overselling*) and in fact, in the Dictionary of the Spanish Academy of Language, is defined as "sale of seats, especially hotel and aircraft, exceeding the available number"[1].

Generally, overbooking is related to some specific activities in the tourist industry such as airlines transport but it extents also to others services like maritime transport, communications by road or by train, as well as hotel services.

Generally, overbooking is related to some specific activities in the tourist industry such as airlines transport but it extents also to others services like maritime transport, communications by road or by train, as well as hotel services.

Overbooking can appear in a variety of forms even in the same area of tourism activities. In this sense, we can say that overbooking not only appears when the client cannot be hosted in the hotel because of overselling but also when the client can be hosted in the room but cannot use some of the services that were contracted by him. This is the matter under decision in the sentence of the Provincial Jury of Madrid, the first of October 2008 [2]. In this case clients of a hotel under an all-inclusive regime were not able to have dinner in specific restaurants of this hotel because the restaurant that worked under a pre-booking system was over-demanded.

On the other hand, it is a topic that overbooking is not a fair practice, even illegal. However, in general overbooking is allowed and regulated in European Law. Also it has a large tradition in some countries like the United States of America where it is considered by scholars and practitioners like a practice that let the tourist business to be more efficient, reducing prices and optimizing resources.

In this sense, it is possible to say that overbooking sometimes can be considered like something positive for consumers when it is possible to obtain some form of compensation like recovering the ticket price or hotel reservation.

Overbooking is relating to massive and impersonal practices and sometimes presents problems in order to be integrated and incardinated with the fundamentals of traditional Contract Law.

Different factors can be drawn up in relation with the increasing significance of overbooking. In countries like Spain, tourism represents one of the most important sectors of its economy. Tourist activities have become not only for elite groups but for consumers in general in the last decades. This generalization, sometimes called democratization of tourism and leisure, is due to different factors like reducing cost of services, better communications and a higher level of life of the general population.

Although this work analyzes overbooking from a legal point of view it also pays attention to the economic aspects of this practice because only by understanding the economical impact it is possible to understand the whole concept.

---

[1] *Vid.* Feal Mariño, (2003), *The overbooking in air transport,* Navarra, pp. 78 and 79 and also, vid. the 20th Edition of the Dictionary of the Royal Academy of the Spanish Language (http://buscon.rae.es/draeI/SrvltConsulta?TIPO_BUS=3&LEMA=overbooking, document recovered the 3th of July, 2009).

[2] Sentence of Provincial Court of Madrid num. 515/2008 (Section 14), 1th of October of 2008.

3. OVERBOOKING TIPOLOGY

Overbooking can be distinguished mainly in the field of transport and hotels. In case of transport, Regulation (EC) No 261/2004 of the European Parliament and of the Council of 11 February 2004 lays down rules on compensation and assistance to passengers in the event of denied boarding, cancellation of flights or long delay. This is probably the best known mode of regulation. Some institutions and doctrine see in the regulation an opportunity to extent it to other types of transport such as bus and boat which are taken into account already in Regulation (CE) Nos *1371/2007* European Parliament and the Council of 23 October 2007 on rights and obligations of rail passengers which recognizes the rights of rail passengers while in the latter there is no specific reference to the *overbooking*.

Article 4 of Regulation (EC) 261/2004 regulates the content of the rights of airplane-passengers in the case of *overbooking* when referring to the denial of boarding. It provides for the scope and procedure to which the airline must comply in order to accredit that denial of boarding of a passenger by lack of places is not unlawful.

Regarding the subjective scope of the Regulation, it basically applies to passengers that have reservations and present themselves on time for checking and boarding. Passengers with free tickets are excluded of this protection unless these free tickets were obtained by a loyalty program.

With respect to the territorial scope, the regulation applies to passengers with departure or destination to countries in EU and operated by community airline[3].

The regulation establishes a procedure for action on the part of the carrier in the event of *overbooking*. In particular, the airline should seek volunteers among the passengers to renounce their reservations and agree the terms with them. If the number of volunteers is not enough, it may deny boarding to passengers against their will resulting in an obligation to financially compensate the passengers affected with a quantity which, according to article 7, varies between 250 to 600 Euros, as well as the obligation to provide adequate assistance (articles 8 and 9), reimbursement of the ticket or obtain re-routing for the passengers as well as providing services of meals and lodging or reimbursing the cost involved.

Regarding *overbooking* in the field of hotels services, unlike the case of air transport, it is not regulated specifically. In any case, there are some regulations that relate to package travel[4], such as those contained in the Royal Decree Legislative 1/2007 of 16 November, approving the consolidated text of the General Law for the Consumer Protection, and other complementary legislation (articles 150 et. seq., and in particular articles 159 et seq. on resolution of the contract and cancellation and derived responsibilities) as well as some decisions of the jurisprudence which can serve as a guide in the resolution of specific problems that may arise in this area.

On this basis, it should be noted that the practice of *overbooking* in hotels services involve a breach of the contract of accommodation in the hotel and therefore it recognizes the right of the customer to demand to be hosted in an establishment of equal or higher category and the right to obtain compensation for the damage suffered. In this sense, different kind of booking by the client can be distinguished. There is a guaranteed reservation if the reservation

---

[3] Vid. in this sense the sentence of Provincial Court of Barcelona SAP No. 210/2008 (part 15), 4th of June of 2008.
[4] Consolidation of Consumer Protection Act of 2007 integrates many regulations. Among others, the previous regulation on "Combined Trips of 1995".

is made through a *tour* operator or via a credit card. There is a non-guaranteed reservation, requiring customer to present himself before the set time, generally 6:00 p.m. The content of rights in the latter case is less than the first.

On the assumption that the *overbooking* occurs as part of a package, the range of responsible subjects is wide according to the rules applicable and cited above. The responsible subjects include the organizer but also the retailer who can be sued under the solidarity rules of liability.

On the other hand, a hotel or other employers offering hosting service can be exonerated from liability by *overbooking* in case the overbooking situation is due to circumstances not attributable to them. This is established in article 162 of the Consolidation of the Consumers Protection Act of 2007, when it refers "... by such force majeure, understanding those circumstances beyond who those invoked, abnormal and unpredictable consequences not have been avoided, in spite of having acted with diligence due..." The jurisprudence is very restrictive in interpreting this assumption of exclusion of liability in the event of *overbooking* in hotels since the *overbooking* "constitutes a normal and foreseeable impact" for this type of entrepreneurs [5].

4. SOME CONSIDERATIONS ON ECONOMIC ASPECTS OF OVERBOOKING

Overbooking has been analyzed in relation with the economic model of "yield management" or "revenue management". This type of model allows benefits and profits for companies by predicting the behavior of consumers. Overbooking in hotels presents, for example, the chance to present flexible prices for clients that made reservations very early[6].

Overbooking is a business strategy also for air carriers. It is essentially used by those companies that have a significant number of passengers with the right to cancel reservations without penalty or with a small economic penalty because of the type of ticket. In consequence, overbooking is a commercial mechanism of the tourist industry in order to be more efficient with their resources and infrastructures.

The negative economical impact for airline companies having to fly with empty seats, due to, for example, "no shows" is big.

At the same time the amount of times overbooking occurs, with passengers being denied boarding, is statistically very low compared to the amount of flights executed.

These two facts are the reason that doctrine considers overbooking as a valuable mechanism but it also considers important a well established system of compensation for consumers that experience overbooking.

Certain economic studies have rated the profitability that this practice represents for airlines, differentiating between airlines that have a significant number of *business class* seats and so called low cost airlines with general no business seats. These studies conclude that the overbooking strategy is only economically beneficial for non low cost airlines with expensive business and first class seats for which the number of cancellations or no-shows is more

---

[5] *Vid.* the sentence of Provincial Court of Baleares Islands num. 475/2007 (Section 3), 4th of December of 2007.
[6] *Vid.* Takeshi Koide y Hiroaki Ishii (2005), The hotel yield management with two types of room prices, overbooking and cancellations, International Journal of Production Economics, Vol. 93-94, pp. 417-428.

frequent and the economic consequence bigger since tickets for these seats are expensive. For low cost airlines, where the planes have a higher occupation grade, the average seat price is low and cancellations are less frequent, an overbooking strategy makes no sense since the cost of compensation will outweigh the benefits[7].

Overbooking has, without doubt, advantages and disadvantages both for the client company. For customer, the advantages that can be highlighted are numerous. It results in cheaper ticket prices since it allows a more efficient employment of aircrafts to the airline[8]. It allows greater flexibility in traveling and it improves availability of seats for potential travelers that otherwise would stay unfilled.

However it is remarkable that in the valuation of pros and contras of overbooking there are some elements that traditionally have not been taken in account. Overbooking can be a profitable activity if it is considered like a mechanism to avoid the "lost of opportunities" representing by no-shows or free space in the plane that could be used and sold to other clients. But there is another kind of potential client called "flight switchers" that is not appreciate by this theory: the client that can't buy a ticket in a plane that is full but could buy another flight in the same company[9].

For airlines, *overbooking* may be a positive commercial strategy that allows to optimize their resources and capabilities, it also presents an economic risk that corresponds to any compensation to be given to passengers which have been denied boarding along with other possible negative effects such as loss of credibility and prestige among its clientele[10]. In this regard, some airline companies, particularly those of *low cost*, publicly rejected fixed compensation schemes established in the Community regulation on rights of passenger aircraft, postulating the necessity that such a system needs to be replaced by another fixed compensation depending on the price of the ticket. In any case it should be taken into account that an approach of this nature could also trigger problems insofar as the choice of the passage to which boarding is denied could not meet criteria entirely righteous depending, precisely, the cost of the ticket and, therefore, possible compensation to meet by that company.

---

[7] Conde Tejón (2008), p. 114.
[8] Conde Tejón (2008), p. 114.
[9] Hereafter we reproduce the interesting considerations of Suzuki (2006), pp. 1-19: "These figures, however, may not reflect the true benefit of overbooking to airlines. Typically, the benefit of overbooking is estimated by taking the difference of the loss of revenues caused by no-shows (and late cancellations) in closed flights when overbooking is not used, and that when overbooking is used (e.g., Alstrup et al., 1989; Nambisan, 2003). Thus, the benefit is typically estimated by simply calculating the amount of revenue gains attained for overbooked (congested) flights, and ignoring the revenue impact of overbooking on other (un-congested) flights. This condition implies that, if overbooking affects revenues of not only congested flights but also un-congested flights, the benefit calculated by the standard approach may be misleading. In theory, overbooking can affect revenues of both congested and un-congested flights. When overbooking is used, reservations for congested flights will increase, as airlines will accept reservations by ''additional'' passengers who demand seats after the flight capacity is reached. Theoretically, these ''additional'' passengers can be classified into two types. The first is ''new customers''. Without overbooking, some travelers demanding seats on an airlines congested flights may have to give up flying or fly by other airlines. The use of overbooking allows an airline to accept reservations by these travelers whose demand would otherwise be lost to other airlines or to other modes of transportation. The second is ''flight switchers''. In the absence of overbooking, travelers who cannot use their preferred flights (congested flights) may reserve seats in other flights of the same airline (e.g., flights with inconvenient arrival times). 2 If overbooking is used, these ''compromising'' travelers may ''come back'' to their preferred flights. Note that if all the ''additional'' passengers are new customers, overbooking has no revenue impact on un-congested flights. If, however, the ''additional'' passengers include flight switchers, overbooking affects revenues of not only congested flights but also un-congested flights, because in this case the revenue gains for the former flights are realized, at least partially, at the expense of reduced revenues for the latter flights."
[10] Vid. Wangenheim & Bayon (2007), p. 36-47. Also Lindenmeier & Tscheulin (2008), pp. 32-43.

In conclusion, the opportunity to regulate the *overbooking* at Community level does not seems to be disputed. However concerning the valuation of certain measures, as the fixing of compensation and extension of the rights of passengers referred to in rules as the 2004 regulation there are many legal and economic issues up for discussion. Issues that must remain subjects of analysis and reflection by scholars in this area to ensure a strengthening of the rights of consumers as part of a contract related to the transport and tourist sector, without prejudice to the economic characteristics within the broader framework of tourist activity.

5. COMPENSATION IN OVERBOOKING CASES: SOME VIEW FROM SPANISH JURISPRUDENCE

Compensation in overbooking can be seen as a mechanism to equilibrate the different interest of airline companies and customers, as it allows the companies to be more efficient with their resources and it ensures customers to obtain a compensation when they are bumped of a plane.

There is an UE Regulation about compensation and assistance to passengers of airlines. This is a fixed compensation (economic compensation of 250 -600 euros (art. 7) and assistance (arts. 8 y 9), refund ticket or alternative transport and care - food, lodging, etc).

In overbooking cases it is very important to determine who is the responsible carrier. In this sense, Regulation (EC) No 261/2004 of the European Parliament and of the Council of 11 February 2004 establishes common rules on compensation and assistance to passengers in the event of denied boarding, cancellation or long delay of flights. It determines as well that it is the carrier who should deal with compensation even if other subjects are involved in the process.

Sometimes there are other types of damaging situations for passengers (like lost of baggage) coexisting with a situation of overbooking. In this sense it is important to mention the case of some travel packages that were regulated by the former "Combined Trips Act" whose text have been refunded and consolidated in the Consolidation Consumers Protection Act of 2007. This law establishes in article 162 the responsibility of organizers and retailers, without fixing amount of money, unlike regulations on overbooking, with compensation limits.

In this respect an interesting ruling from the Provincial Court of Baleares of 23 January 2006 can be mentioned[11], concerning a claim of a married couple to make a honey-moon trip to Dominican Republic. They took a flight from Majorca to Madrid and from there they would take a flight to the final destination. Due to an overbooking problem, the flight that in principle would be made with the company Spanair was finally done with the Air Europe company which had therefore carrier status. The reason behind the claim was due to loss of baggage at the airport of Barajas. Baggage loss brought charges and several disadvantages to the couple that could not enjoy some of the hotel services such as dinners for not having adequate clothing, etc. The question of interest here would be to determine the possible responsible subjects that generated the situation. According to the regulations on "combined trips", the plaintiff could have claimed the organizer or wholesaler and retailer, Viajes Marsans, but they directed their actions against Spanair (which should have conducted the

---

[11] Sentence of Provincial Court of Baleares Islands num. 13/2006 (Section 4), 23 of January of 2006.

transport Palma Mallorca-Madrid) and AirComet (which should have conducted the transport Madrid-Dominican Republic). And also, in these cases there would be a quantitative limit to any compensation.

Spanish jurisprudence treats as well some annexed questions related to overbooking. It relates for example to the determination of the scope of compensation for damages perceived related to overbooking situations. It mentions the possibility that the airline company has the obligation to also repay not only to the passenger but also to other subjects such as travel agencies that have undergone damages by contract cancellations of clients that have suffered overbooking situations. This is a subject not regulated by the Regulation of 2004. This regulation only talks about the compensation of the damage (defining minimum compensation levels) caused to the passengers in situations of overbooking. But the Spanish jurisprudence has been pronounced on the matter through some decisions of the courts like the Sentence of the Provincial Court of Madrid of 4 of February if 2002[12]. It was a decision concerning the claim of a travel agency against Iberia Airlines because of the cancellation of the contract with one client of the travel agency that had contracted hotel and other services in Africa via this agency. This client was a passenger of Iberia and his flight with destination to Johannesburg was overbooked causing the whole trip being cancelled with the consequent losses for travel agency.

REFERENCES

Blanchard, (2004), «Terminal 250: Federal Regulation of Airline Overbooking ", pp. 1799-1832 (document recovered 1th of July of 2009, in http://www1.law.nyu.edu/journals/lawreview/issues/vol79/no5/NYU504.pdf) .

Conde Tejón (2008), El contrato de *charter* aéreo. Especial atención a la responsabilidad en caso de retrasos y cancelaciones, accidentes, daño a los equipajes y *overbooking*, Granada.

Feal Mariño, (2003), *The overbooking in air transport,* Navarra.

IndependentTraveler: http://www.independenttraveler.com/resources/article.cfm?AID=15&category=1.

Klophaus & Pölt, «¿Does Overbooking: A sacred cow ripe for slaughter?», *Aerlines Magazine e-zine edition,* Issue 32, pp. 1-5.

Koide & Ishii (2005), «The hotel yield management with two types of room prices, overbooking and cancellations», *International Journal of Production Economics,* Vol. 93-94, pp. 417-428.

Lindenmeier & Tscheulin (2008), «The effects of inventory control and denied boarding on customer satisfaction: The case of capacity-based airline revenue management, refers to the problem of consumer dissatisfaction in overbooking cases like a drawback of the implementation of revenue management», *en Tourism Management,* 29, pp. 32-43.

Report on the consumer of air transport (June 2009) of the United States Department of transportation: http://airconsumer.ost.dot.gov/reports/2009/June/200906ATCR.PDF.

Suzuki (2006), *The net benefit of airline overbooking,* Transportation Research Part E 42, pp. 1-19.

---

[12] Sentence of Provincial Court of Madrid (Section 13), 4th February of 2002.

The Dictionary of the Royal Academy of the Spanish Language: http://buscon.rae.es/draeI/SrvltConsulta?TIPO_BUS=3&LEMA=overbooking

Wangenheim & Bayon, (2007), «Behavioral consequences of Overbooking Service Capacity», *Journal of Marketing: A quarterly publication of the American Marketing Association.* Vol. 71, no. 4, p. 36.

Chapter 9

HUNTING AND THE ENVIRONMENT

Pilar Domínguez Martínez[*]
University of Castilla-La Mancha, Spain

ABSTRACT

Hunting plays an important role in the conservation of species, and the inter-relation between fauna and flora in the management of natural heritage is aimed at maintaining biodiversity. For this reason the need for protection of wild fauna requires to make the existence of fences compatible with the protection demands on fauna mobility with a view to ensure its conservation and biodiversity. Not surprisingly, in compliance with EIA legislation, conditions relating the so-called impact statements have been imposed on fencing. According to hunting legislation in Spain, fence installation is not mandatory, moreover, a prior authorization is required before installing fencing, being the exclusive obligation of the owner of the hunting ground to place signposts in the property. The aim of this paper is to analyze whether the absence of fencing can be determinant of careless conduct for the purposes of attribution of liability to the holder of the game preserve for damages resulting from traffic accidents caused by collision with animals from the said farm.

Keywords: hunting, environment, fencing, excellent hunting, liability

1. INTRODUCTION

The increasing number of traffic accidents caused by game species-vehicle collisions and the need to attenuate the liability of the owner of the hunting ground as established in Article 33 of the State Hunting Law of 4 April 1970, which consider it as strict liability, gave rise to the publication of Act 19/2001 of 19 December, excluding such liability in the Sixth Additional Provision in case the driver´s noncompliance of the traffic norms can be proved as sufficient cause for the accident. This measure is consistent with the legislation in all

---

[*] Contact author: Pilar.Dominguez@uclm.es

European countries, where the liability for damages caused by collision with animals from a game farm never falls on the preserve or state owner but on the car driver and, exceptionally, on the State (BERNARD, 2004). In Spain, the said legal reform did not have the practical effectiveness expected and this brought about another important modification under Act 17/2005 of 19 July, regulating the permit and driving license using a point system, which modifies the Amended Text of the Law on Traffic, Motor Vehicle Traffic and Road Safety passed by Royal Decree 339/1990 of 2 March. This Act has incorporated the Ninth Additional Provision to this Text, under the heading "Liability for traffic accidents, wildlife collisions" which is applicable in all the state territory and reads as follows:

"In traffic accidents caused by game species-vehicle collision, the vehicle driver shall be considered liable when non-compliance with traffic rules can be proved. The personal and property damages in these claims shall exclusively be charged to the holder of the excellent hunting or, by default, to the owners of the state, in case the accident is the direct result of the action of hunting or there is an absence of diligence in the conservation of the property enclosed. The liability can also be attributable to the holder of the public road where the accident occurs as a result of the liability for road conservation and signposting".

The Provincial Court Decisions (SSAAPP) Lleida (Section 2) of 15 April 2008 (JUR 2008 179 605; FD2) of 31 January 2008 (JUR 2008 138 103) of 25 October, 2007 (JUR 2008, 11 934) and of 20 June 2007 (JUR 2007 300 327) establish that this provision "is similar to the norms contained in the Sixth Additional Provision of Law 19/2001 of 19 December, which partially amended the Law on Traffic, Motor Vehicle Traffic and Road Safety of 2 March 1990, and its interpretation is rather simple in those cases in which a traffic violation by the driver can be duly accredited and that the said offense has causal incidence on the result produced, thus requiring an assessment of the particular circumstances of each case to determine the final attribution of liability to the driver, in whole or in part, with respect to the damages occurred, so that, on the assumption of various causes contributing to the result, fault compensation shall be applied as in any other traffic accident, assessing the relevance of such contributions and the intensity of the negligence attributable to each party; the defendant´s total release from responsibility can only be possible if the exclusive fault of the victim can be demonstrated and presents enough entity to absorb in causal terms the possible incidence of the action attributed to the defendant. "

2. Materials and Field of Study

For treatment and analysis of the scope of responsibility established by the new legislation, the subsequent court decisions ruled by the Superior Courts of Justice and Provincial Courts have been used. Indeed, the interpretation of the said legislation has generated a number of divergences on the scope of the reform and its impact on hunting legislation, at state and regional level. Nevertheless the full applicability of the said Additional Provision should be maintained on the basis of the exclusive competence of the State Administration in the scope of Civil Law on issues not included in regional or special legislation. Indeed, this is not a hunting matter but an issue pertaining traffic legislation. Supporting the previous assertions, the following statements of the Provincial Court Decision "Ourense" of 23 January 2007 (AC 2007, 1613) should be considered, establishing that

"Article 149 of the Spanish Constitution provides as an exclusive competence of the State (with the exceptions stated in the norm itself -none of which are related to the present case-) on civil legislation, field in which that type of liability can be included; likewise, Article 21 establishes traffic and motor vehicles circulation as exclusive competence of the State. After clarifying that the issue regulated under the mentioned Additional Provision here above is the civil liability arising from a traffic accident involving a game species and taking into account the interpretation of the Constitutional Court on the hunting competence under Article 148 of the Constitution, there is an obvious need to opt for the new legislation scheme, which means that the non-application of that provision would rise the question of constitutionality, if the regional legislation prevails.

It is commonly accepted that the said Provision implies a tacit repeal of the prevention under Article 33 of the State Hunting Law and its Regulations of 25 March 1971, which were applicable in all Autonomous Communities which had no specific legislation on hunting, as competencies on hunting issues are transferred, using State Law in substitution for Autonomic Law (Article 149.3 Spanish Constitution, SC).

Article 148.1 (11) SC grants Autonomous Communities the competences on hunting issues and Article 149.1 (23) assigns the State exclusive competence on the basic legislation on woodlands, forestry and inland livestock and environmental protection, without prejudice to the powers of the Autonomous Communities to establish additional standards of protection.

For the same reason there is a repeal of the provisions contained in the regional legislations, due to objective incompatibility according to the norms above mentioned, therefore having preferential applicability the special and latest legislation. This is the doctrine held in the Provincial Court Decisions in Pontevedra of 20 September 2007 (JUR 2008, 42116), in Lugo of 12 January 2007 (JUR 2007 139193) of 4 December 2006 (JUR 2007, 12858), in Orense, 30 March 2007 (JUR 2007 171 991) of 1 October 2007 (JUR 2008, 79 368) of 9 October 2007 (JUR 2008, 79 161) of 23 January 2007 (AC 2007, 1613) of 27 February 2007 (JUR 2007 174 310) and in Pontevedra of 31 January 2008 (AC 2008, 699) of 16 January 2008 (JUR 2008, 87 618), among others. This last decision assigns liability for the wild boar-vehicle collision to the owner of the hunting ground due to the absence of fencing.

However, it should be noted that apart from the State exclusive competence on traffic and road safety under article 149.1.21 EC, State legislation has established the lines to follow in autonomic legislation, as the Autonomous Communities practically transpose State legislation with some variations (Caballero, 2001).

Concerning the driver victim of the accident, the State traffic provision establishes that "the vehicle driver shall be considered liable the non-compliance with the traffic rules can be proved". This new provision is supposed to reduce the number of claims for damages in relation to game species, both concerning hunting ground owners and Road Administration, since road safety legislation is comprehensive and thorough, thus leading to release results for exclusive fault of the driver victim of the collision. To this regard, the Provincial Court Decisions of Lleida of 15 April 2008 (JUR 2008 179 605), of 31 January 2008 (JUR 2008, 138 103 of 20 June 2007 (JUR 2007 300 327) and of 25 October 2007 (JUR 2008, 11 934) make special reference to norms such as "Article 18 of the General Regulations of traffic which obliges the driver to "maintain the necessary visual field and permanent attention to driving" or Article 45 of the said Regulations which imposes the obligation "to respect the speed limits established, taking into account (...) weather conditions, environmental conditions and traffic, and, in general, any other likely condition at any moment, in order to adequate the vehicle speed to such conditions, so the driver can stop the car within the limits of his/her visual field and any obstacles that may appear", thus making the driver find great difficulties to prevent from being charged, in whole or in part, with the liability for the damages caused in case of animal collision.

3. HUNTING GROUNDS: LIABILITY

The new legislation distinguishes two situations in which the hunting ground holder or estate owner shall be liable:

a) "When the accident has been caused directly by the action hunting".

The difficulties to prove the facts and the causal connection of the action of hunting by the driver in case of animal collision would oblige him/her to proceed directly against the holder of the road. The Provincial Court Decision of LLeida (section 2) of 15 April 2008 (JUR 2008, 179605), of 31 January 2008 (JUR 2008, 138103) of 25 October 2007 (JUR 2008, 11934) and of 20 June 2007 (JUR 2007, 300327) reads identical terms on this first criterion on the charge: "A hunting action (article 2 of the Hunting Act of 1970) is defined as the action performed by man using arts, weapons or appropriate means for finding, attracting, tracking or hounding animals mentioned in this Act as game species with the aim of killing, taking them or facilitate its capture by a third person. Only in case the animal collision is a direct, efficient and adequate cause of the action of hunting in the different modalities permitted (awaiting, stalking, beat, hook or *montería*) with the animal irruption - mainly hounded big game species running away – in the road crossing or adjoining the hunting ground, the liability for the damages caused in the collision shall be attributable to the holder of the hunting ground or, by default, to the estate owners. It seems that not much attention has been paid by the legislator to how the vehicle driver can give evidence on the aforementioned facts, just assigning the vehicle driver the burden of proving that the facts occurred as the direct result from the action of hunting, in practice known as *probatio diabolica,* that may be solved if the Law obliged the hunting ground holder to duly notify the competent body on the dates in which the hunting rights are going to be exercised, authorizing those with legitimate interest proved by accreditation (undoubtedly as for the interest of the vehicle driver) the access to such information, as, lacking that, there is an obligation to bring the case to the third route established by the legislator to claim for the damages caused, which is even more complicated and with uncertain outcomes." To this regard, the Judicial Decision of the Contentious-Administrative Court (SJCA) of Oviedo, 21 January 2008 (JUR 2008, 98959) rejects the liability of the estate holder considering that on the date the collision occurs no hunting activity was scheduled and it is impossible to understand that the collision was a direct consequence of the action of hunting.

On the one hand, in order to prevent animal collisions, the exercise of hunting in safety areas or in areas adjoining public roads shall comply with the measures and limits established by the pertaining legislation. Thus, Article 2.6 of the aforementioned Act 6/2006, 23 October, modifying Act 4/1997 of June on Hunting of the region of Galicia, following State Act 17/2005, 19 July, has included Article 25 bis on "hunting on safety areas". Specifically, Section 3 of the said Article establishes that:

"Carrying loaded hunting guns in the car is prohibited as well as using them in highways, motorways, express ways, state, regional and local roads within a distance of fifty meters on both sides of the safety area. In no case may a gun be fired pointing at another public use road or railway. Hunting is permitted in the sides of the roads not included in the previous paragraph, if conditions allow safe hunting, permitting also the placement of hunting posts or snares. Hunting is permitted in low transit paths and rural ways aimed at pedestrians or agricultural or forest use wood if conditions allow safe hunting." In the same restrictive terms we find Article 33.3a 5 of Act 7/1998 on Hunting of Canarias of 6 July, Article 28.1 of Act 4/1996 of 12 July on Hunting of Castilla León. In more general terms, Article 21.2.a of Law 6/2006 on Hunting and River Fishing of Baleares of 12 April considers safe area: "Public roads, ways and railways."

b) "When the accident is the direct consequence of the non compliance with civil standards of care in conserving the enclosed territory".

In such cases, the holders of the hunting exploitation, or by default, the estate owners shall be deemed liable. The point at issue is to decide whether or not this regulation implies a modification with regards, on the one hand, to the liable person and, on the other hand, to the type of strict liability established by the state and regional hunting legislation. Regarding the first issue, the attribution of liability coincides with that under state regulations in Article 33.1 of the Hunting Act of 1970 and in most regional legislative texts. (Among them, see Article 32 of Act 7/1998 on Hunting of the Canary Island of 6 July, Article 34 of Act 8/2003 of the Andalusian Region on wild fauna and flora of 28 October. Article 50 of Act 6/2006 on Hunting and River Fishing of Baleares of 12 April refers to the State civil and traffic legislation)

Nevertheless, it seems relevant that no reference appears regarding the origin of the game animals involved in the vehicle collision, fact which, in contrast, has been considered positive, as it put an end to controversies and "complaints brought by hunting societies and small game hunting grounds holders who had been deemed liable for the damages caused by big game animals that occasionally crossed their hunting exploitations and were hit on their roads." (SOLAZ, 2006 and ORDUÑO, MANZANA, 2007)

With regards to the liability scheme, the legislation categorically establishes a criterion of fault for damage attribution with reversal of the burden of proof, which agrees with numerous decisions of minor case law, in particular in the Autonomous Communities which have amended their respective legislations to adapt them to the new law. The substitution of the concept of strict liability of the hunting ground's holder for the subjective concept of fault is also recognised in the Provincial Court Decisions of A Coruña of 13 February 2008 (JUR 2008, 13710), of 16 January 2008 (JUR 2008, 168116), of 1 February 2008 (JUR 2008, 137968, of 14 February 2008 (JUR 2008, 136573), of 15 February 2008 (JUR 2008, 145487) and of 29 February 2008 (AC 2008, 914), of 18 December 2007 (JUR 2008, 147988), of 28 January 2008 (JUR 2008, 124419), of 13 February 2008 (JUR 2008, 13710), of 14 November 2007 (JUR 77012) and 19 December 2007 (JUR 2008,81407), (JUR 2008,81467). The Provincial Court Decision of Pontevedra of 7 February 2008 (JUR 2008,137286), Provincial Court Decisions of Ourense of 31 March 2008 (JUR 2008, 206972), 27 February 2007 (JUR 2007, 174310), agree in content with the previous decisions. There are also decisions contrary to this liability subjective regime and some adopting an intermediate position. Provincial Court Decision of Salamanca of 21 September 2006 (AC 2006, 2368) affirms that this system may not imply a radical change from strict liability to fault liability "... in which the burden of proof is on the injured claimant; notwithstanding, the new law shall be interpreted as a «softening» of the rigorous system of strict liability previously established by Act 4/1996 on Hunting of Castilla y León (which as for the rest it agrees with regional legislations on Hunting), imposing on the defendant (holder of hunting grounds, or in default, the owners of the ground or the holder of the public road) the burden of proof for failure to comply with traffic norms by the claimant or for sufficient contributory due care in the conservation of the hunting ground of his/her titularity…" The substitution of the strict liability regime of the holder of the hunting ground for the subjective fault liability regime is also acknowledged in the Provincial Court Decision of A Coruña of 13 February 2008 (JUR 2008, 13710).

According to this approach (SOLAZ, 2006) emphasizes the repeal of the previous legal system that imposed a highly objective strict liability, established in the Hunting Law of 1970 and in its Regulation of 25 March 1971, applicable in all Autonomous Communities with no specific legislation on hunting. That system, also contained in the old Article 23 of Act 4/1997 on Hunting of Galicia of 25 July, mentioned in the Provincial Court Decision of Lugo (180/2003) of 21 May, according to which, "on this issue priority should be necessarily be given to the strict liability of the hunting ground's holders or, where appropriate, to the corresponding "Tecor" (organised hunting grounds), therefore, no evidence proving the stay of the animal in the hunting ground with permanent character or because of the geographical origin of its route (impossible in practice) shall necessarily be required, coming this requirement into conflict with the objective nature of attributed to this liability in question by the legal norms on the matter. The said strict liability is recognised in a recent Sentence of the High Court of Justice of Galicia of 13 March 2003 (EDJ 2003/92142) , pointing out, among other issues, that the grounds of the strict liability are based on the fact that those who exploit hunting are liable for the damages caused. The court considers that in these cases the liability becomes objective, so the burden of proof is reversed and passes from the claimant to the defendant."

In fact, the change in the liability regime can be confirmed in Article 23 amended by Act 6/2006 of 23 October. Paragraph 1 thereof establishes that "in traffic accidents caused by wildlife collisions, the personal injuries and property damages shall be deemed in compliance with the existing state regulation on road safety applicable on this issue". The same reference to the state legislation on road safety can be seen in Act 6/2006 on Hunting and River Fishing of Baleares of 12 April and in the Local Law *(Ley Foral)* 17/2005 of 22 December, which expressly recognize a fault liability regime in the Statement of Reasons (whereas).

In spite of this legislative modification, doctrinal interpretation and minor case law, the Supreme Court case law conclusively acknowledges the objective character of this liability, with no mention to this new legislation. **More** specifically, the Supreme Court Decision (Court Room 1) of 23 July 2007 (RJ 2007, 4669), on a case of wild boar-vehicle collision, establishes that the liability is strict and must be tried according to the provisions under Article 33 of the Law on Hunting, which tries a case on the liability to compensate just for the mere production of damages –without claiming any fault on the part of the holder of the hunting ground – based on prior determination of the origin of the animals. (A similar case: see Supreme Court Decision of 22 December 2006 [RJ 2007, 608]

CONCLUSION

Nevertheless, when analyzing the fault of the estate holder, the main obstacle lies on defining the space limit that can be controlled by the estate holder or, by default, by the owner, that is, the standard of care in the conservation of the hunting ground to prevent the risk of animal-vehicle collision. Likewise, the question is to determine the limits of due care required, the control or surveillance measures demandable to the exploitation holder or estate owner. The provincial Court decision of A Coruña of December 19, 2007 (JUR 2007, 81 407) is relevant for this analysis as it sets out the requirements concerning the hunting estate conservation, which should be examined in accordance with the incidence on the said conservation in case the game changed their habitat crossing the road.

The requirements for collision risk prevention can be extracted from the different regional legislations on hunting: signposting regulations (See. Hunting Laws of La Rioja: Articles 23.10 in fine and 83.9; of Aragon: Article 27 and of Extremadura: Article 26, among others), once the resolution for fencing is obtained, the obligation to assume the expenses arising from the practice of hunting, such as surveillance expenses (LAFUENTE, 2006). This author lists the liabilities of the holder of the private hunting ground, affirming that, "except for the hunting legislation of the Region of Valencia where a list of liabilities has been included, liabilities are dispersed in the different legislations of the Autonomous Regions".

The obligation of fencing hunting grounds near roads is not a peaceful issue. Fencing a hunting area consists in enclosing a forest land by placing a metallic fence around the entire estate perimeter. This estate is normally used by its owners as big game territory or other hunting activities. In conclusion, the compliance with the obligations for the improvement of the habitat of the animals in the hunting ground and the consequent rejection to fence the territory would rise the number of accidents and would be contradictory (BERNARD 2004).

The Provincial Court Decision of Burgos of 19 October 2007 (JUR 2008, 58139) rules in favour of fencing in compliance with the obligation of conservation of the hunting ground. On the contrary, the Provincial Court Decision of Lugo of 12 November 2007 (JUR 2008,66681) and the Provincial Court Decision of Orense of 30 March 2007 (JUR 2007,171991) rules against fencing mentioning the incompatibility of the fence with wildlife mobility requirements in order to ensure its conservation and biodiversity. The legal limitations regarding the technical and administrative requirements of fencing (Article 19.9 of the Regulations of the Law on Hunting) are also mentioned. And lastly, the free mobility of non-game wildlife is mentioned (Article 34.f of Act 4/1989 of 27 March on the Conservation of Natural Species and Wildlife Fauna and Flora)". This later was amended by Act 42/2007 of 13 December on Natural Heritage and Biodiversity. On the other hand, the Provincial Court Decision of Lugo of 4 December 2006 (JUR 2007, 12858) mentions the lack of obligation to fence and the right to request fencing to the owners of the grounds. In the same terms the Provincial Court Decision of Ourense of 1 October 2007 (JUR 2008, 79368) reads as follows: "negligence in the conservation of the hunting ground shall not arise in the case that no perimeter fencing exists. Fencing shall not either be imposed to the holders of the land grouping or the integrating plots of land of the enclosed territory." Similar terms are used by the same Court in Decision of 27 February 2007 (JUR 2007, 174310) when referring to the "irrelevance of the absence of closed perimeter: the conservation of the hunting ground shall not necessarily imply fencing"

Although it is true that there is no legal obligation of fencing, it is also true that the need to exhaust the diligence shall require the use of all elements available to prevent damages. Such measures may not include fencing but should be adequate and effective and within the bounds of logic." To this respect we should mention the Provincial Court Decisions of Tarragona of 16 April 2008 (JUR 2008, 179 500) and of Barcelona of 3 June 2008. In the mentioned decisions it was commonly understood that concerning big game exploitations the mere authorization application for fencing would be sufficient to comply with the standard of diligence required. Such requirement would be justified in big game estates but not in small game estates, as affirmed in the mentioned decisions of Lleida of 15 April 2008 (JUR 2008 179 605), of 31 January 2008 (JUR 2008 138 103), of 25 October 2007 (JUR 2008, 11 934) and of 20 June 2007 (JUR 2007 300 327), in the case of animals from small game preserves, as in the cases being tried, "the case involves game species not potentially dangerous or which are not likely to imply any risk for road traffic, therefore in these cases in which the hunting use of the estate is for small game, the lack of fencing cannot be considered (or of the application for fencing) as non compliance with the minimum standard of care requirable to the holder of the said hunting ground in the conservation of the fenced property, as established under the Ninth Additional Provision of Act 17/05, especially considering that, as it is evident in the case of birds, in which fencing is impossible without comparing the risk other small game species (like rabbits) may entail with the risks entailed by big game, in which case the interpretative criteria previously exposed would apply in relation to the standard care requirable to the holder of the hunting ground. For this reason it could be affirmed that in spite of having applied for fencing authorization, in case such permit was not granted, the liability should be transferred to the said Administration responsible for such authorization.

The fencing requirement in turn implies its appropriate conservation, so that there are no holes or tears that facilitate animal exit. The Provincial Court Decision of Badajoz of 13 November 2007 (JUR 2008, 66 602) attributes the liability to estate holder after evidencing the existence of at least one hole through which the animal could escape causing the vehicle collision." In relation to the conservation by the holder of the estate, signposting is acknowledged as subsidiary or complementary to fencing in the Provincial Court Decision of Barcelona (sect. 17) of 3 June 2008 (JUR 2008 266 486).

The decisions of Lleida (sect. 2), of 31 January 2008 (JUR 2008 138 103), of 25 October 2007 (JUR 2008, 11 934) and of 20 June 2007 (JUR 2007 300 327) also agree on the interpretation of the second imputation criterion to the holder and owner of the hunting ground in the following terms:

b) " When the accident is a direct cause from the lack of diligence in the conservation of the property fenced." The norm extends the liability for animal collision to the holder of the hunting ground or, by default, to landowners in cases where the traffic accident is a direct cause from the poor conservation of the estate fenced. That poor conservation of the estate should be understood as the absence of fencing in the areas affecting roads or the non adoption in such areas of any other control or surveillance measure, due to the logical risk implied, mainly in big game estates or those for hunting use of the said species, which would attribute the fault to the property landlord/owner rather than to the estate holder or tenant. However, without attempting to examine the different reasons argued by those who defend that no fencing is part of the best of care in relation to the conservation and use of the game species on the basis of principles such as not discontinuing forest continuity, etc., or preventing endogamy in such way with the unquestionable prejudices this would mean, which should be absolutely respected, the hunting legislation in Spain establishes that fence installation is not mandatory, moreover, a prior authorization is required before installing fencing, being the exclusive obligation of the owner of the hunting ground to place signposts in the property; furthermore, hunting is usually prohibited in fenced properties except in those denominated of "special hunting use", and which is not characteristic in our country. Nevertheless, the fact that the legislation does not require fencing in hunting estates is not an obstacle to understand that hunting estate holders should have at least requested for authorization, as it might seem at first that those estates which had not requested the said authorization shall not be able to allege in their defence that the legislation in force does not require fencing the estate; furthermore, such requirement, in the ordinary sphere of our basic understanding and social respect leads to the consideration that the negligence starts by neglecting those elements which, being accessible and in case of their application, could prevent the generation of damage, although it is not so required by the law. Consequently, in a certain way, those hunting estates are obliged to respond for "not having put forth the minimum diligence when attempting to reduce the production of damages which are predictable; on the contrary, in case the authorization had been requested but is rejected, then this would directly imply the transfer of the liability to the administration which rejected the request, which seems to be the intention of the legislator: transferring the liability to the Administration (whether state, regional or local).

From these arguments, it could be finally deduced that the attribution of liability is transferred from the holder of the estate to the Administration responsible for hunting regulation and for granting the corresponding authorizations for fencing.

In other respects, we should emphasize that despite the scarce case law on the new legislation and its traditional tendency concerning the objective character of this liability, minor case law applies an evidentiary regime in accordance with the culpability character of the estate holder liability. Indeed with no reference to the new road legislation, the Supreme Court Decision (Court Room 1, sect. 1) of 23 July 2007 (RJ 2007, 4669) exclusively requires the proof of the determination of the origin of the animal: "the more or less circumstantial presence in a particular estate is not sufficient", it is necessary to prove that the real place of origin of the animals was where their natural habitat was, with no need to prove any fault of the holder. Similarly, the Supreme Court Decision (Court Room 1, sect.) of 22 December 2006 (RJ 2007, 608) demands "a certain connection between the presence of the animal and the preserve."

Nevertheless as it is clear from the decisions of Provincial Courts and High Courts of Justice, mainly in Galicia, after the new regional legislation, adapted to state legislation and after incorporating the Ninth Additional Provision under Act 17/2005 of 19 July, the accreditation of the origin of the animal is no longer sufficient for imputation of damage to the estate holder. To this respect, see also the Provincial Court Decisions of Burgos (section 2a) of 10 January 2008 (JUR 2008, 125 866) of 12 November 2007 (JUR 2008, 77 309), (sect. 3) of 5 November 2007 (JUR 2008, 67 777), Segovia (sect. 1) of 8 November 2007 (JUR 2008, 77 699), of 30 October 2007 (JUR 2008, 68 417) and of Valladolid (sect. 3) of 18 December 2007 (JUR 2008, 81 568), of Zamora (sect. 1) of 14 November 2007 (JUR 2008, 77 029) and Ciudad Real (sect. 1) of 5 November 2007 (JUR 2008, 67 887). These Decisions make reference to the origin of the animal as a criterion for attribution of liability to the holder of the estate. The actor must allege and prove the lack of diligence of the estate holder in the conservation of the property fenced or that the irruption of the animal was a consequence of the action of hunting as well as the causal relation with the damages caused. Some similar decisions are: A Coruña (section 5) of 16 January 2008 (JUR 2008, 168,116) (section 3) of 1 February 2008 (JUR 2008, 137,968) of 14 February 2008 (JUR 2008, 136,573), of 15 February 2008 (JUR 2008 145 487) and of 29 February 2008 (AC 2008, 914), of 18 December 2007 (JUR 2008 147 988), (Sect. 4) of 28 January 2008 (JUR 2008 124 419), of 13 February 2008 (JUR 2008 , 13 710,) of 14 November 2007 (JUR 2008, 77 012) and 19 December 2007 (JUR 2008, 81 407), (JUR 2008, 81 467). See also Provincial Court Decision of Pontevedra (sect. 1) of 7 February 2008 (JUR 2008 137 286) and of Ourense (sect. 1) of 31 March 2008 (JUR 2008 206 972), .

This means in turn that the defendant must accredit the adoption of the measures required in the conservation and therefore in the prevention of the accident and/or the fault or negligence of the claimant driver. In this respect see the Provincial Court Decisions of A Coruña (section 5) of 25 February 2008 (JUR 2008 166 176), Girona (sect. 1) of 5 March 2008 (JUR 2008 183 001). This decision establishes the liability of the estate holder due to lack of accreditation to prove that the necessary measures to prevent the accident had been taken concerning the diligence on the estate conservation. See also the Court decisions of Pontevedra (sect. 1) of 16 January 2008 (AC 2008, 498) of 23 January 2008 (JUR 2008 124 931), of La Rioja (sect. 1) of 31 October 2007 (JUR 2008, 68 199), of Cantabria (sect. 4) of 14 November 2007 (JUR 2008, 77 012) and of Tarragona (sect. 1) of 6 February 2008 (JUR 2008 146 197) which establish the liability of the holders of the hunting ground for lack of accreditation to prove that the level of diligence requirable in relation to the conservation of the estate was sufficient and decision of 16 April 2008 (JUR 2008, 179500), which affirms: "the evidence of the diligence in the action attempting to avoid traffic accidents must be proved by the estate holder by accrediting the measures adopted to prevent the presence of animals in the road, measures which cannot be limited to fencing and which do not imply fencing as an essential requirement, but which must be adequate and effective within the bounds of logic."

The Provincial Court Decision of 28 November 2008 (JUR 2009, 347 557) establishes the lack of due diligence and the estate holder liability on an accident by collision with a deer during the hunting season, concerning a hunting ground without deer fencing in the area surrounding the road and with no adequate fencing. According to the Court, the joint application of the Additional Provision concerning the norms on the burden of proof as established under the Law of Civil Procedure (article 217) de facto implies that the liability for the accident shall fall on the estate holder/owner, unless the driver liability can be proved for non compliance with traffic regulations which may have a causal incidence in the damage or it is attributable to the holder of the road, since it is extremely difficult, if not impossible, to prove a negative fact, that is, if there is no fault attributable to the driver, there is no lack of diligence in the conservation of the estate, when it is an objective fact and constitutes the presupposition of the action that an accident has occurred.

In a large number of court decisions, as in that of Cantabria of 15 January 2009 (JUR 2009, 277029), the criterion for attribution of liability is the absence of fencing, and furthermore, the liability is established through the lack of accreditation of existence of adequate fencing in accordance with the characteristics of the animals living in the preserve, acknowledging an assumption of the causality relation between hunting and animal stampede.

The existence of fencing in poor conditions is also a determinant criterion in Provincial Court Decisions of Burgos of 4 June 2008 (JUR 2008 302 699) and of 11 September 2008 (JUR 2009, 95046). The lack of fencing or of any other measure of control or surveillance is a determining factor of lliability according to Provincial Court Decision of Lleida of 8 September 2008 (JUR 2009, 124 312) and of Zamora of 22 September 2008 (JUR 2009, 81394). The lack of fencing or substitutory traffic sign is mentioned in the Provincial Court Decision of Barcelona of 3 June 2008 (JUR 2008, 266 486).

The need for protection of wild fauna requires to make the existence of fences compatible with the protection demands on fauna mobility with a view to ensure its conservation and biodiversity. Additionally, in compliance with EIA legislation, conditions relating the so-called impact statements have been imposed on fencing (See the Preamble of Decree 178/2005 of 18 November, of Catalonia establishing the conditions of fencing in wild environment and in hunting enclosed preserves). It is widely believed that fenced hunting grounds have a negative impact on specific species because of the isolation produced by the barrier effect which may bring about the risk of extinction of small populations (SANTIAGO, 1994).

To this respect many recent decisions can be found establishing that the absence of fencing does not imply negligence in the conservation of the hunting estate. Some of which are: Barcelona, of 22 May 2008 (JUR 2008, 267105), Castellón of 15 December 2008 (JUR 2009, 175520), Soria of 30 December 2008 (AC 2008, 835), of 9 November 2009 (JUR 2009, 271302), of 11 December 2009 (JUR 2009, 496551), Albacete of 4 May 2009 (JUR 2009, 271302), A Coruña of 23 June 2009 (JUR 2009, 318983). It is even affirmed that the proximity of the ground does not generate liability and that complete fencing cannot be demandable: Provincial Court Decision of Barcelona of 23 June 2009 (JUR 2009, 318983). Therefore, the absence of fencing does not determine the correct or incorrect constitution of a hunting ground and as a consequence it cannot be considered as a lack of diligence in the conservation of the estate (Prov. Court Decision of Soria, of 30 December 2008 [JUR 2009, 193 632]). The accreditation proving that the hunting ground satisfies all necessary requirements for the exercise of hunting and capture sheets are sufficient proof of the diligence demandable in the conservation of the hunting preserve (Prov. Court Decision of Soria of 20 July 2007 [JUR 2008 103264]).

In summary, there is no uniform interpretation concerning the model of diligence required from the holder of a hunting ground which could be used as exemption of liability, both concerning the conservation of the hunting ground in general and fencing or enclosing the estate in particular. Moreover, it could be affirmed that rather than establishing a model of diligence, we are facing a measure that may be negative for the conservation of species and which requires administrative authorization, thus becoming the contrary conduct a standard of diligence which could rise the number of accidents (VICENTE, 2008). The poor number of studies on this issue claims for the need of research work attempting to compatibilize traffic legislation with environmental legislation on wildlife protection. To this regard, it would be necessary to determine those critical sites for the survival of a particular fauna with conflicting spots for high accident rate caused by wildlife collisions. Furthermore, it would be necessary to develop hunting management models which are compatible with the conservation of endangered species and their habitats, especially in big game fenced hunting farms (García, 1998).

REFERENCES

Agudo González: "La responsabilidad patrimonial de la Administración por daños producidos por animales de caza", *RDUMA*, 1998, p. 113.

Bernard Danzberger (2004): *Atropellos de animales: Problemática en España y Derecho Comparado*. II Jornadas sobre Accidentes de Tráfico, Soria.

Caballero LOZaNO (2001): "Aspectos civiles de la caza desde la perspectiva de la Ley de Castilla y León", Revista General de Legislación y Jurisprudencia, n° 3, Reus. Responsabilidad civil y seguro del cazador, *La Ley*, 21 julio 2000.

Cavanillas Sánchez (2002): "La responsabilidad civil por daños sufridos por los usuarios de autopistas de peaje en la jurisprudencia", *RDP*, 9, Aranzadi.

Cuenca Anaya (1998): *La caza en las Comunidades autónomas. Derecho comparado*, Al-Andalus, Sevilla.

Accidentes provocados por las piezas de caza en la Ley 17/2005 que reforma la de Seguridad Vial", *La Ley*, n° 4, 2005, pp. 1592-1597.

Curatolo y Murphy (1986): "The effects of pipelines, roads, and traffic on the movements of caribou", *Rangifer tarandus. Canadian Field-Naturalist,* 100, pp. 218-224.

Díaz Roldán (1993): "La responsabilidad civil derivada de los daños causados por los animales de caza", *Cuadernos de Derecho Judicial*, CGPJ, Madrid, pp. 429 y ss.

Escribano Collado y Lopez Gonzalez (1980): "El medio ambiente como función administrativa", *Revista Española de Derecho Administrativo*, 26, p. 367.

Fernández Nieto (2006): "Comentario a la STC 210/2005, 18 julio", *Diario de Jurisprudencia El Derecho,* El Derecho Editores, n° 2244, 1de marzo 2006.

Gallego Dominguez (1997): *Responsabilidad civil extracontractual por daños causados por animales*, Bosch, Barcelona.

García y Otros (1998): "Permeabilidad de los vallados cinegéticos de caza mayor. Efecto barrera e implicaciones para la conservación de las especies amenazadas", *Galemys* 10, pp. 109-119.

Lafuente Benaches (2007): *El ejercicio legal de la caza*, Tirant lo blanch, Valencia.

Lopez Menudo (1999): "Responsabilidad Administrativa y exclusión de los riesgos del progreso", *RAAP*, 36. 1999, p. 11.

Martín Aparicio (1996): "Cercados llamados cinegéticos:¿Una laguna o un lapsus jurídico en la política y gestión ambiental sobre recursos y espacios naturales?", *Congreso Nacional de Derecho Ambiental*, Cima Medio Ambiente, Valencia.

Muñoz Machado (1999): *Los animales y el derecho*, Civitas, Madrid.

Ortuño Navalón, Manzana Laguarda (2007): *Régimen de Responsabilidades dimanantes de la caza*. Tirant, monografías, Valencia.

Gálvez Cano (2006): "El derecho de caza en España.- Ed. Comares, Granada.

Ortega Martín, Bernad Danzberguer (2001): *Accidentes causados por animales objeto de caza,* Editorial: Europea de Derecho.

Parra Lucán (1999): "La responsabilidad por daños producidos por animales de caza". *Revista de Derecho Civil Aragonés*, n° 2155, Institución "Fernando el Católico"1999, n° V.

Reglero Campos (2008): *Accidentes de circulación: responsabilidad civil y seguro*, Aranzadi, 2ª ed., enero 2008.

Rodríguez González (2002): "Régimen jurídico de la caza", Obra colectiva *Manual del Cazador*, Marcial Pons.

Santiago (1994): "Influencia de los cercados en la fauna cinegética", *Vallados cinegéticos. Incidencia ambiental, social y económica*, CODA, n° 1, Madrid.

Sánchez Gascón (1988): *El derecho de caza en España. De los terrenos y de las piezas de caza*, Tecnos.

Solaz Solaz (2006): "Responsabilidad en los accidentes de tráfico por atropello de especies cinegéticas", *Revista de Jurisprudencia*, n° 3, abril 2006, El Derecho Editores.

Torres Fernández (1994): "Distribución de competencias en materia de medio ambiente entre el Estado y las Comunidades Autónomas", *Actualidad Administrativa*, n° 13.

Vicente Domingo (2008): "Los daños causados por los animales", *Tratado de Responsabilidad Civil* (coord.) Reglero Campos, Aranzadi, Pamplona.

In: Research Studies on Tourism and Environment
Editors: J. Mondejar-Jimenez et al.
ISBN: 978-1-61209-946-0
© 2012 Nova Science Publishers, Inc.

Chapter 10

MONUMENTAL TREE HERITAGE AS URBAN TOURIST ATTRACTION

Francisca Ramón-Fernández[*], *Lourdes Canós-Darós* *and Cristina Santandreu-Mascarell*
Universidad Politécnica de Valencia, Spain

ABSTRACT

When we think about a city's tourist attraction, we always focus on its architecture and historical heritage. We forget that there is also a living heritage, composed of trees with environmental and historical characteristics, which form the ornamentation of towns and cities, and that, in the case of the region of Valencia (Spain) is especially taken into consideration and legislated. This chapter attempts to demonstrate the importance of trees as living monuments, the monumental tree heritage, in sum, the city's tourist attraction, taking into account its scenic and visual value for potential visitors. This cultural heritage consists in outstanding, monumental or singular trees that are a potential for the city. Their regulation by Law 4/2006 of monumental tree heritage, is the most important tool for their conservation, protection and cataloguing. As we will see, this protection is completely justified by their characteristic exceptions of historical, cultural, scientific and recreational value, since these specimens represent a singular part of the people's environmental and cultural heritage, and, thus, an enormous potential as instrument of reputation and identity of the city as tourist destination.

Keywords: Environment, heritage, law, monument, tree, urban tourism

---

[*] Contact author: frarafer@omp.upv.es

1. INTRODUCTION

City councils are increasingly more aware of the importance of their differentiation to attract new tourists and maintain the loyalty of existing ones. Their aim is to create an attractive image in the mind of current and potential tourists. Therefore, they can count on the value of local heritage, so that it improves cultural tourism, a type of very profitable tourism, formed by people of upper middle-class interested in culture, history or any other heritage resources of a destination.

There are different urban attractions to become a tourist destination and be visited by a certain number of people (Law, 1993). It is usually taken into consideration the location of the city, its climate, communications, infrastructure, in sum, its perspectives for future visitors (Pearce, 2001; Lew, 1987). Nevertheless, there is an aspect that should be considered as a great incentive and with an enormous potential: its heritage. However, when we think of heritage as an element to increase the reputation of a city and so that it is a favourable place and is chosen to be visited, it seems that we concentrate on architecture and historical heritage, but these are not the only ones, since there is another living heritage, the monumental tree heritage. Its importance is such that we shall take it into account as a very important instrument for the consideration as a city's tourist attraction. Moreover, tree heritage can be used as an attractor not only in urban areas, but in rural ones. In this case, entrepreneurs have to designed sustainable objectives (Ferrari et al., 2010) for business, in the line of current legislation in order to protect the environment.

In the recent years, the city council of Valencia and the Government of the Valencia Community, located on the east of Spain bordering the Mediterranean sea, have develop big efforts to promote as well the city, as well de region. For instance, some important events have been organized, for instance, 32th and 33th America's Cup editions (see Canós and Ramón, 2009) and Street Formula 1. In our opinion, visitors can enjoy the touristic resources of the city or the region, not only the activities directly related with the events. One of these resources is the monumental tree heritage. More examples about attractiveness of cities for urban tourism (Turin, Lyon, Lisbon and Rotterdam) can be seen in Russo and van der Borg (2002). Other examples about Manchester and Sheffield are showed in Schofield (1996) and Bramwell (1998), respectively.

Tree heritage is the overall amount of trees whose botanical monumental characteristics or extraordinary circumstances of age, size or other types of historical, cultural, scientific, recreational or environmental events bound to them and their legacy, makes them worthy of protection and conservation. As we have stated previously, it is a living heritage. The trees that form this heritage have peculiar environmental and historical characteristics and are part of the ornamentation of towns and cities. Obviously, the protection of the trees and the environment is good for citizens; if they are committed with the environment, they can improve their quality of life and their richness. A model explaining the environmental attitude and behaviour of Tehran urban residents can be seen in Kalantari and Asadi (2010), which demonstrates the importance of environmental legislation from the residents' point of view.

In this regard, we shall mention one of the first regulations that considered the importance of heritage, although it only makes reference to the existence of trees of singular characteristics. The Preamble of Forest Law 3/1993, of 9 December, of the region of Valencia

(Spain), considered the possibility to establish special regimes to protect endangered forest species and singular trees.

In addition, another regulation that also considered the need of this specific regime was Law 4/2004, of 30 June, on Urban Planning and Landscape Protection, which demands the promulgation of a regulation that increased to the category of true monuments those trees that are landmarks, by their natural features or by historical facts related to them. This possibility is materialized in one of the latest regulations concerning forests promulgated in the Valencia Community: Law 4/2006, of 19 May, of the Autonomous Government, of Monumental Tree Heritage of the region of Valencia. In particular, this Law considers the protection of this cultural heritage that has as main figures outstanding, monumental or singular trees that constitute a potential for the city, in our case, the region of Valencia, through their special protection and cataloguing. As we will see, their protection is completely justified by their characteristic exceptions of historical, cultural, scientific and recreational value, since these specimens (they meet the requirements established by the regulation) represent a singular part of the environmental and cultural heritage of the region of Valencia, and, therefore, their evident public interest of protection and conservation and their enormous potential as instrument of reputation and identity of the region of Valencia.

The main reasons for tree heritage to be object of a specific regulation have been the special environmental and historical characteristics of the region of Valencia, since there exists a great deal of biodiversity of native ligneous vegetal and allochthonous species, that are part of the vegetation of the Valencian forests - being some of them introduced within our spaces in remote times, and others are part of our culture, as they are considered ornamental. All these indicators have determined the existence of botanical groups and specimens that are unique and that have an historical, cultural, scientific and recreational value that constitutes what has been called tree heritage, a singular part of the environmental Valencian heritage and very important to be object of protection and conservation for future generations.

In this chapter we present some regulations for the monumental tree heritage of a city or a region. In particular, we show some laws that are implemented in the region of Valencia (Spain). For this, we analyze the main ideas related contained in Laws referred to monumental trees and the tourism they can attract, and two concrete examples to illustrate them.

2. MATERIAL AND METHODS: THE CATALOGUING OF THE MONUMENTAL TREE HERITAGE: LEGAL INSTRUMENTS IN THE REGULATION

As the Law states, this living tree heritage, formed by trees of spectacular measures, also includes shrubs or other non-tree specimens of remarkable dimensions; those that include an important historical or symbolic meaning and those that represent religious or social traditions or a high ethno-agrarian or ethno-botanical value. Likewise, this section includes specimens of extremely rare ligneous species, whose presence implies an outstanding scientific value and those which society enjoys their contemplation.

The trees that correspond to these characteristics have reached unusual dimensions and forms for their respective species and are indebted to the effort of the human being in their care and multicentennial maintenance; in fact, the great majority of this hoisting corresponds to specimens that have been planted and improved throughout time. Many of them are located in historic gardens, town and city squares, resting areas of cattle routes and other surroundings near rural constructions or farms. Multicentenary specimens of some agricultural species particularly longevous also survive.

The concept of tree covers all the specimens of higher plants, angiospermaes as well as gimnospermaes, autochthonous or allochthons that have one or several trunks sufficiently differentiated. They also include trees of horizontal or crawling growth, palms, certain shrubs and those forms of heavy trunks of lianas or climbing plants, and even isolated specimens, like woods or sets that contain several tree specimens.

For different causes, human as well as environmental, these hoisted spaces are in danger and for this reason, to avoid their degradation and disappearance, different measures are considered to guarantee their viability and survival. Different planning instruments are considered to ensure a suitable management, a supervision of their health conditions, and the application of treatments adapted for their improvement.

The social function of these trees in our surroundings is no longer merely forest, but they are considered as part of the natural and cultural heritage, as they are considered unique species in their kind, with the difference that part of that heritage is a living being that needs specific care to last. Besides that heritage interest, it also stands out the possibilities that offer these species in the scope of environmental education and sustainable development, since they involve a revaluation of the ecological spaces of our geography. It is the monumental tree heritage still has an educational, cultural, social and economic function, which allows to promote sustainable development of the places where it is.

Many of these wooded areas, as stated above, can undergo degradation and disappearance. For this reason, the mentioned Law considers that, in addition to protection, planning instruments are to be established to assure their suitable management, the monitoring of the evolution of their health conditions, the application of the conservation treatments, the restoration of trees and the necessary improvement of the surroundings or other actions.

Since they are living monumental trees, it should be attempted, as the Preamble itself regulates, the best care and attentions, which these works of art, product of nature and culture, deserve. Any type of modification or intervention developed in them, or in their surroundings, can cause serious consequences for their health. Another reason for conservation programs and measures of intervention for each specimen, are adapted based on their vital health conditions, needs and survival.

Therefore, the main objective of this Law is to guarantee the protection, conservation, dissemination, promotion, investigation and increase of the monumental tree heritage of the region of Valencia, understanding by that the set of trees whose botanical monumental characteristics or extraordinary circumstances of age, size or other types of historical, cultural, scientific, recreational or environmental events related to them and their legacy, makes them deserve protection and conservation. This objective can be summarized in an especially significant fragment of the Preamble: "To guarantee that these trees remain alive the longest possible time among us".

The conservation of this monumental tree heritage makes necessary a wide and efficient collaboration and institutional and social coordination, since trees considered as part of this tree heritage, also called outstanding trees, belong to public as well as private owners.

It is necessary a joint action of the Valencian Government, the provincial delegations and the city councils to face the need of economic resources, tangible and human so that the protection and conservation is effective. In this sense, sustainability is promoted. As it is known, sustainable development was defined in the Brundtland report as that which satisfies the needs of the present without jeopardizing the capacity of future generations to satisfy their own. The right to enjoy an adequate environment for the development of people, the right to conserve it and the rational use of natural resources, together with the necessary harmonization of the economic growth, balanced for the improvement of the well-being and quality of life conditions are governing principles of the social and economic policy in the Spanish Constitution - Art. 45- that constitute a basic budget in the urban planning and shall govern the action of the public powers in this matter.

For an effective protection and management of tree heritage it is necessary the establishment of different categories of protection. These categories shall be proportional to the biological, scientific or cultural exceptional nature, establishing, in the same way, different levels of territorial responsibility, to guarantee the participation and main role of the different levels of the public administration. It should also be ensured the capacity of collaboration and participation from different participants in conservation within the framework of their possibilities, duties and rights, to ensure the survival and transmission of this legacy transferred throughout the centuries and millennia, generation after generation.

On one hand, the protection and cataloguing of the tree heritage located in forest land will correspond to the Department of environment. It will also be responsible for the protection and cataloguing of the tree heritage located in non-forest land, regarding trees of generic protection, and those that correspond to city councils considered to be protected. On the other, city councils will be responsible for the protection and/or proposal of cataloguing of the trees of all species that are in forest and non-forest land.

Generally, the protection activities can be distinguished between a generic protection (without needing a solution) to the specimens that surpass one or more of the following parameters (350 years of age; 30 meters of height; 6 meters of perimeter of trunk, measured at a height of 1.30 meters from the base; 25 meters of diameter greater of crown, measured in projection on a horizontal plane), and expressed protection, that could be by the Autonomous Government that will be applied to the specimens declared monumental or singular, and by the city councils.

This expressed protection by the Autonomous Government has two branches:

a) By means of a Council Decree, following a proposal by the competent Department of environment, monumental trees shall be those specimens and tree sets that due to their exceptional characteristics of age, another type of historical, cultural, scientific, recreational or environmental events are deserving protection measures and specific conservation; in particular, the specimens included in this category with a given coefficient of monument whose definition is established by the decree of development of the monumental tree heritage Law.

b) By means of the Order of the Department of environment and following a proposal by the corresponding General Director, singular trees are those tree specimens or sets that without reaching the category of monumental tree according to the decree of development of this Law, are noteworthy due to their remarkable characteristics of age, size or other types of historical, cultural, scientific, recreational or environmental events, that provide them with specific protection and conservation measures; in particular, the specimens included in this category that own a monumental coefficient whose definition shall be established by means of the decree of development of the monumental tree heritage Law. Singular trees are to guarantee the maintenance and extension of monumental tree heritage. This declaration will establish their inclusion in the catalogue of singular trees of the region of Valencia.

Whereas the specific protection by the city councils is carried out:

a) By means of agreement of the plenary session of the corresponding corporation, monumental trees of local interest are those tree specimens or sets that stand out in the local context, by their biological, landscaping, historical, cultural or social characteristics that deserve protection and conservation measures.
b) The declaration will communicate the Department of environment that will proceed to its registration in the corresponding section of the catalogue of monumental trees.
c) The procedure shall be initiated by own initiative or at the request of person or interested organization, which in case of not being the proprietor will not have to provide agreement with the holder.

In the procedure for the specific protection, hearing shall be granted to proprietors and to the city councils in any case and will require a technical report on the values of the trees to protect.

Different activities are considered aimed at the conservation of tree species to be object of agricultural use, since Art. 14 establishes that their conservation includes the development of the slight pruning and fruition, phytosanitary treatments or other traditionally activities for their maintenance and the legitimate extraction of rents from their productions, whenever they do not put in danger the survival of the tree.

This protection includes the technical and normative support, through the Department of agriculture and environment, to facilitate its commercialization of the mentioned productions or its derivatives, such as the oil extracted from multicentennial olive trees or other equivalents. The same regulation states that in the assumptions in which due to the protection indicated there are decreases, losses or damages to the agricultural productions, the competent administration will cooperate in the support of the loads and will compensate the rents not obtained.

3. RESULTS AND DISCUSSION: IMPORTANCE OF THE CATALOGUING OF THE MONUMENTAL TREE HERITAGE FOR THE CITY AND ITS POTENTIAL AS TOURIST ATTRACTION

As mentioned before, the monumental tree heritage is part of the culture of people . For this reason, it is part of their cultural heritage. It is also extended to all graphic, written and oral tradition documents, etc., with the people of Valencia and these trees as main figures. These reasons motivate that it is also necessary a recovery of the documental, ethno-botanic and bibliographical legacy, of the historical-cultural value group that gathers these outstanding trees (an example can be found in Morton (1998)).

The monumental tree heritage Law specifically considers the creation of the catalogue of monumental and singular trees of the region of Valencia where tree specimens or sets will be registered with a generic or specific protection on the part of the Government as well as from the city councils. This catalogue will be managed by the competent Department of environment and each city council will manage its corresponding catalogue of monumental trees of local interest. The competent General Director in the management of natural means will proceed to the subsequent inscription in the catalogue of declarations communicated by the corresponding administrations.

The cataloguing of a tree will take place by means of the corresponding inscription that will detail the characteristics of the unit, the species at hand, the reasons for their cataloguing, the proprietor and the surroundings of protection that, as minimum, will include a circle around the base of the tree where their roots are extended. The discontinuation or loss of the condition of catalogued tree comes by the death or disappearance of the specimen. Transplant to a new location, the decrease in the height, diameter of crown or other dimensions, does not imply discontinuation.

As the mass media have shown, 840 specimens have been included so far in the Technical Catalogue of Monumental and Singular Trees in progress. This catalogue will be the seed of the future Catalogue of Monumental Trees that considers the Law object of our chapter. It is expected to include a total of 1200 specimens. The cataloguing by means of the automatic incorporation of 206 Monumental specimens has taken place (12 specimens in compliance with the requirement of height – being the highest a white eucalyptus of 38.46 meters, and 56 by the requirement of perimeter, - being the larger an oak of 28 meters; 50 specimens with a greater diameter of crown - being the major a microcarp rubber plant, and 114 of an age older than 350 years - being the oldest an olive tree of 1500 years, and 17 older than 1000 years) of the 840 catalogued; 18 monumental trees of local interest and 616 as singular ones with a provisional incorporation, at the expense of what is established later on as monumental, singular or monumental of local interest. A total of 98 different species are included (56 of Valencian flora and 42 allochthonous). However, although the works have begun in this respect and it is advancing in the cataloguing of tree heritage, the complete catalogue has still not been finished.

3.1. Specific Cases in the Region of Valencia: The Centennial and Millenarian Olive Trees and the Palm Grove of Elche

3.1.1. Millenarian Olive Trees

Within these trees, and as an example, are the centennial and multicentennial olive trees located in the region of Valencia and compliant with the Law under consideration. The existence of magnificent specimens older than 500 years motivated their protection by means of the Law under study, in article 4 of Law/2006, to the trees that exceed 30 years of age.

The presence of the olive grove in the region of Valencia goes back to the time of the Romans, more than two thousand years ago. There is certainty of its culture in zones of the province of Valencia, in the regions of the Maestrazgo of Castellon, in areas near the August Route, that communicated the capital of the Roman Empire, with Cadiz. It is in that region where numerous specimens older than two thousand years, being easy to verify the presence of millenarian olive trees thanks to the effort of the cooperatives and their relaunch as the so-called routes of tourist interest of the "millenarian oil".

Favourable decision to the request of registration in the community registry of protected denominations of origin and protected geographic indications is adopted, foreseen in Regulation (EC) 510/2006, of the Council, 20 March, Protected Denomination of Origin Oil of the region of Valencia and its conditions are published, according to Resolution of 24 June, 2008, of the Department of Agriculture, Fishing and Food.

3.1.2. Palm Grove in Elche

(http://www.turismedelx.com/en/palmeral/;http://whc.unesco.org/en/list/930;http://www.fodors.com/world/europe/spain/the-southeast/review-450159.html).

The transition from the 19th to the 20th century marked a dramatic inflection in the evolution of the Palm grove. The impact of the industrial and city-planning revolutions put under serious risk its historical continuity. The railroad, inaugurated in 1884, sectioned in two the great periurban Palm grove and stimulated the occupation of the adjacent orchards by the incipient Elche industry of footwear. New districts grew on the orchards of palms to lodge the workers attracted by the demand of manufacturers. The destiny of the Palm grove seemed fixed, even more if it is considered that agriculture of oases lost part of its reason of being due to the establishment, between 1915 and 1923, of the companies "Nuevos Riegos El Progreso" and "Riegos de Levante", that gave a bonus to the city of Elche of leftovers of the Segura river and of some channels of water-drainage of the Vega Baja, pumped from the lagoon of the Hondo. Since 1979, Elche's citizens also have access to waters of the Tajo-Segura transfer. Luckily, the citizens of Elche reacted and in the 1920's qualified voices were raised in defence of the Palm grove, headed by the municipal archivist, Pedro Ibarra-Ruiz. Thanks to his perseverant campaign, the value of the Palm grove obtained recognition at local and national level. The 2nd Republic promulgated laws in defence of the Palm grove in 1933, and general Franco's regime declared it artistic garden in 1943. Between 1930's and 1980's, the City council of Elche promulgated a long city-planning regulation aimed at a more effective protection of the Palm grove.

The urban growth was concentrated on the right shore of the Vinalopo river, far from the orchards, and the intrusions in the Palm grove, a minority, had as premise the respect to the palms and the watered structure of the land. For this reason, the Palm grove is still recognizable even where the cultivable land was adapted to new functions, such as parks, schools or hotel facilities. At present, the Palm grove enjoys the maximum level of protection, under the protection of the Law for the Regulation of the Trusteeship of the Palm grove of Elche endorsed by the Valencian Government in 1986, and of the General Plan of Urban Planning agreed upon in 1997 by the city council of Elche.

The future of the Palm grove is assured. The merit corresponds to the people of Elche, whose early vindications led back the process of modernization and economic growth towards a true sustainable development. Elche has known to preserve the great mass of the urban Palm grove, the most threatened, with 61,454 palms in 1997, adapting the uses of the orchards to the requirements of the modern society, and developing an imaginative public management, that contemplates the acquisition of orchards, the concession of grants to private proprietors, their exchange by other plots of urban ground, the reforestation, and a global maintenance of the periurban orchards, where the workers of the Board of the Palm grove compensate the abandonment of the agricultural activity by individuals.

In 1991, the research centre Estacion Phoenix was settled down for the obtaining of resistant and economically profitable palms. At present, the public management of the urban Palm grove is aimed at the representation of the orchards in its original agrarian functions, often as breeding grounds for the obtaining of young palms of different species. On the other hand, the dispersed Palm grove in the field of Elche, alien to the tribulations of industrialization and modern urbanism, maintains its original structure and agrarian functions. In spite of the improvements in the irrigated land network, water continues being scarce, which has prevented monoculture of speculative species, and has allowed the continuity of the oasis landscape.

Unavoidable mixed farming has turned the field of Elche into one of the most beautiful Spanish landscapes of irrigated land. There, traditional cultures of dry land and irrigated land are juxtaposed, next to modern plantations of ornamental horticulture: fig trees and carob trees, olive trees and almond trees are next to with pomegranate trees, lemon trees and orange trees; wheat and cotton grow among roses and carnations.

The social function of these trees in our surroundings is no longer merely forest in itself, but they are considered as part of the natural and cultural heritage, as they are considered unique species in their kind, with the difference that they comprise that living heritage that needs specific care to last (Abella, 2007). Besides that heritage interest, it is also pointed out the possibilities that these species offer in the field of environmental education and sustainable development, since they involve a revaluation of the ecological spaces of our geography. As indicated in the own Preamble of the Law, monumental tree heritage at present still has an educational, social and economic function, which allows us to promote sustainable development of the places where it is.

These monumental trees are like olive trees and carob trees that ennoble our mistreated rural landscape. They are also the guardians of the soul of the Mediterranean culture, sheltering celebrations, processions and love encounters. Others are witnesses of a civilized bourgeoisie that lived not long ago in our land, creating beauty with their small palaces and woods. They are the great palms (starting from 12-meter-wide trunks) that accompany us in our day to day, and which we have mentioned previously an example.

Conclusion

The city's tourist attraction is to be bound to the interest that it arises, since, if we observe the programs of events, for example, always a part of them is aimed at social programs. In the election of a city as a tourist destination is going heritage is going to have an influence. The importance of monumental tree heritage of the region of Valencia as instrument to increase the reputation of the city that lodges the catalogued specimens is pointed out, since it involves a very important incentive at the time of the design of tourist activities and the influence of the surroundings at the time of selecting the place.

The extraordinary importance of the behaviour of the tourist to increase the reputation through living monuments -the trees- since they reflect the traditions and the idiosyncrasy of a place, and will increase the interest of visitors. In sum, the protection and cataloguing of the monumental tree heritage of the region of Valencia (Spain) is subject to regulation by means of the current Law 4/2006, of 19 May, which promotes the consideration of the natural heritage and culture of unique trees that increase the perspective of a city as the venue of an event.

In this chapter we show the most important ideas expressed in regional Laws about monumental tree heritage in Valencia (Spain) related with tourism attraction. Then, we develop some examples of the application of these regulations.

ACKNOWLEDGMENT

The translation of this chapter was funded by the Universidad Politécnica de Valencia, Spain.

Partially supported with projects TIN2008-06872-C04-02, PAID-06-08-2431 and GV/2009/020.

REFERENCES

Abella, I. (2007). Monumental trees: an essential heritage. *Seminario sobre Convergencia del Mundo Rural y la Biodiversidad,* Santo Domingo de Silos, Burgos (Spain).

Bramwell, B. (1998). User satisfaction and product development in urban tourism. *Tourism Management,* v. 19, No 1, pp. 35-47.

Canós, L. & Ramón, F. (2009). Economic effects of the 32nd America's Cup in Valencia. *4th Tourism Industry and Education Symposium. Innovative and Sustainable Products in the Tourism and Hospitality Business. Jyvaskyla* (Finland).

Ferrari, G.; Mondéjar-Jiménez, J & Vargas-Vargas, M. (2010). Environmental Sustainable Management of Small Rural Tourism Enterprises. *International Journal of Environmental Research* (IJER), v. 4, No 3, pp. 407-414.

Kalantari, Kh. & Asadi, A. (2010). Designing a Structural Model for Explaining Environmental Attitude and Behavior of Urban Residents (Case of Tehran). *International Journal of Environmental Research* (IJER), v. 4, No 2, pp. 309-320.

Law, C.M. (1993). *Urban tourism: attracting visitors to large cities.* Mansell Publishing Limited. London/New York.

Lew, A.A. (1987). A framework of tourist attraction research. *Annals of Tourism Research,* v. 14, No 4, pp. 553-575.

Morton, A. (1998). *Tree heritage of Britain and Ireland: a guide to the famous trees of Britain and Ireland.* Swan Hill Press.

Pearce, D.G. (2001). An integrative framework for urban tourism research. *Annals of Tourism Research,* v. 28, No 4, pp.926-946.

Russo A.P. & van der Borg, J. (2002). Planning considerations for cultural tourism: a case study of four European cities. *Tourism Management,* v. 23, pp. 631-637.

Schofield, P. (1996). Cinematographic images of a city: alternative heritage tourism in Manchester. *Tourism Management,* v. 17, No 5, pp. 333-340.

http://www.turismedelx.com/en/palmeral/

http://whc.unesco.org/en/list/930
http://www.fodors.com/world/europe/spain/the-southeast/review-450159.html

In: Research Studies on Tourism and Environment
Editors: J. Mondejar-Jimenez et al.
ISBN: 978-1-61209-946-0
© 2012 Nova Science Publishers, Inc.

Chapter 11

MODELING DIMENSIONS OF THE TOURIST SATISFACTION: THE CULTURAL/HERITAGE CASE OF TOLEDO

Gema Fernández-Avilés[*1], *José-María Montero Lorenzo*[1] *and Jean Pierre Lévy-Mangin*[2]
[1]Universidad de Castilla-La Mancha, Spain
[2]Université du Québec en Outaouais, Canada

ABSTRACT

Because of people's inclination to look for new attractive activities including traditional cultures, the heritage tourism has become a major "new" area of tourism demand and almost all policy–makers are now aware of and anxious to develop. Despite the emergence of new strategies (those based on equity, norm, or perceived overall performance, those where emotions play an important role in satisfaction-formation, and those with a cognitive affective view) the models of expectation are still very popular in tourist research. In this context, we pursuit to measure tourist satisfaction through three first-order factor variables: accommodation, offered services and tourist attractiveness by using a causal model among other three second-order factor variables: importance given to tourist attributes, satisfaction and expectations.

This research relies on the information provided by 1,500 respondents who were given a small questionnaire specially designed to measure tourist satisfaction in the emblematic part of Toledo, Spain (a UNESCO World Heritage City).

Keywords: Tourism, satisfaction, expectations, structural equation modelling

---

[*] Contact author: Gema.faviles@uclm.es

1. INTRODUCTION

In 2006, the World Travel and Tourism Council (WTTC) showed the tremendous scale of the world's tourism sector (WTTC, 2006) and pointed out that the travel and tourism industry accounted for the 13.2% of the world GDP and had a turnover of US$6,477.2 billion, and supported 234 million jobs (8.7% of the total world employment). It is therefore clear that tourism is the major force in the economy of the world, an activity of global importance and significance. Moreover, as pointed out by Cooper et al. (2008), tourism has been remarkable in its resistance to adverse economic and political conditions. Events such as the 9/11 terrorist bombing, and the 2004 Boxing Day Tsunami clearly demonstrates the sector's ability to regroup and place emphasis on a new vocabulary including words as "safety", "security", "risk management", "crisis" and "recovery". Inevitably though, growth is slowing as the market matures and, as the nature of the tourist and his or her demands change, the sector will need to be creative in supplying products to satisfy the "new tourist".

A particular case of tourism is cultural/heritage tourism, where "cultural/heritage" can be defined as the visit of all set of monuments, buildings and archaeological sites of outstanding universal value from the point of view of history, art or science.

Cultural/heritage tourism is the fastest growing segment of the tourism industry, and obviously, the growth in the cultural/heritage tourism market may provide several benefits to cultural/heritage destinations. Because of people's inclination to seek out novelty, including that of traditional cultures, heritage tourism has become a major "new" area of tourism demand, of which almost all policy–makers are now aware of and anxious to develop. Heritage tourism, as a part of the broader category of "cultural tourism" is now a major pillar of the nascent tourism strategy of many countries. Cultural/heritage tourism strategies in various countries have in common that they are a major growth area, that they can be used to boost local culture, and that they can aid the seasonal and geographic spread of tourism (Richards, 1996).

According to Huh (2002), if the cultural/heritage tourism market can be segmented so that planners can easily understand market niches, the contribution to the field is three-fold. First, comprehending what tourists seek at cultural/heritage attractions may help tourism marketers better understand their customers. Second, identifying which attributes satisfy tourists who visit cultural/heritage destinations could help tourism planners develop strategies to attract customers. Third, knowing who the satisfied tourists are may reduce marketing costs and maintain the cultural/heritage destination's sustainability.

Many models have been developed to explain consumer satisfaction; we are here more interested in the global tourist satisfaction in relation with touristic services; some of these papers are very update (Wood 2008, Huang et al., 2006, Ciavolino and Dahlgaard, 2007, Cronin et al. 2000), also many other papers have been written on the selection of tourist destination and trip experiences (Crompton 1992, Fesenmaier, 1990, Zeithaml et al. 2002, Yang and Fang, 2004).

In general, making customers satisfied is important from different perspectives. Research has shown that satisfaction can affect customer retention and also it can make them recommend the goods or services to others; this can also be applied to tourism. Thus, it is crucial to investigate which factors are important for tourists bearing in mind this. Providing that tourism is an experience made up of many different interdependent parts, some more tangible than others, tourist satisfaction may be treated as a cumulative measure of total purchase and consumption experience over time (Haber and Lerner, 1998). At this point it is also crucial to take into account that according to quality management theories many key product and service attributes have a non linear relationship with satisfaction (Conklin et al. 2002).

Then, this research will introduce a causal model based on a relationship between three second order factor variables: *Importance* which will define the importance of accommodation and lodging, the city tourism attractions and the offered services of the city of Toledo; *Satisfaction* will measure the satisfaction towards accommodation and lodging, the city tourism attractions and the offered services; finally *Expectations* will also measure the tourist expectations related to accommodation and lodging, the city tourism attractions and the offered services of the city of Toledo.

2. CASE STUDY

2.1. The Study Site

The study area for this research is the emblematic old part of the city of Toledo in Spain. Toledo is a UNESCO World Heritage City with an economy driven to a significant extent by commercial activities deriving from tourism.

The city of Toledo (pop. 76,618 in 2010) is located in central Spain, about 71 km from the capital city of Madrid. Toledo was imperial capital until 1563 when the court moved to Madrid, and is currently the seat of regional government of Castilla-La Mancha, one of the most extensive regions in the country. As a medieval city known today as the city the three cultures due to its Islamic, Christian, and Jew heritages, Toledo has been highly successful in the task of preserving its historical and architectural character. This accomplishment has been recognized by the UNESCO, which has granted the denomination of World Heritage City. The preservation of centuries-old city walls has meant that Toledo has experienced relatively little expansion, with virtually all recent growth taking place beyond the perimeter of the old city (see Figure 1). Given its historical and cultural interest, its proximity to Madrid, and excellent connections with the capital and other regions, which include highways and a new High Speed Train service, Toledo has stood to benefit from a very active tourist sector, and related commercial and service activities. The importance for the economy of the city of the service sector in general, and tourism in particular is very high, with about 84.5% of Toledo's gross product coming from that sector, and about half of it is due to tourism. The importance of these activities is also reflected in the distribution of the economically active population in the municipality, 86.5% of which is associated with the service sector. According to Esteban et al. (2005), some 45% of visitors to the city are international travellers. In terms expenditures while in the city, about 80% of all visitors (domestic and international) have a daily budget-person of between 50 and 100 Euros, and about 14% have a budget of between 100 and 200 Euros. Besides lodging, other significant expenses include food and beverages, clothing and shoes, and souvenirs. There are about 11,000 establishments in the city that cater to the demand for these services.

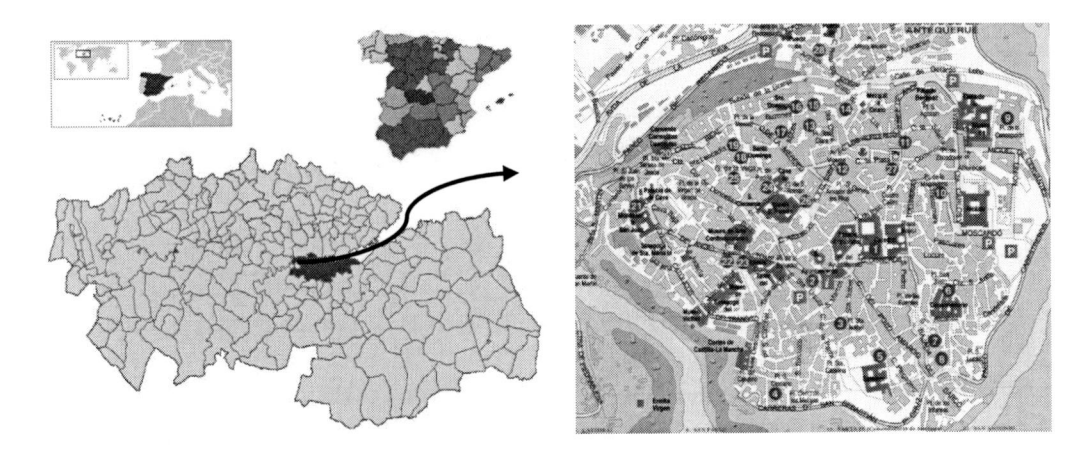

Figure 1. Toledo: a UNESCO World Heritage City.

2.2. The Dataset

The information has been collected through a specially developed questionnaire, the type of survey being a personal interview. Data were collected during the months of April, May and June 2009. The number of questionnaires was 1.500. Statistical analysis of questionnaires revealed no significant differences among them, this being the reason why the set of polled tourists can be considered as a unique sample of tourists. Tourists were interviewed in different locations of the old part of Toledo city (monuments, squares, restaurants, hotels, etc.) at different hours, to try to collect a range, as wider as possible, of people and situations.

The questionnaire included the following tourist attributes: accommodation, tourist attractions and or/ walks, entertainment/ cultural and recreational activities, gastronomy, shopping, cleanness, hygiene, public toilets, Internet/ communications/ phone, public lighting, hospitality, information, security, tourist information offices, providers of tourist services, public transport connections, signposting, banks/ ATM, local transportation, environmental care, accessibility, and ability to settle problems. The non-response rate was 6.8%, and most of the non-respondents were Asian people.

Although the population of visitors is unknown, we can infer the appropriateness of the sample from indirect indicators as percentage of male and female, percentage per age strata, percentage per nationality, etc., The structure of these percentages is very similar to the official data (if available).

Table 1. Technical Data

Population	About 600,000 tourists
Geographical Area :	Toledo City
Sample size:	1,500
Sample error:	± 5.1%
Significance level:	5%
Sample period:	April, May and June 2009

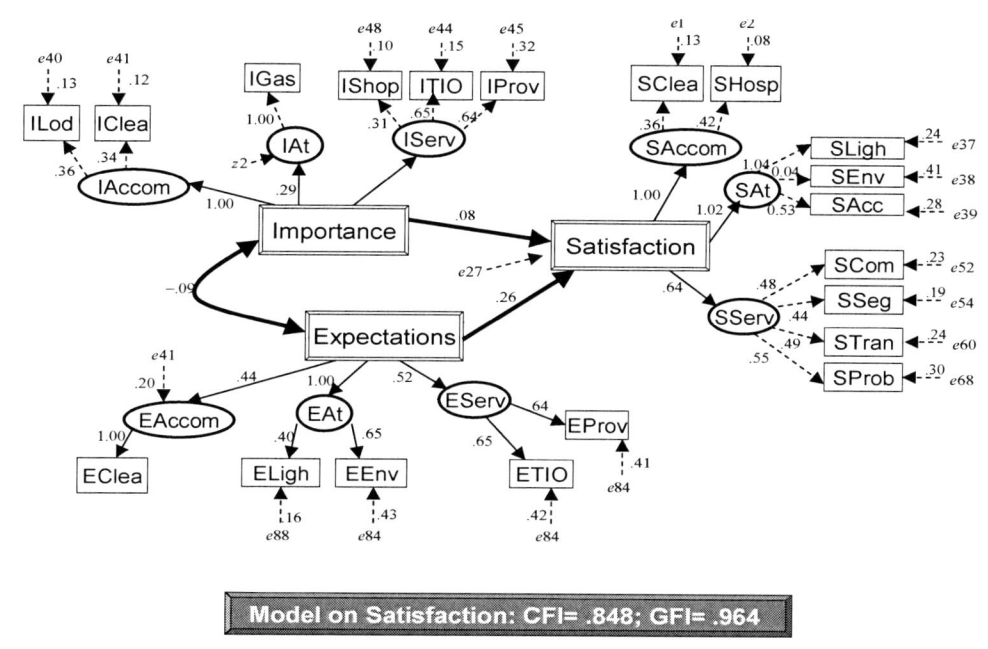

Source: own elaboration.

Figure 2. Second order factor causal model on tourist satisfaction.

3. METHOD

To analyze the tourist global satisfaction in a cultural heritage context like the Toledo city, we have chosen to present a regression causal model of second order factor variables.

This model is based on Joreskog- Sorbom second order approach model and consists on a regression of three second order CFA (confirmatory factor analysis) latent variables, *Importance, Satisfaction,* and *Expectations*. Each second order latent variable is defined by three first order latent variables, all of them measuring the *Accommodation (accommodation, lodging)*, the *Tourist attraction* and the *Offered services*.

The questionnaire is divided in 17 questions; the first 14 are general questions or classification questions. Questions 15 to 17 ask the interviewee to rate the different items in a five point scale from 1 (less important) to 5 (very important) for a total of 60 items.

3.1. The Structural Model (Or Second Order Factor Variable Causal Model)

Figure 2 presents the global theoretical causal model relating *Importance, Satisfaction* and *Expectations* latent variables and theirs relationships.

This model is issued from an original model accounting for 60 variables, which has been purified or re-specified by erasing all multi-collinear variables. The model we work with (Figure 2) is the optimized and re-specified model.

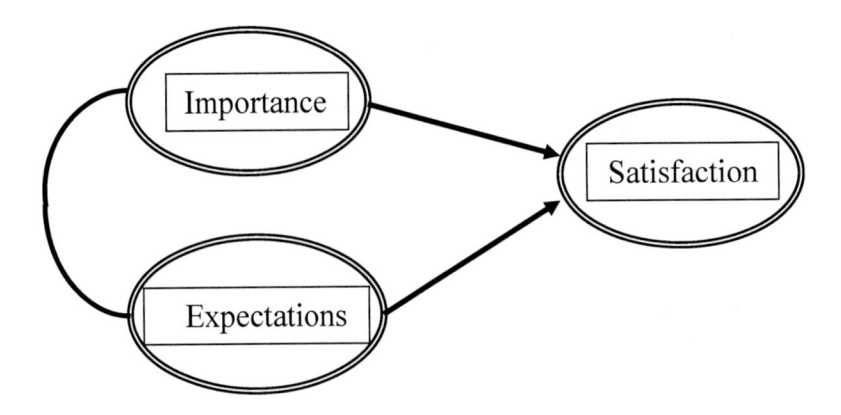

Figure 3. The Structural Model.

The first latent variable, *Importance*, is a second order factor variable that describes the importance of all items for the tourist in the city of Toledo. This variable plays an important role as an independent latent variable because the tourist has a pre-idea of what is important for him or for her to have a good stay in a cultural and heritage city like Toledo. *Importance* evaluates through a first order factor variables (*accommodation, IAccom, attractions, IAt, services, IServ*) the 20 items that represent these three latent variables.

Satisfaction is a positive affective satisfactory state; more precisely satisfaction is considered an emotional state that results from a specific transaction That is to say, it is an emotional reaction following a disconfirmation experience (Oliver 1981) or a summary psychological state resulting when confirmed or disconfirmed expectations exist with respect to the service experience. *Satisfaction* is represented by another second order factor latent variable which consists in evaluating through a first order factor variables (*accommodation, SAccom, attractions, Sat, services, Sserv*), the tourist satisfaction on the same items than before for latent variable *Importance*. This second order factor latent variable is a dependent variable from *Importance* to *Expectations*.

Expectations measure the tourist expectations in the city of Toledo. This latent variable is the third second order factor latent variable which consists in evaluating through a first order factor variables (*accommodation, EAccom, attractions, EAt, services, EServ*) the tourist expectations on all the same items than before.

Figure 3 represents the theoretical model relating the causal relationship of the three second order latent variables. *Importance* and *Expectations* have a direct relation with *Satisfaction* while *Importance* and *Expectations* are correlated each other.

3.2. The Measurement Model

The measurement model depicts three first order latent variables for each second order latent variable. *Importance* is defined by first order latent variables *IAccom, IAt* and *IServ*, *Satisfaction* by *SAccom, SAt* and *SServ* and *Expectations* by *EAccom, EAt* and *EServ*. It can be noticed that these first latent variables measure the same concepts for each second order latent variable.

Table 2. First order standardized factor loadings

Observed variable or item		First order factor	Loading
SCommunications (SCom)	←	SServ	.492
SCleaness (SClea)	←	SAccom	.378
SSecurity (SSec)	←	SServ	.429
SHospitalility (SHosp)	←	SAccom	.418
SLighting (SLigh)	←	SAt	.508
SEnvironmental care (SEnv)	←	SAt	.647
SAccesibility (SAcc)	←	SAt	.535
SLocal transoportation (SLt)	←	SServ	.496
SProblem resolution (SProb)	←	SServ	.526
ILodging (ILod)	←	IAccom	.357
ICleaness (IClea)	←	IAccom	.343
IGastronomy (IGas)	←	IAt	1.000
IShopping (IShop)	←	IServ	.314
ITourist information offices (ITIO)	←	IServ	.648
IProvider of tourist services (IProv)	←	IServ	.563
EEnvironmental care (EEnv)	←	EAt	.653
ELighting (ELigh)	←	EAt	.405
EProvider of tourist services (EProv)	←	EServ	.644
ETourist information offices (ETIO)	←	EServ	.651
ECleaness (EClea)	←	EAccom	1.000

The model presented in Figure 2 has been re-specified using a structural covariance program (AMOS); we used Modification Indices to eliminate those multi-collinear variables that reduce the global fit of the model. Each second latent variable is not defined anymore by all the original first order indicators; those which have had any multi-collinear effect have been suppressed. The final model for each second order latent variables is like presented in Figure 2.

As general considerations we can say that we have accepted loadings superior to 0.30 which are generally accepted in social sciences (Hair et al. 2000). We could increase the acceptable factor loading order but as a matter of fact we could lose some interesting items for this research and some branches of this first order factor model. In another round of re-specification we could eliminate those items but the model could lose some explanation capabilities. A second round of re-specification will certainly improve the general fit of the model.

Table 3. Second order factor loadings

First order factor		Second order factor	Loading
SAccom	←	Satisfaction	1.000
SServ	←	Satisfaction	1.000
SAt	←	Satisfaction	1.000
IAccom	←	Satisfaction	1.000
IAt	←	Importance	.291
IServ	←	Importance	.796
EAccom	←	Expectations	.444
EAt	←	Expectations	1.000
EServ	←	Expectations	.523

All second order regression coefficients are significant for P<0.05.

Reliability Measures and AVE

The reliability measures internal coherence of all indicators in relation with the latent variables. It is determined by the measure of Composite Reliability for which 0.7 should be an acceptable threshold (Nunnally, 1978).

The convergent validity represents the common variance between observed variables or items and the latent variable or factor measured by the Average Variance Extracted (AVE); the acceptable threshold should be superior to 0.5 (Fornell and Larker 1981, Vinzi 2003). It can be observed that the second order latent variable *Importance* has very low composite reliability and AVE. *Satisfaction* and *Expectations* seem to have more regular levels.

Table 4. Composite reliability and AVE

Latent Variable	Composite Reliability	AVE
IAccom	0.743887	0.683044
IAt	1.000000	1.000000
IServ	0.371149	0.503392
Importance	0276498	0.282493
SAccom	0.719927	0.562411
SAt	0.761074	0.516859
SServ	0.709932	0.381722
Satisfaction	0.838570	0.637820
EAccom	1.000000	1.000000
EAt	0.774195	0.631581
EServ	0.881479	0.799355
Expectations	0.742983	0.494076

Table 5. Correlation of latent variables-1st order and AVE square roots

	IAccom	IAt	IServ	SAccom	SAt	SServ	EAccom	EAt	EServ
IAccom	0.826								
IAt	0.113	1.000							
IServ	0.148	0.166	0.709						
SAccom	-0.037	0.080	0.030	0.750					
SAt	-0.006	0.050	0.002	0.391	0.718				
SServ	-0.002	0.076	-0.081	0.393	0.594	0.617			
EAccom	0.028	-0.007	-0.023	0.097	0.069	0.094	1.000		
EAt	0.011	-0.040	-0.019	0.064	0.164	0.120	0.317	0.794	
EServ	0.078	0.020	0.118	-0.005	-0.067	-0.108	-0.130	-0.311	0.894

Discriminant Validity

To evaluate any presence of discriminant validity among first order factors it is necessary that the AVE square root be superior to the correlation between factors (Fornell and Laker 1981). Table 5 shows that this should be the case (the table presents correlations between latent variables and AVE square roots in diagonal).

4. RESULTS

To evaluate the model pertinence and suitability we have computed the model fit indices; among them we have identified the most common ones (see Lévy Mangin and Varela 2006, or Hair et al. 2000). The GFI and AGFI are close to 1, RMSEA and RMR are very good to good (the residuals have to be inferior to 0.06), CFI is may be not so strong (it goes from 0 to 1). In general the model fits reasonably well.

The standardized regression weights are significant for a P<0.10 for *Importance* to *Satisfaction*, and for a P<0.05 for *Expectations* to *Satisfaction*.

This is a core result. On the one hand, the more important is the service for the tourist, the more satisfied is the tourist with the service. On the other hand, the relation *Expectations-Satisfaction* suggests the confirmation of the Assimilation Theory in cultural heritage contexts. According to the Assimilation Theory (Sherif and Hovland, 1961), individuals suffer a psychological conflict when they perceive discrepancies between performance and prior beliefs and they tend to adjust perceptions to their expectations in order to minimise, or even remove, that tension (Oliver, 1997). In these circumstances, expectations are a driver of satisfaction.

Fit Indices	RMR	GFI	AGFI	RMSEA	CFI
Second order factor model	.057	.946	.932	.046	.848

In the context of the expectation/disconfirmation model, if the overall performance, while or after visiting a destination, exceeds or meets initial expectations, then the tourist is considered satisfied (positive disconfirmation). Otherwise, the tourist may be dissatisfied (negative disconfirmation). An alternative way of checking the Assimilation Theory consists of deriving a Chi-square test to determine association between expectation (low, high) and disconfirmation (negative, positive) in tourist attributes. Positive association implies the verification of the theory.

Standardized Regression Weights			Estimate
Satisfaction	←	Importance	.080
Satisfaction	←	Expectations	.259

The correlation between *Importance* and *Expectations* is not significant for a P<0.05 but it is for a P<0.10. This could mean that an increase in *Importance* should result in a decrease in *Expectations* and vice-versa.

This result suggests that in cultural heritage cities tourist are not excessively exigent with the more important services for them, which along with the verification of the Assimilation theory suggest that in questionnaires the variable Satisfaction could be overestimated; or in other words: cases of slight positive disconfirmation could not be, in reality, positive disconfirmation but negative one.

Thus, the interpretation of measurements of satisfaction in cultural heritage cities such as Toledo must be interpreted with caution, because to the extent that the Assimilation Theory is at work, satisfaction, and positive disconfirmation, could be overestimated.

Correlations			Estimate
Importance	↔	Expectations	-.095

The model predicts with just one significant second order factor (*Expectations*, the relationship between *Importance* and *Satisfaction* is just significant for a P<0.10) and this should be the reason why the R^2 is not very significant. Obviously, this makes that the model predictability is very low.

Explained Variance of the Endogenous Factor *Satisfaction*	Estimate
R^2 for *Satisfaction*	.07

CONCLUSION

It is clear that prior to their visit, tourists give importance to lodging and accommodation, and to cleanness and gastronomy for touristic attraction. For services the importance goes to shopping, to different tourist services and good tourist information (*Importance*).

The tourist expectations (*Expectations*) are important for cleanness, lodging and hygiene, public lighting and illumination; environmental care for touristic attractions and for services, offices of tourist information and good providers of tourist services. The final tourist satisfaction (*Satisfaction*) is based on lodging cleanness and hospitality; for tourist attraction, good lighting and illumination in the heritage city, environmental care and accessibility; for services, good communications, internet and telephone, security, local transportation and ability to resolve problems.

The research does not presents a very strong relationship between *Importance* and *Satisfaction* (but significant for a P<0.10), that means that there is a mild relationship between the prior importance given to all services that could be offered in the city of Toledo and the satisfaction (*Satisfaction*) with the offered services.

It is our understanding that some institutional advertising of the city of Toledo could change these perceptions and even correct them. The advertising should be directed to the same dimensions than satisfaction.

Expectations measure the tourist expectations. Here, there is a stronger relationship between these two latent variables; an increase in *Expectations* has a positive impact on tourist *Satisfaction*.

The main conclusion of this paper is that taking into account that a) the more important the service for the tourist is, the more satisfied the tourist is with the service, b) the more important the service is for the tourist, the lower are the tourist expectations, and c) an increase in *Importance* should result in a decrease in *Expectations*, and vice versa, the interpretation of measurements of satisfaction in cultural heritage cities as Toledo must be interpreted with caution, because to the extent that the Assimilation Theory is at work, satisfaction, and positive disconfirmation, could be overestimated.

Finally, from the predictive point of view, there is not the first time that a research made to analyze the tourist satisfaction does not suit a complex model. The low explained variance of *Satisfaction* tells us that there are other variables that predict the tourist satisfaction. But this question goes beyond the scope of this research.

REFERENCES

Ciavolino E., Dahlgaard, J. (2007): "ECSI Customer Satisfaction Modelling and Analysis: A Case Study". *Total Quality Management*, 18 (5), 545-554.

Conklin, P., Powaga, K., Lipovetsky, S. (2002): "Customer satisfaction analysis: Identification of key drivers". *European Journal of Operational Research*, 154, 819- 827.

Cooper, C., Fletcher, J., Wanhill, S., Gilbert, D. (2008): *Tourism: Principles and Practice*. (3rd ed.). Harllow: Pearson Education.

Crompton, J. (1992): "Structure of Vacation Destination Choice Sets". *Annals of Tourism Research*, 19, 420-434.

Cronin J, J, Brady, M.K., Hult, G.T.M. (2000): "Assessing the Effects of Quality, Value, Value and Customer Satisfaction on Consumer Behavioural Intentions in Service Environments". *Journal of Retailing*, 76 (2), 193-218.

Esteban, A., Martín- Consuegra, D., Molina, A., Díaz, E. (2005): Turismo y consumo: El caso de Toledo. *Centro de Estudios de Consumo de la Universidad de Castilla- La Mancha y de la Junta de Comunidades de Castilla- La Mancha*. Working paper. Toledo.

Fesenheimer, D.R. (2005): "Theoretical and Methodological Issues in Behavioural Modelling, Introductory Comments". *Leisure Sciences*, 20, 175-191.

Fornell, C., Larcker, D. F., (1981): "Evaluating Structural Equation Models with Unobservable Variables and Measurement Errors". *Journal of Marketing Research*, 18 (1), 39-50.

Haber, S., Lerner, M. (1998): "Correlates of Tourist Satisfaction". *Annals of Tourism Research*, 25(4), 197-201.

Hair, J.F., Anderson, R.E., Tatham, R.L., Black, W.C. (2000): *Multivariate Analysis*, (5th edition). Englewood Cliffs: Prentice Hall

Huang, H.H., Chiu, C.K., Kuo, C. (2006):"Exploring Customer Satisfaction, Trust and Destination Loyalty in Tourism". *The Journal of American Academy of Business*, 10 (1), 156-159.

Huh, J. (2002): Tourist Satisfaction with Cultural/Heritage Sites: The Virginia Historic Triangle. Doctoral Thesis. Virginia Polytechnic Institute, State University, Virginia.

Lévy Mangin, J. P., Varela Mallou, J. (2006): *Modelización con Estructuras de Covarianzas en Ciencias Sociales*. A Coruña: Netbiblo.

Ndubisi, N. O. (2006): "A Structural Equation Modelling of the Antecedents of Relationship Quality in the Malaysia Banking Sector". *Journal of Financial Services Marketing*, 11 (2), 131-141.

Nunnally, J.C. (1978): *Psychometric Theory*, 2nd edition, NewYork: McGraw-Hill.

Oliver, R, L. (1981): "Measurement and Evaluation of Satisfaction Process in Retail Settings". *Journal of Retailing, 57,* 25-48.

Oliver, R.L. (1997): *Satisfaction: A Behavioral Perspective on the Consumer*. New York: Mc Graw- Hill.

Richards, G. (1996): Production and consumption of cultural tourism in Europe. *Annals of Tourism Research, 23*(2), 261-283.

Sherif, M., Hovland, C. I. (1961): *Social judgment: Assimilation and contrast effects in communication and attitude change*. New Haven: Yale University Press.

Wood, J. (2008): "The Effect of Buyer´sPerception of Environmental Uncertainty on Satisfaction and Loyalty". *Journal of Marketing Theory and Practice*, 16 (4), 309-320.

Yang, Z., Fang, X. (2004): "On line Service Quality Dimensions and their Relationship with Satisfaction". *International Journal of Service Industry Management*, 15 (3), 302-326.

Zeithaml, V. A, Parasuraman, A., Malhotra, A. (2002): *An Empirical Examination of the Service Quality Value Royalty Chain in an Electronic Channel. Working Paper, University of North Carolina*, Chapel Hill, NC.

WTTC (2006): http://www.wttc.org.

In: Research Studies on Tourism and Environment
Editors: J. Mondejar-Jimenez et al.

ISBN: 978-1-61209-946-0
© 2012 Nova Science Publishers, Inc.

Chapter 12

LUXURY RESORTS AND SUSTAINABLE TOURISM IN THE MALDIVES

Blanca de-Miguel-Molina, María de-Miguel-Molina[*] *and Mariela Rumiche-Sosa*
Universidad Politécnica de Valencia, Spain

ABSTRACT

Small islands are favoured locations for luxury resorts. However, this type of tourism, which aids economic development in some countries, can cause a threat to these islands' local environment. In the case of the Maldives, the country's government has established specific regulations to prevent this from happening but these are not sufficient to ensure the future conservation of local ecosystems. Therefore, this chapter examines, from a conceptual sustainable tourism framework, whether luxury resorts in the Maldives have implemented measures to ensure sustainable tourism and whether they have communicated them to their visitors in order to establish whether luxury and sustainable tourism are compatible. To carry out our study, we analysed around 40 luxury resorts in the Maldives using a content analysis methodology and we examined those which were supposed to be more eco-friendly.

Keywords: Small islands, Sustainable Tourism, Luxury Tourism, Corporate Social Responsibility, Recreational aspects of natural resources

1. INTRODUCTION

Tourism is a major economic driver in many small islands (Shareef and McAleer, 2005; Nurse and Moore, 2005; Belle and Bramwell, 2005; McElroy, 2006; Clampling and Rosalie, 2006). However, the economic and environmental aspects of tourism need to be balanced (Henderson, 2001) to guarantee long-term benefits to communities (UNWTO, 2004). While

---

[*] Contact author: mademi@omp.upv.es

tourism can bring many economic advantages to small islands, there are many examples of rapid, unplanned tourist development which have produced over-reliance on this one industry, environmental degradation and excessive concentration at the lower quality end of the mass tourism market. As a result, in the 1990s many islands started to remedy this situation by showing greater commitment to planning, upgrading their facilities and developing new markets (Bull and Weed, 1999). This included promoting ecotourism. The popularity of ecotourism as a conservation and development tool largely stems from the associated theoretical benefits and the perception that it is a viable alternative to more traditional, destructive and extractive forms of rural development (Powell and Ham, 2008).

Both internal (tourism impact: Zubair et al., 2010; Belle and Bramwell, 2005; Georges, 2006) and external (climate change: Briguglio, 1995; Belle and Bramwell, 2005; Roper 2005) factors can have an impact on the environment of small islands, which can reduce the attractiveness of these coastal tourism destinations and may reduce the number of people who want to visit small islands in tropical and subtropical regions (Nurse and Moore, 2005). In an attempt to preserve its local ecosystem, the Maldives signed all the major international agreements promoted by the UN Environment Programme, and the Maldives Government established specific regulations to develop sustainable tourism: the Environmental Protection and Preservation Act of Maldives (Maldives Government, 1993), the Tourism Act of Maldives (Maldives Government, 1999) and the Regulation on the Protection and Conservation of Environment in the Tourism Industry (Maldives Government, 2006).

Small island states should readily accept on one hand that they are unlikely to be in a position to access substantial external resources to adapt their model of tourism to an eco-tourism model and on the other, that their strategies to combat climate change should be integrated into existing plans and programmes (Nurse and Moore, 2005). For example, municipal solid waste is the most significant waste stream in many small islands (Georges, 2006). In this sense, small islands could set an example for the rest of the world (Roper, 2005).

With respect to the characteristics of eco-friendly tourists (EFTs), few personal characteristics have been examined in the research published to date, which leads us to the conclusion that all we really know about EFTs at the moment is that they are more educated, earn more money and are interested in learning. Many of the socio-demographic descriptors which could readily be used by destination managers to actively target EFTs have either produced inconsistent results (e.g. age) or have only been included in a small number of studies (e.g. gender) (Donicar S. et al., 2008).

Up until now, studies about small islands, including the Maldives, have not analysed every island separately. When studying sustainable tourism, indicator analyses are based on national data so as to compare different countries (Buzzigoli, 2009). Thus, we have not found any studies that focus on the eco-friendly image that the resorts in the Maldives give to tourists or on whether luxury and sustainable tourism are compatible in the Maldives.

Thus, our chapter's goal is to examine, from a conceptual sustainable tourism framework, the RQ1: whether the luxury resorts of the Maldives have implemented measures to ensure sustainable tourism and whether they have communicated them to their visitors. Moreover, we also analysed RQ2: Are luxury and sustainable tourism compatible in the Maldives?

2. MATERIALS AND METHOD

In 1987, the UN World Commission on Environment and Development created the term Sustainable Development to refer to development that meets the needs of the present without compromising the ability of future generations to meet their own needs (Brundtland Report, 1987). Later, in 2004, the UNWTO also established the definition of Sustainable Tourism as an enterprise that achieves an effective balance among environmental, economic, and socio-cultural aspects of tourism to guarantee long-term benefits to communities. Recently, climate change (including global warming) has been added to these definitions of sustainable tourism which now has a "quadruple bottom line" (Buzzigoli, 2009). The Third National Environment Action Plan of the Maldives (2009-2013) includes ensuring a balance between resource protection and resource usage, and develops environmental guidelines for the selection of islands to be developed as resorts. Furthermore, new resorts are subject to rigorous environmental impact studies and developers and planners cannot build on over 20% of the islands (Maldives Government, 2006).

Studies about small islands (Sathiendrakumar and Tisdell, 1989; McElroy and Albuquerque, 1998; Nurse and Moore, 2005; Roper, 2005; Fry, 2005; Belle and Bramwell, 2005; Shareef and McAleer, 2005; Yahya, Parameswaran and Sebastian, 2005; Campling and Rosalie, 2006; McElroy, 2006; Van der Velde et al., 2007) have pointed out the islands' economic and environmental vulnerabilities (Figure 1). For example, Shareef and McAleer (2005), Nurse and Moore (2005), Belle and Bramwell (2005), McElroy (2006) and Campling and Rosalie (2006) have underlined the value of tourism as an economic activity in small islands. The Maldives is a group of small isolated islands that have become an attractive tourist destination for their exoticism (Scheyvens and Momsen, 2008) and biodiversity (Campling and Rosalie, 2006). The islands' characteristics offer different income-generating activities, such as water sports (diving, yachting, windsurfing, dinghy sailing, water skiing and scuba-diving) (Bull & Weed, 1999). However, some of these activities may have negative effects on the local ecosystem, for example, they may damage coral reefs.

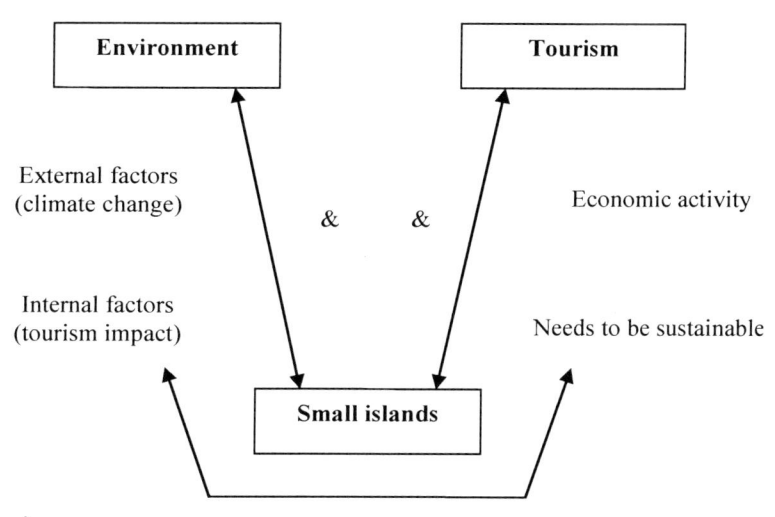

Source: Author's own.

Figure 1. Small islands, tourism and environment.

Worachananant et al. (2008) show that coral reefs which attract scuba divers are those with the most sensitive species and this leads to their rapid destruction. However, if access to these places is limited, tourists are less likely to visit them. The solution is to provide better information to scuba divers so that they are more careful. Along the same lines, Nurse and Moore (2005) and Belle and Bramwell (2005) consider that coral reefs are one of the attractions of small islands whose loss can affect income-generating activities such as diving and snorkelling. Hence, the relationship between protecting biodiversity and sustainable tourism is a major issue in small islands (Fry, 2005).

As represented in Figure 1, studies about small islands include the external and internal factors that produce environmental damage. The first are related to climate change (Briguglio, 1995; Belle and Bramwell, 2005) while the latter are related to the economic influence of tourism on small islands.

The disadvantage of tourist resources that are concentrated on the coast is that these islands are particularly vulnerable to climate change impacts such as beach erosion, salinity of lagoons and reef damage (Belle and Bramwell, 2005). Aware of their vulnerability to climate change, some small island states created a coalition in 1990 called AOSIS, the Alliance of Small Island States, to establish a work schedule in the planning and implementation of sustainable development. Tourist resources were among their priority areas (Chasek, 2005).

Thus, a new focus of study on the relationship between biodiversity protection and sustainable tourism (Zubair et al., 2010; Fry, 2005; Campling and Rosalie 2006) emerged. Many studies have centred on the protection of coral reefs (Nurse and Moore, 2005; Worachananant et al., 2008), developing waste management (Sullivan-Sealey & Cushion, 2009; Georges, 2006) and water management (Sullivan-Sealey & Cushion, 2009; Belle and Bramwell, 2005) and using alternative energies (Roper, 2005; Stuart, 2006). For example, tourists staying in hotels have a high consumption of drinking water (Belle and Bramwell, 2005) and generate more waste (Georges, 2006). In the case of water, one solution would be to desalinate seawater (Stuart, 2006). Specifically, Roper (2005) considers the problem of global warming in the case of the Maldives and emphasises the use of renewable energy.

Furthermore, Fotiu et al. (2002) believe that tourism may be a solution for funding the protection of marine and coastal areas in small islands. However, this would require the involvement of both public and private organisations to protect the natural environment which is in fact the main attraction for many tourists. One way to encourage hotels to promote protection is the use of an ecolabel or an environmental certification that also warns customers about what to expect before booking. It should not be forgotten that tourists are prepared to participate in responsible tourism, and thus omitting them from this debate overlooks any positive contributions they can make whilst on holiday (Standford, 2008).

We have taken the Republic of Maldives as the focal point of our chapter. The Maldives are an island country in the Indian Ocean formed by a double chain of twenty-six atolls, 21 of which have resorts. In 2008, there were 94 resorts, with a total of 19,860 beds (Ministry of Tourism, Arts and Culture, 2009). 44% of the total number of resorts are luxury class (5 stars or more) and it is these resorts which were the focus of our analysis. We considered whether they have a specific environmental certification and also looked into which are the leisure activities they offer that could be potentially negative for the local ecosystem. The remaining data were obtained by the content analysis methodology from these resorts' websites.

The type of content analysis used was conceptual analysis, which studies the presence of concepts and not the frequency with which they appear (Cohn, 2009). The selection of concepts was done before the websites were searched, based on existing theories about sustainable tourism and small islands. Subsequently, the variables related to these concepts were looked up on the resorts' websites, taking into account Weber's criteria (1990) that "a variable is valid to the extent that it measures or represents what the investigator intends it to measure". Therefore, we did not use the standard indicators proposed by various organizations as a reference because they did not fit our objectives (Buzzigoli, 2009).

The chapter's goal was to examine whether luxury and sustainable tourism are compatible in the Maldives. The selected variables are shown in Table 1. These variables were then searched on the resorts' websites from October 2009 to March 2010 to find out what information they give out to current and potential tourists. Moreover, we carried out a cluster analysis to determine whether luxury and sustainable tourism are compatible.

The first group of variables includes water sports that could damage the ecosystems of the lagoons and coral reefs. These water sports are offered by all the resorts. However, it has been shown that scuba diving and thus human contact with the coral reefs can damage them (Saphier and Hoffmann, 2005), as can boat anchors and chains (Saphier and Hoffman, 2005). Moreover, these water sports require the construction of a jetty, which can also affect lagoons and coral reefs (Zubair et al., 2010).

The second group of variables relates to the resort construction. For example, installing water bungalows can damage lagoons and coral reefs (Zubair et al., 2010).

The third group variables are those directly related to the resort's development of an environmental policy. This includes whether the resort has an ecolabel (Fotiu et al., 2002), whether it has a water management policy (Sullivan-Sealey & Cushion, 2009; Belle and Bramwell, 2005), a waste management policy (Sullivan-Sealey & Cushion, 2009; Georges, 2006) and whether it uses renewable energies (Roper, 2005; Stuart, 2006).

Table 1. Definition of variables and categories

Concept	Variables	Categories
Water sports which are potentially negative for the ecosystem	1. Parasailing 2. Water skiing or wakeboarding 3. Ringo riding or banana riding 4. Jet skiing 5. Submarine diving (using an engine) 6. Scuba diving 7. Catamaran sailing	If the resort offers the sport: 1 If does not offers it: 0
Services related to the construction of the resort which are potentially negative for the ecosystem	1. Tennis or badminton 2. Swimming pool 3. Water bungalows 4. Average number of beds 5. A/C rooms	If the resort has it: 1 If the resort doesn't have it: 0
Environmental policy	1. Ecolabel 2. Water management 3. Waste management 4. Alternative energy	If the resort has it: 1 If the resort doesn't have it: 0

Source: Author's own.

Table 2. Results of the content analysis and conglomerates

Resort	Water sports	Construction	Environmental Policy	Conglomerate
RL1	5	3	0	1
RL2	4	3	2	2
RL3	3	2	4	2
RL4	3	3	0	2
RL5	2	3	0	2
RL6	2	2	2	2
RL7	3	4	2	2
RL8	2	3	0	2
RL9	5	5	0	1
RL10	5	4	0	1
RL11	6	5	0	1
RL12	2	5	0	2
RL13	5	3	0	1
RL14	5	4	0	1
RL15	4	2	0	1
RL16	3	5	0	1
RL17	4	3	0	1
RL18	4	4	0	1
RL19	4	4	0	1
RL20	4	4	0	1
RL21	1	3	0	2
RL22	3	4	0	1
RL23	3	4	0	1
RL24	4	4	0	1
RL25	2	3	0	2
RL26	3	2	0	2
RL27	6	5	3	1
RL28	7	5	0	1
RL29	7	4	0	1
RL30	2	5	0	2
RL31	3	3	4	2
RL32	3	4	3	2
RL33	6	5	0	1
RL34	5	4	0	1
RL35	4	4	4	2
RL36	2	3	0	2
RL37	2	3	4	2
RL38	4	4	0	1
RL39	4	3	0	1
RL40	5	4	0	1

Source: Author's own.

3. RESULTS AND DISCUSSION

In order to answer our RQ1 as to whether luxury resorts in the Maldives have implemented measures to ensure sustainable tourism and whether they have communicated them to their visitors, we analysed the information taken from the content analysis of the resorts' websites from October 2009 to March 2010 which is accessible to current and potential tourists. The results on environmental policies are shown in Table 2, and demonstrate that some luxury resorts in the Maldives are supposed to have implemented measures to ensure sustainable tourism (ecolabel, water management, waste management and use of alternative energies) and that these measures are communicated through their websites. However, the number of these resorts is low. Although it must be said that these eco-friendly resorts in the Maldives tend to have implemented more than one of these measures.

In order to answer our RQ2: Are luxury and sustainable tourism compatible? we conducted a K-means conglomerate study using SPSS software. The K-means analysis was proved for two and three conglomerates. The ANOVA results in the case of three conglomerates proved that two were the same, so the results for two were used. Two cases were verified in the conglomerates according to whether or not the resorts were eco-friendly (Tables 2 and 3). The first group (1) was made up of those that we called *less eco-friendly* and the second group (2) of those that were *more eco-friendly* resorts. This second conglomerate includes resorts which have an ecolabel (8 resorts), waste management and water management policies, and/or use alternative energies.

The relevant results that can be observed in Table 4 are that more eco-friendly resorts have a lower mean of water sports and services related to the construction of the resort which are potentially negative for the ecosystem. Moreover, the mean of resorts which have some kind of environmental policy is higher in more eco-friendly resorts. Therefore, we can conclude that, in more eco-friendly resorts, luxury goes hand in hand with sustainable tourism.

This type of tourism could be a solution to balancing the economic and environmental aspects of tourism (Henderson, 2001) and guaranteeing long-term benefits to communities (UNWTO, 2004) such as the Maldives. The relationship between biodiversity protection and sustainable tourism is a major issue in small islands (Fry, 2005), thus ecotourism is a viable alternative to more traditional, destructive and extractive forms of rural development (Powell and Ham, 2008). However, eco-friendly tourists (EFTs) are difficult to reach (Donicar S. et al., 2008) if resorts do not communicate their environmental policies. As pointed out by Worachananant et al. (2008), tourists need information in order to be forewarned about the consequences of their actions and thus be more careful.

Hence, as to whether luxury and sustainable tourism are compatible in the Maldives, we can conclude, like Standford (2008), that tourists who go to the Maldives are prepared to participate in responsible tourism and that omitting them from this debate overlooks any positive contributions they can make whilst on holiday.

On the other hand, as our empirical results show, eco-friendly luxury resorts subscribe to more than one environmental policy measure. Thus, policymakers should encourage resorts to obtain an environmental certification or ecolabel. As pointed out by Fotiu et al. (2002), resorts and the Maldives Government will need to work jointly to preserve the ecosystem of these small islands.

Table 3. Conglomerates

| 1. Less eco-friendly | 23 resorts |
| 2. More eco-friendly | 17 resorts |

Source: Author's own.

CONCLUSION

The environment of small islands is affected by external (climate change: Briguglio, 1995; Belle and Bramwell, 2005) and internal (tourism impact: Zubair et al., 2010) factors. Tourism is an important source of revenue for small islands (Shareef and McAleer, 2005; Nurse and Moore, 2005; Belle and Bramwell, 2005; McElroy, 2006; Campling and Rosalie, 2006), but the impact of tourist activities differs according to the islands' ecosystems (Saphier and Hoffmann, 2005; Saphier and Hoffman, 2005). Thus, maintaining this source of income also requires protecting the ecosystem of these islands, including mainly coral reefs.

Measures that small islands could implement to ensure sustainable tourism include water management (Sullivan-Sealey & Cushion, 2009; Belle and Bramwell, 2005), waste management (Sullivan-Sealey & Cushion 2009; Georges, 2006), and the use of alternative energies (Roper, 2005; Stuart, 2006). Eco-friendly resorts in Maldives tend to have implemented more than one of these measures which could also be reinforced by obtaining an ecolabel.

The empirical results of our chapter show that some luxury resorts in the Maldives have implemented measures to ensure sustainable tourism, which are communicated through their websites to current and potential visitors. These resorts have implemented a variety of eco-friendly measures, and thus we can conclude that luxury and sustainable tourism are compatible in the Maldives. Although this situation is not widespread, it may become a future opportunity to link tourism to protecting local ecosystems.

Table 4. Conglomerates and Descriptives (concepts and type of resort)

Concepts and type of resort		F and p-value	N	Mean	S. deviation
Water sports	1 Less eco-friendly	44.574 $p<0.01$	23	4.70	1.146
	2 More eco-friendly		17	**2.53**	.800
	Total		40	3.78	1.476
Construction	1 Less eco-friendly	8.028 $p<0.01$	23	4.00	.798
	2 More eco-friendly		17	**3.24**	.903
	Total		40	3.68	.917
Environmental policy	1 Less eco-friendly	11.736 $p<0.01$	23	.13	.626
	2 More eco-friendly		17	**1.47**	1.736
	Total		40	.70	1.381

Source: Author's own.

The main limitation of our study is that our content analysis was restricted to the resorts' websites information retrieved between October 2009 and March 2010. But the information of those websites seems to have changed at present, especially in the point of eco-labels due to the Green Globe certification is being replace by EarthCheck. This means that if we require more information we will have to develop our own surveys. In a previous round, we found that resorts often have more information than they are communicating. However, this is the information which is accessible to tourists so resorts that do not communicate these values may be losing eco-friendly visitors.

ACKNOWLEDGMENT

This chapter is part of a research project which is supported by the Valencian Regional Government (Spain). Reference: GV/2009/020.

REFERENCES

Belle, N. & Bramwell, B. (2005). Climate change and small island tourism: policy maker and industry perspectives in Barbados. Journal of Travel Research, No. 44, pp. 32-41.

Briguglio, L. (1995). Small Island Developing Status and Their Economic Vulnerabilities. World Development, v. 23, no. 9, pp. 1615-1632.

Bruntland, G. H. (1987). Our Common Future: The World Commission on Environment and Development. (Oxford University Press).

Bull, C. & Weed, M. (1999). Niche markets and small island tourism: the development of sports tourism in Malta. Managing Leisure, No. 4, pp. 142-155.

Buzzigoli, L. (2009). Tourism Sustainability: conceptual issues, data and indicators. (In Ferrari, Mondéjar, Mondéjar & Vargas (Eds.), Principales tendencias de investigación en turismo (pp. 135-157), Oviedo: Septem ediciones, Spain).

Chasek, P.S. (2005). Margins of Power: Coalition Building and Coalition Maintenance of the South Pacific Island States and the Alliance of Small Island States. Review of European Community & International Environment Law, v. 14, No. 2, pp. 125-137.

Clampling, L. & Rosalie, M. (2006). Sustaining social development in a Small Island Developing State? The case of Seychelles. Sustainable Development, No. 14, pp. 115-125.

Cohn, Ellen G. (2009). Citation and Content Analysis. 21st Century Criminology: A Reference Handbook. SAGE Publications. Retrieved 2009, from http://www.sage-ereference.com/criminology/ Article n45 .html

Donicar S., Crouch G., & Long P. (2008). Environment-friendly Tourists: What Do We Really Know About Them? Journal of Sustainable Tourism, v. 16, No. 2, pp. 197-210.

Fotiu, S., Buhalis, D. & Vereczi, G. (2002). Sustainable development of ecotourism in small islands developing states (SIDS) and other small islands. Tourism and Hospitality Research, v. 4, No. 1, pp. 79-88.

Fry, I. (2005). Small Island Developing States: Becalmed in a Sea of Soft Law. Review of European Community & International Environment Law, v. 14, No. 2, pp. 89-99.

Georges, N.M. (2006). Solid Waste as an indicator of Sustainable Development in Tortola, British Virgin Islands. Sustainable Development, No. 14, pp. 126-138.

Henderson J.C. (2001). Developing and managing small islands as tourist attractions. Tourism and Hospitality Research, v. 3, No. 2, pp. 120-131.

Maldives Government (1993). Environmental Protection and Preservation Act Of Maldives. Act No. 4/1993.

Maldives Government (1999). Maldives Tourism Act. The Tourism Act of Maldives 1999.

Maldives Government (2006). Regulation on the Protection and Conservation of Environment in the Tourism Industry.

Maldives Government (2009). "Aneh Dhivehi Raajje": The Strategic Action Plan National Framework for Development 2009 – 2013.

McElroy, J.L. & Albuquerque, K. (1998). Tourism penetration index in small Caribbean islands. Annals of Tourism Research, v. 25, No. 1, pp. 145-168.

McElroy, J.L. (2006). Small island tourist economies across the life cycle. Asia Pacific Viewpoint, v. 47, No. 1, pp. 61-77.

Ministry of Tourism, Arts and Culture of Maldives (2009) Tourism Yearbook 2009. Republic of Maldives.

Nurse L. & Moore R. (2005). Adaptation to Global Climate Change: An Urgent Requirement for Small Island Developing States. RECIEL, v. 14, No. 2, pp. 100-107.

Powell, R.B. & Ham S. (2008). Can Ecotourism Interpretation Really Lead to Pro-Conservation Knowledge, Attitudes and Behavior? Evidence from the Galapagos Islands. Journal of Sustainable Tourism, v. 16, No. 4, pp. 467 – 489.

Roper T. (2005). Small Island States – Setting an Example on Green Energy Use. RECIEL, v. 14, No. 2, pp. 108-116.

Sathiendrakumar, R. & Tisdell, C. (1989). Tourism and the economic development of the Maldives. Annals of Tourism Research, v. 16, No. 2, pp. 254-269.

Scheyvens, R. & Momsen, J. (2008). Tourism in Small Island States: From Vulnerability to Strengths. Journal of Sustainable Tourism, v. 16, No. 5, pp. 491-510.

Shareef, R. & McAleer, M. (2005). Modelling International Tourism Demand and Volatility in Small Island Tourism Economies. International Journal of Tourism Research, No. 7, pp. 313-333.

Saphier, A.D. and Hoffmann, T.C. (2005). Forecasting models to quantify three anthropogenic stresses on coral reefs from marine recreation: Anchor damage, diver contact and copper emission from antifouling paint. Marine Pollution Bulletin, No. 51, pp. 590-598.

Standford, D. (2008). Exceptional Visitors: Dimensions of Tourist Responsibility in the Context of New Zealand. Journal of Sustainable Tourism, v. 16, No. 3, pp. 258-275.

Stuart, E.K. (2006). Energizing the Island Community: a Review of Policy Standpoints for Energy in Small Island States and Territories. Sustainable Development, No. 14, pp. 139-147.

Sullivan-Sealey, K. & Cushion, N. (2009). Efforts, resources and costs required for long term environmental management of a resort development: the case of Baker`s Bay Golf and Ocean Club, The Bahamas. Journal of Sustainable Tourism, v. 17, No. 3, pp. 375-395.

UNWTO (2004). Sustainable Development of Tourism, Mission statement, Conceptual Definition. Retrieved 2010, from http://www.unwto.org/sdt/mission/en/mission.php?op=1

Van der Velde, M., Green, S.R., Vanclooster, M. & Clothier, B.E. (2007). Sustainable development in small island developing states: Agricultural intensification, economic development, and freshwater resources management on the coral atoll of Tongatapu. Ecological Economics, No. 61, pp. 456-468.

Worachananant, S.; Carter, R.W.; Hockings, M. & Reopanichkul, P. (2008). Managing the Impacts of SCUBA Divers on Thailand's Coral Reefs. Journal of Sustainable Tourism, v. 16, No. 6, pp. 645-663.

Yahya, F., Parameswaran, A., & Sebastian, R. (2005). Tourism and the South Asia littoral: Voices from the Maldives. South Asia-Journal of South Asian Studies, v. 28, No. 3, pp. 457-480.

Zubair, S., Bowen, D. & Elwin, J. (2010). Not quite paradise: Inadequacies of environmental impact assessment in the Maldives. Tourism Management (in press).

In: Research Studies on Tourism and Environment
Editors: J. Mondejar-Jimenez et al.

ISBN: 978-1-61209-946-0
© 2012 Nova Science Publishers, Inc.

Chapter 13

TOWARDS AN EXPLANATORY MODEL OF INNOVATION IN THE CULTURAL TOURISM DISTRICTS

Pedro M. García-Villaverde, Dioni Elche-Hortelano, María J. Ruiz-Ortega, Gloria Parra-Requena, Pilar Valencia-De Lara, Ángela Martínez-Pérez, Job Rodrigo-Alarcón and Miguel Toledo-Picazo
University of Castilla-La Mancha, Spain

ABSTRACT

In the last decade there has been an increasing political, economic and scientific interest for the dynamic of the urban and cultural tourism as a basis for sustainable local development, especially in small and medium size historical cities. Cultural touristic cluster remain common features of industrial districts: interdependence between firms, flexible boundaries of companies, partnerships and competition, confidence through sustained collaboration among stakeholders and a cultural community with the support of public policies. We try to advance into this field to analyse in depth the heterogeneity in the behaviour and innovation of firms belonging to cultural tourism districts. We propose an integrative model to analyse the role that two key factors, such as relationships and a higher level of knowledge, has on the innovation developed by firms in the cultural tourism districts. In this sense, we have established several theoretical propositions based on different approaches to the literature. We have improved the previous models by introducing an intermediary effect, which provides a better explanation of the causal process that leads to the generation of a greater performance of innovation. Thus, we highlight the role of knowledge acquisition, as a factor that mediates between the combination of internal and external relationships and the development of innovation by firms in the cultural tourism districts. This research contributes to integrate three theoretical perspectives such as network and relationships approach, knowledge-based approach and territorial perspective to the study of innovation.

Keywords: cultural tourism districts, cross institutional relationship, knowledge management, innovation, explanatory model

1. INTRODUCTION

During the last decades, the studies about firms' agglutinations show that membership in a particular district or agglomeration affects the performance obtained by these firms (Signorini, 1994; Paniccia, 1999, Hernandez and Soler, 2003). The competitive advantages of these firms are due to the externalities generated by the spatial concentration and interrelationships that are created between players. Traditionally it has been established the homogeneity of performance between firms in a district -district effect-. However, the literature reveals that there are major differences among the performance of these firms because of their internal characteristics (Lazerson and Lorenzoni, 1999; Rabellotti and Schmitz, 1999). Firms, as independent entities, have different levels of resources and capabilities, for example, access and control of shared resources, links with institutions and other companies, absorption capacity, skills, among others, which will affect their performance in terms of innovation, efficiency, profitability, growth, etc. In this regard, there are still few studies on clusters that have focused on the firm as unit of analysis and deepened in the heterogeneity in behaviour and firms' performance.

In the last decades there has been a growing interest among politicians, economists and scholars in urban cultural tourism as a basic instrument for sustainable local development, especially in the historic small and medium size towns (Troitiño, 1998). Nowadays cultural tourism districts are gaining importance due to the continued growth of the culture industry in modern economies and the increasing incorporation of cultural elements in traditional tourism products to increase their interest (Scott, 2000). Specific features of historic cities, whose touristic development is mainly based on a cultural heritage located geographically, form a suitable environment to develop studies from the viewpoint of the districts / clusters. In this sense, the cultural tourism districts share common features with industrial districts: interdependence between firms, flexible boundaries of firms, partnerships and competition, trust sustained through collaboration between the actors and a cultural community with the support of public policies (Hjalager, 2000). Thus, in the historic cities there is a predominance of small and medium tourism firms, the dependence of multinational firms is not very high, the collaborative behaviour predominates above the opportunistic behaviour and it tends to be some stable structures for supporting and cooperation between firms (Prats, Guía and Molina, 2008).

On the other hand, knowledge is recognized as a key element in the firm because of its contribution to the generation of innovations and competitive advantages, which receives high attention in the literature of strategic management (Grant, 1991; Peteraf, 1993). It is also highlighted that social networks are key elements for the development of the tourism districts, because in this area these networks are considered more cohesive than in other sectors and introduce the complex cultural, economic and social connections (Merinero, 2008).

Several studies suggest that knowledge transfer is facilitated by social interactions among organizations (Zahra, Ireland and Hitt, 2000; Yli-Renko, Autio and Sapienza, 2001). Consequently, interactive learning and better performance are not caused by a firm belonging to a district, but rather on the relationships with the various stakeholders of the district (Bathelt et al., 2004). However, some companies are not able to take advantage of the potential available knowledge in the tourism districts. Therefore, only when firms in the tourism districts are able to acquire and generate knowledge from internal and external relationships that have the other players in the districts, may be more innovative. In addition, it is important to supplement the relationships with internal agents to the district with external agents to obtain a new and valuable knowledge, avoiding problems of redundancy of information (Tiwana, 2008).

The main aim of this chapter is to propose a model to analyze the direct and mediated effect (by the acquisition of knowledge), of the relationships with internal and external agents to the district on the innovation of firms belonging to a cultural touristic district. In order to achieve this aim, in the next section we review the concept and characteristics of the cultural tourism districts. We analyze the networks of relationships, both with internal and external agents to the district, and its importance for innovation in the field of cultural tourism districts. We also analyze the role of the acquisition of knowledge in the relationship between contacts with internal and external actors to the district and firms' innovation in a cultural district. In the last section, we expose the discussion of the proposed model a series of conclusions.

2. THE CULTURAL TOURISM DISTRICTS

One of the pioneer economists in the field of endogenous development and the spatial location was Marshall (1920) who used the concept of "district", understood as a territorial-based entity that serves as the unit of analysis of economic development to study the geographical organization. These agglomeration economies, which encourage all firms in the cluster, arising from the proximity of firms, especially when their expertise is developed along the value chain. Subsequently, Becattini (1989) proposes the concept of industrial district, based on the idea that external economies are derived from the merger of firms. These firms, usually small in size, are specialized in different activities of the same production process and are an important driver of local development. In this sense, the advantages that firms achieve for being located in a district are due to external economies to the firm, perhaps even outside the industry, but internal to the district in which they operate.

This system of local development that based on the concentration of related firms, in whose relationships play an important role the social and cultural characteristics of the territory, can be applied to the touristic production. Tourism encompasses very different economic activities, closely interrelated with activities of other sectors. Generally, the supply of a touristic product includes a number of private services, public infrastructure and natural resources. Therefore, tourism is structured as a multidimensional network of horizontal, vertical and diagonal connections between undertakings (Huybers and Bennett, 2000; Michael, 2007). The relationships established between firms in the agglomeration imply that these actions are conditioned by the actions of the other firms. This explains the coexistence of rivalry and cooperation between firms in the district. In this context, the key factor for firms to improve their competitiveness lies in competitive advantages that go beyond sectoral boundaries extending its coverage to more sectors. Thus, companies must find their competitive advantage through specialization, rather than trying to exactly imitate the actions of other districts. In this sense, firms must implement differentiation strategies, and develop resources that are unique to the agglomeration to achieve and maintain competitive advantages (Porter, 1998).

Although traditionally linked to the concept of industrial district, in the most recent literature there are several studies that address the agglomerations in the tourist business (Michael, 2007; Lazzaretti and Petrillo, 2006, Novelli, Schmitz and Spencer, 2006; Merinero, 2008; Aurioles, Fernandez and Manzanera, 2008, Weidenfeld, Butler and Williams, 2010). This chapter focuses on the notions of district or cluster of tourism, which are based on several key elements, the territory, the different actors in the territory and the productive relationships that are established between them. We consider that a touristic cluster is configured around a group of firms from various sectors and other entities associated with the competition, which offer complementary products and services as a holistic touristic experience (Wang and Fesenmaier, 2007).

Therefore, the concept of district traditionally used in the industrial sector fits easily with the idea of merger that arises by a common interest in exploiting a resource located in the territory. In the case of tourism, it seems clear that the merger raises the possibility of exploiting external economies and domestic firms to the district, which are often the result of cooperation between these firms and the government for external promotion in the area. This cooperation is established in order to boost infrastructure necessary for all the member or institutions related with the activity, such as school catering, information offices or the tourist boards (Aureoles et al., 2008).

Considering the similarity between a touristic district and an industrial district, Hjalager (2000) identifies the main features of industrial districts in the tourist districts. These include items such as: a) The interdependence between the firms. Firms operate independently but are linked through a series of actions or processes performed together (Jackson and Murphy, 2002). b) The existence of flexible limits of the firm. Hjalager (2000) refers to the flexibility in time, place and functional flexibility, and the firms' ability to avoid discontinuities. He claims that many touristic destinations are temporary flexibility with regard to employment, creating a possible limit to the qualified employees. c) Competitive and cooperative relationships. This is one of the paradoxical characteristics of the districts, since there is a set of companies that target the same market, but are able to share access to different opportunities in different ways. In tourism districts further cooperative efforts are institutionalized so that it could compensate for the lack of trust. d) Confidence sustained through collaboration between the actors. This represents the tacit knowledge between businesses through relationships that go beyond a contract. This tacit knowledge is improved by the geographical and cultural proximity, reducing transaction costs. e) A 'cultural community' with the support of public policies. In the touristic sector, public policies are involved in the generation of infrastructure and the planning of the use of land. The role of public administration in the district is of vital importance as it provides a stable legislative system, provides transportation and infrastructure, communications and relationships with market forces to establish linkages and externalities (Jackson and Murphy, 2002).

In recent years, urban cultural tourism is creating a great economic, social and political interest by its potential for sustainable local development, especially in the historic towns of small and medium size (Troitiño, 1998). Culture is becoming an innovative and consistent new platform for the development of the current knowledge society (Sacco, Blessi and Nuccio, 2009). It is also used as a means of social and economic regeneration. The cultural touristic market is the need to provide new cultural attractions and heritage centers in cities in order to be competitive. Therefore, that creativity is especially relevant in developing new products for urban tourism. Despite the globalizing nature of contemporary cultural elements, in many cases creativity seems to be still linked to local contexts based on geographic, social and economic benefits (Scott, 2000; Mizzau and Montanari, 2008). Thus, certain cities are taking shape as solid cultural districts, from an equity holder, incorporating a wide range of creative activities, cultural and artistic activities, museums, art galleries, utilities, food, etc. and integrate a competitive touristic offer.

In general, the evolution of touristic demand raises further guidance on the development of cultural touristic districts, as the complementarity of internal and external networks, the generation of new knowledge, development of local entrepreneurs, attraction of external firms, expanding cultural offerings, the development of local talent, attract foreign talent, the ability to build and educate the local community, local community involvement and improvement in local government (Sacco et al., 2009).

3. INTERNAL AND EXTERNAL RELATIONSHIPS AND INNOVATION IN CULTURAL TOURISM DISTRICTS

The concept of social networks has acquired great importance in recent years in the field of social sciences. At the district level, firms have substantial differences in terms of their potential to discover and exploit the competitive capabilities across their networks. Thus, it is important to understand how the distinctive patterns that firms follow in the establishment of relationships are linked to their competitiveness. In this sense, the relationships generated between firms and other agents of the district are very important to the achievement and maintenance of competitive advantages (Gulati, 1999).

One of the distinctive elements of the districts is the productive specialization of firms operating within it, which involves a high degree of dependence between them. This dependence favours the establishment of relationships between firms and between them and the other agents. These relationships can be vertical and horizontal (Sorensen, 2007). But besides the integration of experiences of different firms, it may happen diagonally, forming what Michael (2003) called 'diagonal clustering', where the co-location of SMEs directly and indirectly, not only add value to the network and the experience of the members of the district, but also to the touristic experience (Novelli et al., 2006).

In the cultural tourism districts, cooperative relationship and the process of coordination connect different players, processes and products related to tourism and cultural production. For example, travel agents, facilities, administration, conferences, education related to tourism, R & D institutions, libraries, urban management and services, public transport, telecommunications, advertising, strategic media, cultural industries (art, architecture, cultural heritage), the new guidelines of culture (high-tech and high touch, subculture, pop), industry events, sports, arts, etc. (Murphy and Boyle, 2005).

On the other hand, we emphasize the need for firms within the district to establish contacts with actors outside the district, so as to obtain new and valuable knowledge, avoiding problems of redundancy of information (Tiwana, 2008).

In the last decades, the literature on strategic management has considered innovation as a critical factor for economic growth. The interest on innovation among academics, economists and politicians is due to its influence on the firms' competitiveness. It becomes a key factor in achieving and maintaining competitive advantage and therefore a key determinant of firms' success (Zahra and Covin, 1995, Teece et al., 1997, Covin & Miles, 1999; Zaheer and Bell, 2005). In addition, innovation has become an engine of social and economic development in all the modern economies.

Although firms are the major innovative players, they do not act in isolation. The innovations are the result of social interactions between different economic actors. In this sense, the ability of the firm and the way it interacts with its environment to leverage the experience and knowledge of other agents are key determinants of innovation (Fernández, 2005). From this approach, the districts are raised as social structures, where the proximity of the agents facilitates the exchange of information and, therefore, the development of innovation in firms belonging to the district (Jensen, 2001, Alvarez and Gonzalez, 2006; Aureoles et al., 2008, Mattsson, Sundbo and Fussing-Jensen, 2005, Novelli et al., 2006; Sorensen, 2007; Prats et al., 2008). This approach establishes that innovation is a spatial phenomenon that takes place primarily within a limited geographical area, where industries are concentrated and there are knowledge spillovers.

Prats et al. (2008) establish a model called Touristic Local Innovation System, based on territorial agglomerations of touristic business from a network of relationships. This local innovation system focuses on an area where a group of agents that interact with each other, are supported by ancillary industries and external agents. All of them generate relational assets and establish links with their macro-environments that enable collective learning and common understanding, both critical in determining the innovation capacity of the system. In other words, relational assets allow the creation and diffusion of innovations and may explain the competitive advantage of a tourist destination. This generates flows of knowledge throughout the system that encourage collective learning and the creation of new knowledge, continuously developing innovative capacity.

Because of its specific characteristics the touristic sector can be an example of geographically organized networks that incorporate the benefits of both, local networks, generating benefits in terms of learning and innovation (Maskell and Malmberg 1999) and non-local networks, providing relevant external information along with the benefits of learning (Oinas and Malecki 1999). Therefore, through these networks, the development of continuous innovations in the district is enhanced.

3.1. Combining Internal and External Relationships to the District and Innovation

In recent years the concepts of social networks and social capital have been used extensively in the literature on innovation. Since the social network approach holds that within the networks it is generated information necessary and suitable for innovation (Ahuja, 2000). In this sense, innovation is conceived, not as specific events from isolated inventors, but as the result of interactions and exchanges of knowledge that involve a variety of actors and elements in situations of interdependence (Landry, Amara and Lamari, 2002). Within a cultural district, the relationships that are created around these networks are the basic factor driving innovation, so it is necessary to identify and characterize the relationships that occur between the territory touristic stakeholders with the outside agents (Merinero, 2008).

The relationships that firms develop with agents inside the district have several benefits that directly affect the ability of firms to innovate. In this respect, trust and social relationships in the district are important mechanisms that promote innovation (Amin and Cohendet, 2005). Thus, social networks have a positive effect on firm's performance, and the most important mechanism underlying them is the trust (Jones, Hesterly and Borgatti, 1997). In fact, many studies highlights trust as the main mechanism of the dynamics of networks (Othman, 2009). In this sense, the strong ties between firms, which are typically firms that are located near each other, are valuable tools for business competitiveness, especially when firms are faced with uncertainty (Keister, 1999). In fact, they can provide stable flows of new ideas, technological innovations and operational support. The strength of strong ties based on mutual understanding, social content and the specific investments in relationships, helps keep the leadership and the firms' ability to innovate (Capaldo, 2007).

In cultural tourism firms, the group of firms and sectorial context can create a localized network through which the producers of global touristic product are linked, there are also complementarities in production and geographical proximity creates networks of production, which can lead to foster cooperation and innovations (De Propis, 2002). Therefore, a key characteristic of the cultural tourism districts is the localized nature of the relationships, which contributes to innovations (Saxena, 2005). We believe that, in this context, those companies that exploit the potential of social networks in the districts to develop frequent, narrow, of trust and face to face interactions, while sharing a common culture and language throughout of time (Wolfe, 2002; Prats et al., 2008), will develop more innovations.

In spite of being widely held that geographical proximity, by creating localized learning, increases innovation, recent contributions are questioning that geographical proximity is sufficient for learning and innovation (Othman, 2009). In this sense, we must consider the negative effects of information redundancy in homogeneous networks, since, although strong links between individuals facilitate the flow of information, individuals may possess highly redundant information (Tiwana, 2008). A high level of redundancy in relationships causes the inefficient exchange of information that is not in the pursuit of competitive advantage or the development of innovations. In dynamic environments with changing competitive conditions, the access to new information is a key element to generate innovations and thereby improve firms' performance. Thus, the district will benefit from frequent updating of people, ideas and cultural and creative styles, resulting in greater innovations, and help the district to avoid the typical problem of internal blocking of the districts (Mizzau and Montanari, 2008).

When the heterogeneity of the members in a relationship increases through weak ties and the "structural holes", the ideas, perspectives, and information are expanded and diversified (Reagan and Zuckerman, 2001). This approach extends the repertoire of available solutions, and the likelihood of novelty arising from the recombination of previously isolated perspectives (Lapre and Wassenhove, 2001). Therefore, the relationships of the firms belonging to a cultural district with firms outside the district, increase the potential of innovation of these firms. Therefore, by means of the addition of weak ties and "structural holes" in their network of relationships, firms will probably add non-redundant contacts - network diversification (Burt, 1992)-, increasing innovations (Rowley et al. 2000, Capaldo, 2007).

Therefore, relationships allow the creation and diffusion of innovation and may explain the competitive advantage of a touristic destination in particular (Prats et at. 2007). Thus, we consider that the internal and external relationships to the district are complementary rather than substitutes ones and, therefore, generate a greater number of innovations when acting in combination (Ruef, 2002; Boschma and ter Wal, 2007; Sorensen, 2007). From the above arguments we can establish the following proposition:

P1: In the cultural tourism districts, the combination of relationships of the firms with internal and external agents to the district has a positive influence on firms' innovation.

4. THE ROLE OF KNOWLEDGE ACQUISITION

In recent decades the interest of scholars to analyze organizational knowledge is continuously growing. Knowledge management has become in a key element to the success of firms, since knowledge is a determinant factor of competitive advantage of firms. In this regard, the first studies developed from the Resources-Based View (RBV) reveal the importance of knowledge as one of the most strategic resources because it is unique, inimitable and valuable (Wernerfelt 1984; Prahalad and Hamel 1990, Barney, 1991).

The typology of knowledge usually used in the literature is the one established by Polanyi (1967), which identifies two types of knowledge: tacit and explicit. Explicit knowledge is the knowledge that can be expressed in words, numbers, symbols, and can be easily shared and transferred, taking the consideration of public good. While tacit knowledge involves abstract and complex knowledge that is difficult to codify -technical knowledge, attitudes and skills-, and therefore, its transmission is much more complex. Such knowledge is especially "sticky" and it is learned through experience, so that requires intense interaction to be codified and transferred (Dyer and Nobeoka, 2000). Therefore, this type of knowledge has a greater potential as a source of sustainable competitive advantages in time (Grant, 1996). On the other hand, a part of tacit knowledge is collective, that is. it is distributed and shared among members of the firm, as it arises from the interaction between them (Spender, 1996). Part of this tacit and collective knowledge is semipublic in the district level. In this sense, the mechanisms for the transfer of knowledge are essentially social, because knowledge (whatever its source) is transferred between other agents by interactive mechanisms based on rules norms, information flows and common channels (Saxena, 2005).

Hjalager (2002) also analyzes the channels of acquisition of collective knowledge in the tourism industry, and detects that there exists heterogeneity among firms in the access of this knowledge. This can be explained because of, among other reasons, the relevance of human resources for the transmission of tacit knowledge, and the difference in training and employee turnover in the tourism sector.

4.1. Combination of Internal and External Relationships and Knowledge Acquisition

In the literature, the effects of internal and external relationships to the district have generally been studied separately. However, we believe that the combination of internal and external relationships generate greater knowledge, which, in turn, leads to the development of further innovations.

Traditionally, it has been defended the importance of geographical location and internal relationships for the acquisition of tacit knowledge, especially under the close personal contact (Lorentzen, 2008). As a result, flows of knowledge of the firm are assumed to be largely localized (Waterings and Ponds, 2009). Considering the differentiation of tacit and codified knowledge, both affect and are affected by the learning mechanisms and the division of labor (Cainelli and De Liso, 2005). As it has been extensively highlighted in the literature on industrial districts, a great part of the competitive advantage of districts can be attributed to forms of tacit knowledge overflow between the business relationships internal to the district. However, it has not been given enough importance to the intentional efforts of knowledge acquisition, where the novel knowledge are particularly important, and with them, the relationships with the agents external to the district become very important.

On the other hand, we believe that the combination of internal and external relationships to the district is required to explore and exploit knowledge. Strong ties that are very important in internal relations to the district, are usually associated with the exchange of effective and efficient information, and the transfer of tacit knowledge. Moreover, these links denote trust between the partners. In this context, strong ties serve to a better exploitation of the existing knowledge (Rowley et al., 2000). On the other hand, weak ties, which are mostly in relationships outside the district, are associated with exploration, which is the access to new areas of knowledge (Rowley et al., 2000). Therefore, if we only consider one type of link - strong or weak-, we can access to an insufficient knowledge for its efficient exploration and exploitation. Therefore, a network of relationships that combines internal and external links to the district could provide an ideal configuration of knowledge (Burt, 1992, Capaldo, 2007; Tiwana, 2008).

The arguments raised on the need to combine the internal and external relationships to the cultural touristic district to generate completed knowledge are reinforced in the context of strong globalization process in which we are immersed. Some recent contributions specifically discuss the role of the local and global level in innovation and learning (Lorentzen, 2008). Firms are able to develop linkages of knowledge in different spatial systems, not just regional but global networks of relationships are essential in this regard (Lorentzen, 2008). Several authors have stressed that local and global flows of knowledge may indeed be a complement in the innovation process (Wolfe and Gertler, 2004, Asheim and Gertler, 2005). Therefore, the networks of firms of cultural tourism are an example of networks geographically organized (Sorensen, 2007), which incorporate the supposed benefits of local networks of the agglomerations, providing great benefits in terms of learning and innovation (Maskell and Malmberg 1999). These networks also incorporate non-local networks, which provide additional external information and learning, providing major benefits in terms of knowledge (Oinas and Malecki, 1999). We have found several studies in the literature that demonstrate the need for a combination of internal and external relationships to the district to generate valid knowledge (Freel and Harrison, 2006; Waxell and Malmberg, 2007; Vang and Chaminade, 2007). One of the most important studies that combines internal and external relationships to the district to generate a high level of knowledge has been the metaphor of "local buzz vs. global pipelines" from Bathelt et al. (2004). These authors explain the advantage conferred by the coexistence of local integration in the highest level and well-developed network of partners worldwide –in innovative firms and groups-. From these arguments we raise the following proposition:

P2: In the cultural tourism districts, the combination of relationships of the firms with internal and external agents to the district has a positive influence on firms' knowledge acquisition.

4.2. Knowledge Acquisition and Innovation

The organizational knowledge is considered as a basic resource of strategic importance for the production system (Drucker, 1996). Firms, through their activity create value for society and for this purpose, require relevant knowledge, which then becomes innovation. Therefore, the development of knowledge from existing knowledge in the firm, together with that acquired from external sources, is the key to generate innovations and improve firms' competitiveness (Nonaka and Takeuchi, 1995). The new knowledge can be incorporated in the form of new equipment, integrated knowledge, and the incorporation of technology with origin in the information technology and communications, "new skills, human capital, or new organizational routines (Alvarez and Gonzalez, 2006).

Although for a touristic firm it is important the capability to produce new knowledge derived from its activities, -learning by doing- it also requires the capability to acquire and absorb the innovations of other agents, -internalize the knowledge that is codified and make it tacit knowledge, and to distribute their innovations –outsourcing- in the form of codified knowledge to distribute it more easily. This capacity depends on structural factors such as firm size or type of firm. However, there are certain cognitive factors such as proactive or reactive nature of routines and knowledge management capabilities that are of particular relevance in this process

The innovation process requires external knowledge flows to boost its development (Dyer and Singh, 1998, Lane and Lubatkin, 1998). Depending on the degree to which a firm has access to external sources of knowledge, it can take advantage of its resources in a greater way to generate innovations (Kogut and Zander, 1992; Decarolis and Deeds, 1999). Therefore, the ability of a firm to innovate may be associated with their ability to obtain and exchange knowledge (Kogut and Zander, 1992). Thus, the acquisition of external knowledge becomes a fundamental element for the development of innovative activity of the firm.

Thus, the mechanisms of cooperation and dissemination of knowledge that characterize the cultural tourism districts have been considered as key tools for the development of innovations of the firms in the district (Albors and Molina, 2001, Tallman et al., 2004). As a result, several recent studies about territorial agglomerations have emphasized the district's ability to support processes of knowledge acquisition and innovation as the basis for creating competitive advantage (Inkpen and Tsang, 2005). In this regard, studies like Yli-Renko et al. (2001), Ahuja and Katila (2001), Quintana and Benavides (2007), Chen and Huang (2008), among others, verified the positive effect of the acquisition of knowledge on firms' innovation.

The increase of competition as well as the difficulty to protect innovation in tourism firms leads to a greater importance of knowledge acquisition and continuous learning, linked to the supply, the management and demand of specific markets, in the development of new products and services of cultural touristic firms (Tremblay, 1998). The above arguments lead us to propose the following proposition:

P3: In the cultural tourism districts, the firms' knowledge acquisition has a positive influence on firms' innovation.

4.3. The Intermediary Role of Knowledge Acquisition

Social networks in the cultural tourism districts are key elements in the generation of knowledge. These networks establish complex connections of cultural, economic and social character between firms which can boost up innovation (Merinero, 2008). Therefore, the cultural tourism district is an appropriate context that promotes innovation due to the creation of connections with every agent of district, by sharing cooperation and competition relationships.

However, we consider that the establishment of relationships with internal and external agents in the district is not enough to enhance the innovation process in firms belonging to a cultural tourist district. This is because all firms in the district do not have the same access to such knowledge, neither they are able to use it to develop innovations. Therefore, although the innovation is the result of social interaction between different economic actors, the individual capability of a firm and its interaction with the environment in order to take advantage of the experience and knowledge of other agents these are key determinants of the innovation process (Fernández, 2005). In the field of cultural tourism districts there is tacit knowledge, the exchange of this knowledge occurs through technological spillovers, informal relationships and movements of people. This tacit knowledge is generated through interdependence between firms, the creation of a cultural community and the intervention of government in infrastructure development, training, information offices, etc…(Hjalager, 2000; Aurioles, et al., 2008). But all firms in the district do not have the same access to this knowledge and not making the most of it is to innovate.

From this perspective, the acquisition of knowledge allows the connection of the relationship between internal and external agents to the district with the development of innovations. A cultural tourism district is generated in a particular geographical area where a group of agents interact with each other with the support of auxiliary industries and agents. All of them generate relational assets that facilitate collective learning and common knowledge, which is critical in determining the innovation capacity of the system (Prats et al., 2007). Therefore, there are factors such as density, quality and frequency of contacts, reliance and commitment in their relationships, as well the similarity in culture, values and goals with the agents, which are important sources of innovation for firms in the cultural tourism districts, while these are focused on acquiring relevant and complementary knowledge to development and exploitation of innovations.

From this perspective, the acquisition of knowledge allows the connection of the relationship between internal and external agents to the district with the development of innovations. A cultural tourism district is generated in a particular geographical area where a group of agents interact with each other with the support of auxiliary industries and agents. All of them generate relational assets that facilitate collective learning and common knowledge, which is critical in determining the innovation capacity of the system (Prats et al., 2007).

Since there is a connection between the internal and external relationships to the district, knowledge and innovation, we propose that this effect is essential for understanding the origin of the innovations developed by firms belonging to a cultural tourism district. Thus, only those firms of the district which lead their internal and external relations -correctly combined- to generate more knowledge, may reach the potential to generate more innovation. According to these arguments we suggest the fourth proposition:

P4: In the cultural tourism districts, the firms' knowledge acquisition mediates the relationship between the combination of internal and external relationships between agents of district and firms' innovation.

CONCLUSION

We propose a model that explains the link between the internal and external relationships that firms establish and the innovation that these firms develop in a cultural tourism district. The model is shown in Figure 1. In this model, we gather the contributions of several theoretical and empirical studies on districts, relationships with internal and external agents, the acquisition of knowledge and innovation (Becattini, 1990; Signorini, 1994; Nahapiet and Ghoshal, 1998; Tsai and Ghoshal, 1998; Dyer and Singh, 1998; Soler, 2000; Yli-Renko et al., 2001; Molina, 2005). We have also analysed several studies focusing on cultural tourism districts (Michael, 2003; Saxena, 2005; Novelli et al, 2006; Aurioles et al., 2008; Merinero, 2008).

One of the main contributions of this research is the integration of three theoretical perspectives such as network and relationships approach, knowledge-based approach and territorial perspective to the study of firms' innovation. These three perspectives have tried to explain the performance of innovation, but separately. In this sense, we try to fill this gap in the literature in the intersection between key concepts such as the internal and external relationships and the acquisition of knowledge. In this study we try to provide linkages between theoretical and empirical studies to fill this gap. In this sense, we have developed an integrative model that has allowed us to analyse the determinants of the innovation developed by firms in a particular type of network, this is the cultural tourism districts.

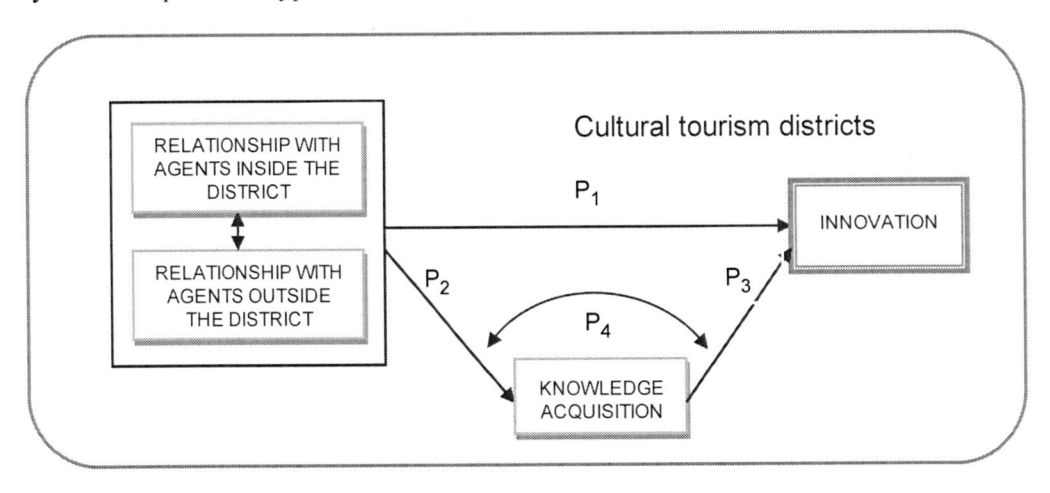

Figure 1. Theorethical model.

Several studies have confirmed the existence of a "district effect" linked to superior innovation and performance of firms belonging to the districts with regard to non-members. We have examined the causes of the heterogeneity on innovation among firms in cultural tourism districts.

We consider that there are several determining factors of relationships established between firms in the cultural tourism districts that significantly influence the firms' knowledge acquisition. On the other hand, we consider that the firms' knowledge acquisition becomes a key determinant of the innovation of firms belonging to a district. Therefore, there is an indirect effect of the combination of internal and external relationships on firms' innovation through the level of knowledge generated by these firms.

We consider that the integrative model contributes to justify the role that two key factors, such as firms' relationships and a higher level of knowledge, has on the innovation developed by firms in the cultural tourism districts. The proposed model has allowed us to establish several theoretical propositions basing on different literature approaches. In this sense, we highlight the integrative character of the proposed model. Therefore, this study contributes to fill a theoretical gap in the research on districts introducing relevant concepts -internal and external relationships to the district, knowledge and innovation-, also enquiring into the mechanisms that connect them to explain why some firms are more innovative than others. On the other hand, we focus on the cultural tourism districts, whose study has been addressed recently. Furthermore, we have established the firm as the unit of analysis, being this approach not very common in the literature about districts.

Therefore we have improved the previous models by introducing an intermediary effect, which provides a better explanation of the causal process that leads to the generation of greater firms' innovation performance. Thus, we highlight the role of knowledge acquisition, as a factor that mediates the relationship between the combination of internal and external relationships of the firms and the development of innovations. In this sense, we strengthen the link among internal and external relationships to the district and the knowledge acquisition, and among the knowledge acquisition and innovation. The main contribution of this study is to propose and justify the intermediary role of knowledge acquisition between the firms' relationships and innovation in the cultural tourism districts.

We highlight as a limitation of this research, its theoretical approach. However, we consider that the wide review of the literature and the strength of the provided arguments have allowed us to establish the propositions. We must also clarify that the proposed model does not try to provide a comprehensive explanation of innovation developed by firms in the cultural tourism districts. We assume the partial character of the model; while we consider that the factors and effects proposed in the model contribute significantly to explain the variability of innovation in the districts. Furthermore, we consider it necessary to add new explanatory variables that help us to clarify the relationship between internal and external relations and innovation in the district.

According to this approach, it might be interesting to analyse the role of other actors in the cultural tourism districts such as large firms. Although Becattini (1990) emphasized small firms as a feature in the definition of the district, more recently, the role that large firms may have in the district have been highlighted (Molina, 2005a). In this sense, Lorennzoni and Lazerson (1999) point out that large firms often organize production activities in groups of smaller firms, also developing innovations and expanding to existing markets. Therefore it would be interesting to analyse the role played by these large firms in the cultural tourism district.

Moreover, we consider it would be interesting to examine how internal and external relationships have an influence on the transfer of different types of knowledge, since several studies reveal that each type of knowledge can have a different impact on organizational processes and competitive advantage. In this sense, we believe it is essential to distinguish between tacit knowledge -it is very important in internal relations in the district- and novel codified knowledge, coming mostly from established relationships with external actors to the district.

On the other hand, it would be interesting the consideration of different types of innovation to analyse which one predominates in the districts, and also how the acquisition of knowledge influences the diverse types of innovation. Because of the characteristics of the district we expect that from internal relations the innovation developed by firms is more incremental than radical (Molina and Martínez, 2003).

Among the recommendations for firms belonging to cultural tourism districts that can be derived from this analysis, we highlight the following ones. Firms must combine local relationships, characterized by creating a dense network with strong links -acquisition of tacit knowledge- with external relationships, characterized by being disperse networks with structural holes, -acquisition of new and exclusive information-. Thus, they may develop jointly strategies of exploitation and exploration, since the combination of both can be crucial for the firms' innovation (Lant, Milliken and Batra, 1992).

In addition, it would be advisable for these firms to interact with local institutions or even institutions in other levels, which may play an important role, since the firms could avoid the disadvantages from being in a network with redundant links (McEvily and Zaheer, 1999). Therefore, it would be desirable for firms to interact with institutions to develop strategies for exploration, since they can carry out the function of connection between the district and outside, by providing the firms new and no redundant information.

Similarly, it would be desirable that the firms led and built their relationships in a proactive manner to acquire relevant knowledge effectively and efficiently. Thus, firms should enhance the confidence in their relationships and develop rules, values and a culture shared with their contacts, since they appear as the most influential elements in the generation of knowledge and innovation.

REFERENCES

Ahuja, G. (2000a): "Collaboration networks, structural holes, and innovation: a longitudinal study", *Adm Sci Q*, 45(3), pp. 425–55.

Ahuja, G. and Katila, R. (2001): "Technological acquisitions and the innovation performance of acquiring firms: A longitudinal study", *Strategic Management Journal*, 22, pp. 197-220.

Albors, J. and Molina, X. (2001): "La difusión de la innovación, factor competitivo en redes interorganizativas. El caso de la cerámica valenciana". *Economía Industrial*, 339, pp. 167-186.

Álvarez, J.A. and González Morales, M.O. (2006): "Base de conocimientos y capacidad innovadora de los sistemas locales de producción turística españoles", XII Jornadas AEDE, Donosti-San Sebastián.

Amin, A. and Cohendet, P. (2005): "Geographies of Knowledge Formation in Firms", *Industry and Innovation*, 12 (4).

Asheim B. and Gertler, M. S. (2005): "Regional Innovation Systems and the Geographical Foundations of Innovation", en: Fagerberg J, Mowery D, Nelson R, *The Oxford Handbook of Innovation* (Oxford, Oxford University Press), pp. 291-317.

Aurioles, J.; Fernández, M.C. and Manzanera, E. (2008): "El distrito turístico", *Mediterráneo Económico*, 13, pp. 299-326.

Barney, J. (1991): "Firm resources and sustained competitive advantage", *Journal of Management*, 17(1), pp. 99-120.

Bathelt, H.; Malmberg, A. and Maskell, P. (2004): "Cluster and knowledge: local buzz, global pipelines and the process of knowledge creation", *Progress in Human Geography*, 28 (1), pp. 31-56.

Becattini, G. (1989): "Sectors and/or districts: Some remarks on the conceptual foundation of industrial economics?" en E. Goodman and J. Bamford (Eds.): *Small firms and industrial districts in Italy*. Routledge. London.

Becattini, G. (1990): "The marshallian industrial district as a socio-economic notion" en Pyke, F., Becattini, G. and Sengenberger, W. (Eds.): *Industrial districts and local economic regeneration*. International Institute for Labor Studies, Geneva.

Boschma, R.A. and ter Wal, A.L.J. (2007): "Knowledge Networks and Innovative Performance in an Industrial District: The case of a footwear district in the south of Italy", *Industry and Innovation*, 14(2), pp.177-199.

Buhalis, D. (1998): "Strategic Use of Information Technologies in the Tourism Industry", *Tourism Management*, 19(5), pp.409-422.

Buhalis, D. (2001): "The Tourism Phenomenon: The New Tourist and Consumer", en: S. Wahab & C.Cooper (Eds), *Tourism in the Age of Globalisation*, pp. 69-96 (London:Routledge).

Burt, R. (1992): *Structural Holes: The Social Structure of Competition*, Harvard University Press: Boston, MA.

Cainelli, G. and De Liso, N. (2005): "Innovation in Industrial Districts: Evidence from Italy", Industry & *Innovation*, 12 (3), pp. 383-398.

Capaldo A. (2007): "Network structure and innovation: The leveraging of a dual network as a distinctive relational capability", *Strategic Management Journal*, 28, pp. 585-608.

Chen, C. and Huang, J. (2008): "Strategic Human resource practices and innovation performance , The mediating role of knowledge management capacity", *Journal of Business Research*.

Covin, J.G. and Miles, M.P. (1999): "Corporate entrepreneurship and the pursuit of competitive advantage". *Entrepreneurship Theory and Practice*, 23 (3), pp. 47-63.

De Propris L., (2002): "Types of Innovation and Inter-firm Cooperation", *Entrepreneurship and Regional Development*, 14.

DeCarolis, D.M. and Deeds, D.L. (1999): "The impact of stocks and flows of organizational knowledge on firm performance: an empirical investigation of the biotechnology industry", *Strategic Management Journal*, 20(10), pp. 953-968.

De Brabander, G. and Gijsbrechts, E. (1994): "Cultural Policy and Urban Marketing, a General Framework and some Antwerp Experiences", en *Urban Marketing in Europe*, ed. by G. Ave and F. Corsico, Turin, Torino Incontra, pp. 814-841.

Drucker, P. (1996): "The effective executive", *HarperCollins Publishers Inc.*, Nueva York.

Dunning, J.H. and McQueen, M. (1982): "The Electric Theory of the Multinational Enterprise and the International Hotel Industry, en A.M. Rugman (Ed.) *New theories of the Multinational Entreprise*, pp. 79-106 (London: Croom Helm).

Dyer, J. and Nobeoka, K. (2000): "Creating and managing a high-performance knowledge-sharing network: the toyota case", *Strategic Management Journal*, 21, pp. 345-367.

Dyer, J. and Singh, H. (1998): "The relational view: cooperative strategy and sources of interorganizational competitive advantage", *Academy of Management Review*, 23(4), pp. 660-679.

Fernández, E. (2005): "Estrategia de Innovación". Thomson, Madrid.

Freel, M. S. and Harrison R. T. (2006): "Innovation in the small firm sector: Evidenced from 'Northern Britain'", *Regional Studies*, 40, pp. 289-305.

Gollub, J.; Hosier, A. and Woo, G. (2003): "Using cluster-based economic strategy to minimize tourism leakages". San Francisco, California.

Grant, R.M. (1991): "The resource-based theory of competitive advantage: implications for strategy formulation", *California Management Review*, 13 (33), pp. 114-135.

Grant, R.M. (1996): *Dirección estratégica: Conceptos, técnicas y aplicaciones*. Editorial Civitas. Madrid.

Gulati, R. (1999): "Network Location and Learning: The Influence of Network Resources and Firm Capabilities on Alliance Formation", *Strategic Management Journal*, 20 (5), pp. 397-420.

Hernández, F. and Soler V. (2003): "Cuantificación del efecto distrito a través de medidas no radiales de eficiencia técnica", *Investigaciones Regionales*, 3, pp. 25-39.

Hjalager, A.M. (2000): "Tourism destinations and the concept of industrial districts", *Tourism and Hospitality Research*, nº 3, pp.199-213.

Hjalager, A.M. (2002): "Repairing innovation defectiveness in tourism", *Tourism Management*, 23 (5), pp. 465-474.

Huybers, T. and Bennett, J. (2000): "Impact of the environment on holiday destination choices of prospective UK tourists: implications for Tropical North Queensland". *Tourism Economics*, 6 (1), pp. 21-46.

Inkpen, A. and Tsang, E. (2005): "Social capital, networks, and knowledge transfer", *Academy of Management Review*, 30(1), pp. 146-165.

Ioannides, D. (1998): "Tour-operators: The Gatekeepers of Tourism", en D. Ioannides & K. G. Debbage (Eds), *The Economic Geography of the Tourist Industry*, pp.139-158 (London: Routledge).

Jackson, J. and Murphy, P. (2002): "Tourism destinations as clusters: Analytical experiences from the New World", *Tourism and Hospitality Research*, 1 (4), pp. 36-52.

Jensen, R.A. (2001): "Strategic intrafirm innovation adoption and diffusion", *Southern Economic Journal*, 68 (1), pp. 120-132.

Jones, C.; Hesterly, W. and Borgatti, S. (1997): "A General Theory of Network Governance. Exchange Conditions and Social Mechanisms", *Academy of Management Review*, 22 (4), 911-945.

Keister, L. A. (1999): "Where do strong ties come from? A dyad analysis of the strength of interfirm exchange relations during China's economic transition", *International Journal of Organizational Analysis*, 7, pp. 5-24.

Kogut, B. and Zander, U. (1992): "Knowledge of the firm, combinative capabilities, and the replication of technology", *Organization Science*, 3 (3), pp. 383-397.

Landry, R., Amara, N. and Lamari, M. (2002): "Does social capital determine innovation? To what extent?", *Technological Forecasting and Social Change*, 69, pp. 681-701.

Lane, P. J. and Lubatkin, M. (1998): "Relative absorptive capacity and interorganizational learning", *Strategic Management Journal*, 19 (5), pp. 461-477.

Lant, T.K.; Milliken, F.J. and Batra, B. (1992): "The role of managerial learning and interpretation in strategic persistence and reorientation: An empirical exploration", *Strategic Management Journal*, 13(8), pp. 585-608.

Lapre, M. and Wassenhove L. (2001): "Creating and transferring knowledge for productivity improvement in factories", *Management Science*, 47(10), pp. 1311-1325.

Lazerson, M.H. and Lorenzoni, G. (1999): "The firms that feed industrial districts: A return to the italian source". *Industrial and Corporate Change*, 8, pp. 235-266.

Lazzaretti, M. and Petrillo, E. (2006): *Tourism Local Systems and Networking*, Elsevier, Oxford.

Lazzeretti, L., Boix, R. and Capone, F. (2008): "Do Creative Industries Cluster? Mapping Creative Local Production Systems in Italy and Spain", *Industry and Innovation*, 15 (5), pp. 549-567.

Lorentzen, A. (2008): "Knowledge networks in the experience economy: An analysis of four flagship projects in Frederikshavn", *Center for Regional Development & Department of Development and Planning Working paper Series*, 321.

McEvily, B. and Zaheer, A. (1999): "Bridging ties: A source of firm heterogeneity in competitive capabilities", *Strategic Management Journal*, 20 (12), pp. 1133-1156.

Marshall, A. (1920: *Principles of Economics*, Novena edición Variorum, C. W. Guillebaud (ed.), Vol II: Notes, Londres, Macmillan.

Maskell, P. and Malmberg, A. (1999): "Bridging ties: a source of firm heterogeneity in competitive capabilities", *Strategic Management Journal*, 20, pp. 1133-1156.

Maskell, P.; Bathelt, H. and Malmberg, A. (2006): "Building Global Knowledge Pipelines: The Role of Temporary Clusters", *European Planning Studies*, 14, pp. 997-1013.

Mattsson, J., Sundbo, J. and Fussing-Jensen, C. (2005): "Innovation Systems in tourism: The roles of attractors and Scene-Takers". *Industry and Innovation*, 12 (3), pp. 357-381.

Merinero, R. (2008): "Micro-cluster turísticos: El papel del capital social en el desarrollo económico local". *Revista de Estudios Empresariales*, 2, pp. 67-92.

Michael, E.J. (2003): "Tourism micro-clusters". *Tourism Economics*, 9 (2), pp. 133-145.

Michael, E.J. (2007): "Micro-clusters and networks: The growth of Tourism", Elsevier, Oxford.

Mizzau, L. and Montanari, F. (2008): "Cultural districts and the challenge of authenticity: the case of Piedmont, Italy", *Journal of Economic Geography*, 8, pp.651-673.

Molina F.X. (2005): "The territorial agglomerations of firms: a social capital perspective from the Spanish tile industry", *Growth and Change,* nº 36(1), pp. 74-99.

Molina, F.X. and Martínez, T. (2003): "The impact of industrial district affiliation on firm value creation", *European Planning Studies,* nº 11(2), pp. 155-170.

Murphy, C. and Boyle, E.: "Testing a conceptual model of cultural tourism development in the post-industrial city: A case study of Glasgow", *Tourism and Hospitality Research*, 6 (2), pp.111-128.

Nahapiet, J. and Ghoshal, S. (1998): "Social capital, intellectual capital, and the organizational advantage", *Academy of Management Review*, nº 23(2), pp. 242-266.

Nonaka, I. and Takeuchi, H. (1995): *The Knowledge Creating Company: How Japanese Companies Create the Dynamics of Innovation*, Oxford university press, New York.

Novelli, M., Schmitz, B. and Spencer (2006): "Networks, clusters and innovation in tourism: a UK experience", *Tourism management*, 27: 1141-1152.

Oinas, P. and Malecki, E.J. (1999): "Spatial Innovation Systems", en Oinas P. and Malecki E.J. (eds.) 1999, *Making Connections: technological learning and regional economic Change*, pp. 7-34, Ashgate, Aldershot, UK.

Ozman, M. (2009): "Inter-firm networks and innovation: a survey of literature", *Economics of Innovation and New Technology*, 18 (1), pp. 39-67.

Paniccia, I. (1999): "The performance of IDs. Some insights from the Italian Case", *Human Systems Management*, 18(2), pp. 141-160.

Peteraf, M.A. (1993): "The cornerstones of competitive advantage: a resourcebased view", *Strategic Management Journal*, 14 (3), pp. 179-191.

Polanyi, M. (1967): *The tacit dimension.* Londres: Routledge and Kegan Paul.

Porter, M. (1990): *The Competitive Advantage of Nations.* The Free Press, New York. Versión española: *La ventaja Competitiva de las Naciones.* Plaza & Janés Editores, Barcelona, 1991.

Porter, M. (1998): "Cluster and the Economics of Competition", *Harvard Business Review*, 76 (6), pp.77-91.

Prahalad, C.K. and Hamel, G. (1990): "The core competence the corporation", *Harvard Business Review*, 3, pp.79-91.

Prats, L; Guía, J. and Molina, F.X. (2008): "How tourism destinations evolve: The notion of Tourism Local Innovation System". *Tourism and Hospitality Research*, 3, pp. 178-191.

Quintana, C. and Benavides, C.A., (2007): "Concentraciones territoriales, alianzas estratégicas e innovación: un enfoque de capacidades dinámicas", *Cuadernos de Economía y Dirección de la Empresa*, 30, pp. 5-38.

Rabellotti, R. and Schmitz, H. (1999): "The internal heterogeneity of industrial districs in Italy, Brazil and Mexico", *Regional Studies*, 33(2), pp 97-108.

Reagans, R. and Zuckerman, E. (2001): "Networks, diversity, and productivity: The social capital of corporate R&D teams", *Organization Science*, 12 (4), pp. 502-517.

Rowley, T.; Behrens, D. and Krackhardt, D. (2000): "Redundant goverance structures: An analysis of structural and relational embeddedness in the steel and semiconductor industries", *Strategic Management Journal*, 21(3), pp. 369-386.

Ruef, M. (2002): "Strong ties, weak ties, and islands: Structural and cultural predictors of organizational innovation", *Industrial and Corporate Change*, 11, pp.427-449.

Sacco, P.; Blessi, G. and Nuccio, M. (2009): "Cultural Policies and Local Planning Strategies: What Is the Role of Culture in Local Sustainable Development?", *The Journal of Arts Management, Law, and Society*, 39 (1), pp.45-63.

Saxena, G. (2005): "Relationships, Networks and the Learning Regions: Case Evidence from the Peak District National Park", *Tourism Management*, 26, pp. 277-289.

Scott, A.J. (2000): "The Cultural Economy of Cities-Essays on the Geography of Image-Producing Industries", *London: Sage Publications*.

Signorini, L.F. (1994): "The price of Prato, or measuring the industrial district effect", *Paper in Regional Science*, 73(4). pp. 369-392.

Soler, V. (2000): "Verificación de las hipótesis del distrito industrial. Una aplicación al caso valenciano", *Economía Industrial*, 334, pp. 13-23.

Sorensen, F. (2007): "The geographies of social networks and innovation in tourism". *Tourism Geographies*, 1 (9), pp. 22-48.

Spender, J.C. (1996): "Organizational knowledge, learning and memory: three concepts in search of a theory", *Journal of Organizational Change Management*, 9(1), pp. 63-78.

Tallman, S., Jenkins, M., Henry, N. and Pinch, S. (2004): "Knowledge, clusters, and competitive advantage", *Academy of Management Review*, 29(2), pp. 258-271.

Teece, D.J.; Pisano, G. and Shuen, A. (1997): "Dynamic capabilities and strategic management", *Strategic Management Journal*, 18 (7), pp. 509-533.

Tiwana, A. (2008): "Do bridging ties complement strong ties? An empirical examination of alliance ambidexterity", *Strategic Management Journal*, 29, pp. 251-272.

Tremblay, P. (1998): "The Economic Organization of Tourism". *Annals of Tourism Research*, 25 (4), pp. 837-859.

Troitiño, M.A. (1998): "Turismo y desarrollo sostenible en ciudades históricas", *Ería*, 47, pp. 211-227.

Trullén, J. and Boix, V. (2008): "Knowledge externalities and networks of cities in creative metropolis", en P.Cooke and L.Lazzeretti (eds), *Creative cities, cultural clusters and local economic development*, Edward Elgar.

Tsai, W. and Ghoshal, S. (1998): "Social capital, and value creation: the role of intrafirm networks", *Academy of Management Journal*, 41(4), pp.464-478.

Vang, J. and Chaminade, C. (2007): "Cultural clusters, global–local linkages and spillovers: theoretical and empirical insights from an exploratory study of Toronto's film cluster", *Industry and Innovation*, 14(4), pp. 401-420.

Wahab, S. and Cooper, C. (2001): Tourism in the Age of Globalisation, (eds) London: Routledge.

Wang, Y. and Fesenmaier, D.R. (2007): "Collaborative destination marketing: A case study of Elkhart county, Indiana". *Tourism Management*, 28 (3), pp. 863-875.

Waxell, A. and Malmberg, A. (2007): "What is global and what is local in knowledge-creating interaction? The case of the biotech cluster in Uppsala, Sweden", *Entrepreneurship & Regional Development*, 19, pp. 137-159.

Weidenfeld, A., Butler R.W. and Williams, A.M. (2009): "Clustering and compatibility between tourism attractions", *International Journal of Tourism Research*, 12 (1), pp. 1-16.

Wernerfelt, B. (1984): "A Resource-based view of the firm", *Strategic Management Journal*, 5, pp.171-180.

Weterings, A. and Ponds, R. (2010): "Do Regional and Non-regional Knowledge Flows Differ? An Empirical Study on Clustered Firms in the Dutch Life Sciences and Computing Services Industry", *Industry & Innovation*, 16 (1), pp. 11-31.

Wolfe, D. (2002): "Social capital and cluster development in learning regions", in Holbrook, J.A. and Wolfe, D. (eds). *Knowledge, Clusters and Learning Regions*. School of Policy, Queen's University Kingston.

Wolfe, D. and Gertler, M. (2004): "Clusters from the inside out: local dynamics and global linkages", *Urban Studies*, 41(5/6), pp. 1071–1094.

Yli-Renko, H., Autio, E. and Sapienza, H. (2001): "Social capital, knowledge acquisition, and knowledge exploitation in young technology-based firm", *Strategic Management Journal*, 22, pp. 587-613.

Zaheer, A. and Bell, G.G. (2005): "Benfiting from network position: Firm capabilities, structural holes, and performance", *Strategic Management Journal*, 9 (26).

Zahra, S.A. and Covin, J.G. (1995): "Contextual influences on the Corporate Entrepreneurship – Performance relationship: a longitudinal analysis", *Journal of Business Venturing*, 10, pp. 43-58.

Zahra, S., Ireland, R. and Hitt, M. (2000): "International expansion by new venture firms: International diversity, mode of market entry, technological learning, and performance", *Academy of Management Journal*, 43(5), pp. 925-950.

In: Research Studies on Tourism and Environment
Editors: J. Mondejar-Jimenez et al.
ISBN: 978-1-61209-946-0
© 2012 Nova Science Publishers, Inc.

Chapter 14

YOU ARE AN ORDER CITIZEN, ARE YOU? THEN, WHY DON'T YOU SEPARATE SELECTIVELY YOUR TRASH? AN EXPERIMENTAL TEST

Marta Magadán and Jesús Rivas[*]
Universidad de Oviedo, Spain

ABSTRACT

Yellow, green, blue and black trash cubes can be found by any street, in front of each block of houses. Usually, our municipalities are which determine the day of the week for throw away plastics, glasses, cardboard and paper. All in favor of recycling. All for recycling. But, How many people believe in recycling? More even, how many people are recycling practicants? To answer these questions we need to dive into the most unknown motivations of human behavior. Not many people face voluntarily to a recycling routine. Then, which are the variables that give us a good explanation to understand a non recycling attitude? Our experiment tries to throw some light to this respect.

Keywords: Experiment, trash, recycling attitude

1. INTRODUCTION

The relation between Experimental Economics and Environmental Economics starts from the experimental analysis of externalities. Davis and Holt (1993) illustrate this aspect analyzing, firstly, the common-pool-resource mechanism [Gordon (1954), Gardner, Ostrom and Walker (1990) and Walker, Gardner and Ostrom (1990)]; secondly, the problem of double-auction in the presence of externalities –in the form of a 'damage schedule' [Plott (1983) and Harrison et al. (1987)]; and thirdly, they do a review on experiments related with corrective policies [Ostrom and Walker (1991), Plott (1983), Walker and Gardner (1990) and Harrison et al. (1987)].

---

[*] Contact author: rivasjesus@uniovi.es

In this chapter, starting from a general framework of experimental analysis, the authors will try to measure, at least, the weight of three dimensions –let us say *attitudes*- in relation with an specific behaviour: separation of trash in origin. As seen, this problem will have relation, not only with environment but with expectations, too. And the main aim of authors will consist on clearing up the structure of individual behavior when face on this question.

Yellow, green, blue and black trash cubes can be found by any street, in front of each block of houses. Usually, our municipality are which determine the day of the week for throw away plastics, glasses, cardboard and paper. All in favor of recycling. All for recycling. But, How many people believe in recycling? Moreover, how many people are recycling practicants? To answer these questions we need to dive into the most unknown motivations of human behavior. Not many people face voluntarily to a recycling routine. Then, which are the variables that give us a good explanation to understand a non recycling attitude? Our experiment tries to throw some light to this respect.

The dimensions considered in this work are the following: i) attitude to risk (individual damage in terms of taxes and penalties), ii) attitude to environment (individual perspective of social damage), and finally iii) attitude faced to cost-benefit analysis (individual potential revenues).

2. EXPERIMENTAL FRAMEWORK

Experiments were conducted during october, november and december of 2009 and march of 2010. Students were recruited from several universities: Oviedo, Spain (hereinafter, OVD); Guadalajara (hereinafter, GDL) and Colima (hereinafter, COL), both in Mexico (see Table 1). After recruitment, students were divided in three groups, taking into account individual's response to risk.

Table 1. Characteristics of samples

Place	Individuals	Men	Women
OVD	24	12	12
GDL	24	12	12
COL	24	12	12

Table 2. Risk attitude in samples

Place	Individuals	Risk propense	Risk neutral	Risk averse
OVD	24	8	8	8
GDL	24	8	8	8
COL	24	8	8	8

The size of these samples was the same: twenty four individuals, twelve of whom –in each sample of research- were women and the rest, men. The range of ages was from eighteen to twenty, and they were selected from faculties of Tourism. Inside of each sample it is determined three cathegories of response to risk: i) risk propense, ii) risk neutral and iii) risk averse (see Table 2).

3. LOTTERY AND RISK

A rational behavior is considered traditionally in any economic analysis. But real world teach us that not always individuals are absolute rational. Even more, a limited rationality can increase potentially the degree of individuals' risk attitude [Allais (1953 and 1979)].

As Davis and Holt (1993) stablish:

> 'Many individual-choice experiments involve situations in which a subject must choose between probability distribution of payoffs, that is, between lotteries. Such experiments may be used to evaluate assumptions and theories, about how decisions are made under uncertainty[1].'

And, in relation with risk preferences and economic models, Davis and Holt (1993) emphasize that:

> 'Although many economic models are quite general, the predictions of these models often depend critically on assumptions about agents' risk preferences. Typically, agents are assumed to be risk neutral, or at the least to harbor homogeneous risk attitudes. Absent controls for risk preferences, laboratory tests of such models are really tests of the joint hypothesis that risk attitudes are as assumed, and that the model is a good one[2].'

In this case, it is necessary to divide in three groups of 'risk-attitudes' to understand, not only their effects on experimental results but to 'translate' appropiately what there is behind these different behaviors from a psicological perspective. Quoting again to Davis and Holt (1993):

> 'If a subject is risk averse, then their preferences cannot be represented solely in terms of monetary payoffs. Rather, such preferences mus be represented in terms of the utility of a lottery[3]'.

The concept of lottery –for experimental purposes- is simple: a list of prices and associated probabilities[4]. Each lottery pairs must be labelled with R or S (where R means 'risky lottery', and S means 'secure lottery'.

In the experiments conducted to analyze individuals' attitude to trash selective separation, a lottery test has been developed previously and subsequently. This test allowed to assign an specific individual reaction to risk (see Table 3).

---

[1] See Davis and Holt (1993), p. 68.
[2] See Davis and Holt (1993), p. 472.
[3] See Davis and Holt (1993), p. 79.
[4] See Davis and Holt (1993), p. 442.

Table 3. Lotteries to control risk attitude

Lottery	S, secure	N, neutral	R, risky
L1	(1.000$, 1; 0$, 0)	(1.500$, 0,5; 500$, 0,5)	(2.000$, 0,5; 0$, 0,5)
L2	(2.000$, 1; 0$, 0)	(3.000$, 0,5; 1.000$, 0,5)	(4.000$, 0,5; 0$, 0,5)
L3	(4.000$, 1; 0$, 0)	(6.000$, 0,5; 2.000$, 0,5)	(8.000$, 0,5; 0$, 0,5)

4. EXPERIMENTAL STRUCTURE

As said above, three groups were considered in this experimental analysis. Both GDL and COL groups acted as 'control' groups of OVD group, in order to contrast and strengthen the results obtained. The reason to do that is simple: in GDL and COL there is not a deep culture of trash selective separation. Rather, in OVD *does* exist this kind of culture or social attitude.

The questionaire considered the following variables: time, individual income, individual taxes, individual revenues and a dummy variable reflecting individual behavior in relation to the selective separation of trash (see Table 4).

The experiments takes 13 periods of time or sessions: from t to $t+12$. The personal income was the same for each individual, in order to avoid undesirable effects caused by different levels of income. Session after session, the personal income can be the same as in the previous period if individual is not audited. Rather, if audited and found that individual behaves appropriately in relation with selective separation of trash, only will pay the stablished tax rate. On the other hand, if audited and found 'guilty' of bad environmental behavior, then individual will face the individual tax. Plus a penalty of the same rate (for instance, if tax rate is 10% and individual is found guilty, then the penalty added is 10% calculated over his/her personal income before taxes). 'Trash selective separation' worked as a dummy variable (yes=1, no=0) reflecting, as said above, individual behavior in relation to the *key* problem analyzed. A positive attitude to selective separation of trash implies –in these experiments- a certain individual revenue in terms of public goods and services.

Table 4. Structure of questionaire

Time period	Individual income	Trash selective separation (yes=1, no=0)	Probability of detection	Individual revenue in terms of public goods and services	Individual Taxes and penalties
t					
$t+1$					
...					
$t+12$					

There is an inmediate explanation: every money saved in waste proccessing is directed to reinforce other public programs (health, education, *environment*, security, ...) which are individually valued in positive terms. The variable "Individual taxes" reflects the monetary consequences of increasing environmental costs from an "avoidance of responsability" in relation with trash selective separation. If individual does not separate his/her trash, more public money will be needed to trash proccessing. People who do not separate selectively their trash will face a proportional increasing tax with a probability of detection associated to this kind of behavior.

5. ATTITUDE TO TAXES

The tax rate applied to all groups was proportional –initially, from 10%- and, period by period, it is increased one percentual point, reaching up to 22%. Table 5 resumes the main results of different groups, considering their structure of risk:

Table 5. Attitude to taxes

Group	Attitude to taxes			
	Risk Averse	Risk Neutral	Risk Propense	Globally
OVD	Pay less	Pay less	Pay more	Pay less
GDL	Pay less	Pay more	Pay more	Pay more
COL	Pay less	Pay more	Pay more	Pay more

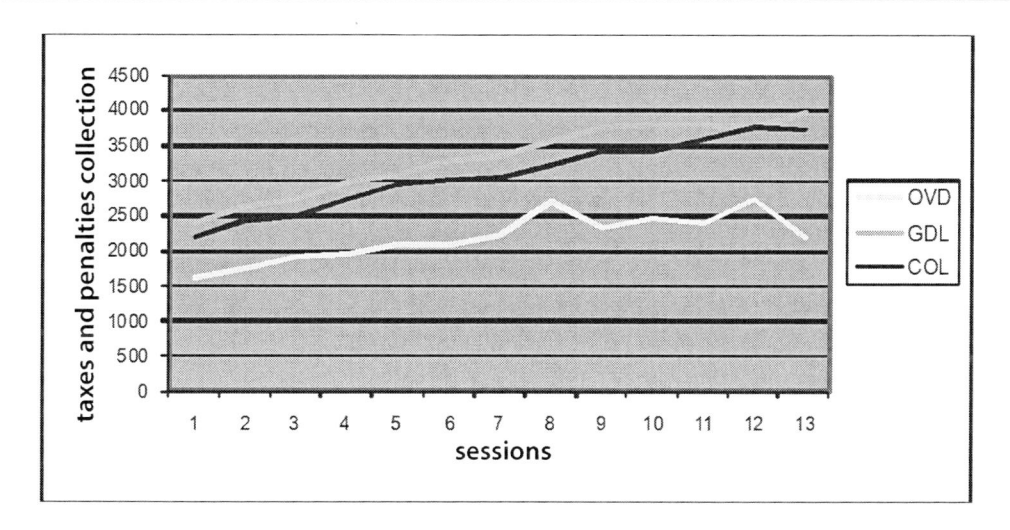

Figure 1. Aggregate tax revenue in each experimental group (cohort).

From Table 5 it can be deduced that risk-averse individuals, in all groups, pay less taxes because there is a conservative attitude to loose individual o personal incomes. Rather, in the case of risk-propense individuals, in all groups, pay more taxes as a consequence of their attitude to avoid their personal responsability to separate selectively their trash. Differences begin to appear with risk-neutral individuals in these three groups: both in GDL and COL groups can be observed how risk-neutral individuals pay more taxes. This is probably explained by a lack of sensitivity faced to trash selective separation. In relation to gender, there are not significative differences.

An importan aspect is the evolution of tax collection in each experimental group (see Figure 1).

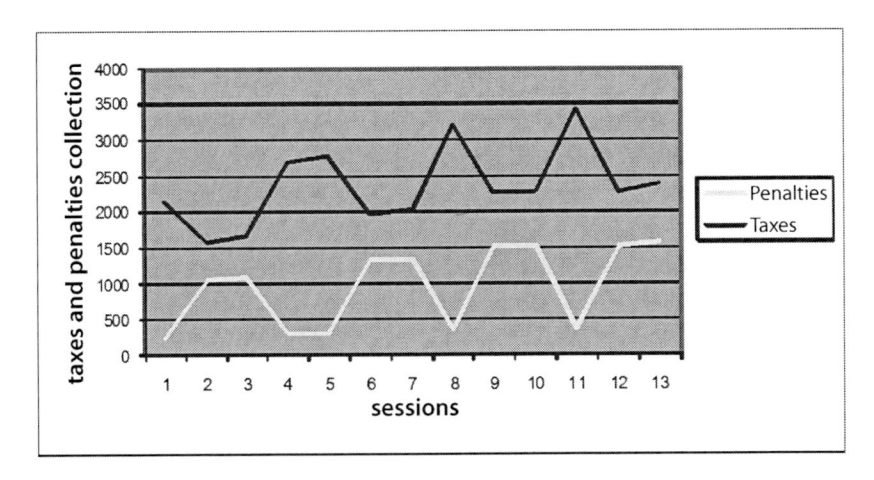

Figure 2. Disaggregated tax collection (tax and penalties) in GDL.

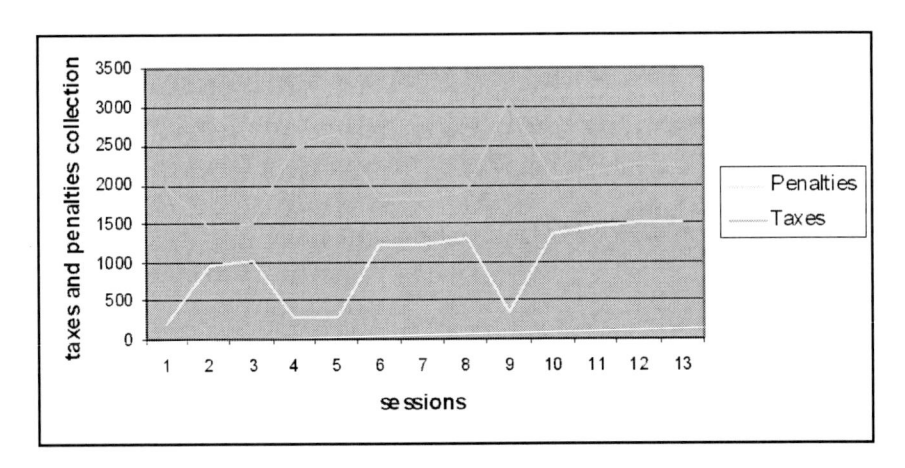

Figure 3. Disaggregated tax collection (tax and penalties) in COL.

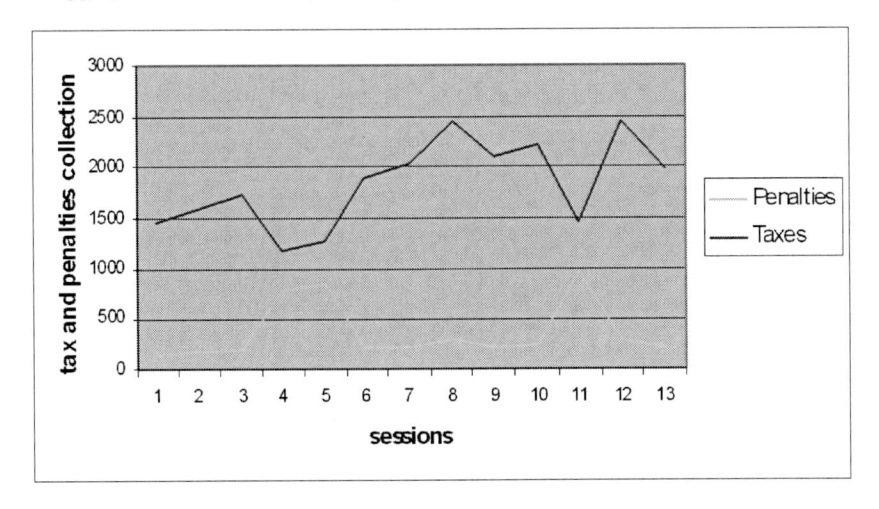

Figure 4. Disaggregated tax collection (tax and penalties) in OVD.

The OVD group generates less tax collection as a consequence of a certain social sensitivity to environmental problems that affects to individual behavior. Beside this, can be considered, of course, the fact that individuals try to save personal income for the next period, in function of his/her risk preferences. Rather, the control groups –GDL and COL- shows more tax payoffs.

Figure 2, Figure 3 and Figure 4 show, more specifically, the experimental dynamics of taxes and penalties in cohorts of control, GDL, COL and OVD.

The GDL cohort shows a very active behavior trying to maximize individually their net income, after taxes and penalties.

Table 6. Experimental tax scheme

Session	Tax rate	Penalty	Probabilities of detection
t	10%	10%	
t+1	11%	11%	
t+2	12%	12%	
t+3	13%	13%	
t+4	14%	14%	
t+5	15%	15%	$P_1=0,1$
t+6	16%	16%	$P_2=0,4$
t+7	17%	17%	
t+8	18%	18%	
t+9	19%	19%	
t+10	20%	20%	
t+11	21%	21%	
t+12	22%	22%	

Table 7. Attitude to environment

Group	Risk Averse	Risk Neutral	Risk Propense	Globally
OVD	Sensititvity	Sensitivity	Insensitivity	Sensitivity
GDL	Sensitivity	Insensitivity	Insensitivity	Insensitivity
COL	Sensitivity	Insensitivity	Insensitivity	Insensitivity

The structure of probabilities of detection follows an scheme developed by Landsberger and Meilijson (1982) [Magadán and Rivas(2010)]. As seen above, this strategical structure put under control the tendencies of GDL and COL to behave against social interest. For our experimental purposes $p_1=0,10$ and $p_2=0,40$. In Table 6 it is possible to evaluate the evolution of our experimental tax scheme. Must be taken into account that the differences between both probabilities are quite evident in order to generate an 'ex ante' attitude to avoid fulfilling with the stablished pro-environmental behavior.

6. ATTITUDE TO ENVIRONMENT

As seen in section 5, there is a close relation between payment of taxes and attitude to enviroment (see Table 7).

Among the three cohorts, OVD stands out for its environmental sensitivity. Sociologically is a real fact. And experimentally can be observed the important influence of environmental education at this respect. The other two cohorts tend to behave without much sensitivity to that problem, but solely to their own personal income.

7. ATTITUDE TO COST-BENEFIT ANALYSIS

On considering the individual revenue in terms of public goods and services, the results are resumed in Table 8:

The idea underlined by the results resumed in Table 8 fits with the other results obtained above. Individuals in control cohorts tends not to consideration the social benefit. Only centers their interest in the final net income, after all tax payoffs. Do not forget that OVD cohort is selected from a region that in the last year –from october, 2009 to october, 2010- the selective trash collection has increased up to 5%[5].

Table 8. Attitude to cost-benefit analysis

Group	Attitude to cost-benefit analysis			
	Risk Averse	Risk Neutral	Risk Propense	Globally
OVD	Conscious	Conscious/Unconscious	Unconscious	Conscious
GDL	Conscious	Conscious/Unconscious	Unconscious	Unconscious
COL	Conscious	Conscious/Unconscious	Unconscious	Unconscious

Table 9. Main standard nonparametric tests for experimental purposes

	Classificatory data	Numerical data
One-sample design	Binomial χ^2	Kolmogorov-Smirnov
Two-sample design (related samples)	Binomial χ^2	Randomization Wilcoxon
Two-sample design (independent samples)	Fisher Exact Probability χ^2	Randomization Mann-Whitney Kolmogorov-Smirnov

Source: Davis and Holt (1993), p. 552.

---

[5] La Nueva España, 22/11/2010 (http://www.lne.es/asturias/2010/11/22/aumenta-5-recogida-selectiva-basura/997858.html)

8. NONPARAMETRIC TESTS

The main standard nonparametric tests applied to experimental designs are summarized in Table 9.

But our experimental analysis is considering three samples: OVD, GDL and COL. The Kruskal-Wallis test is appropiate for evaluating the hypothesis that the medians of *S* independent samples are equal, against the general alternative that they are not equal. The mathematical formulation of this test is summarized in Table 10.

Table 10. Kruskal-Wallis test

The Kruskal-Wallis test is represented by H:

$$H = \frac{12}{n(n+1)} \sum_{i=1}^{c} \frac{R_i^2}{n_i} - 3(n+1)$$

where:
  c = number of samples,
  R_i = sum of ranks of n_i observations forming the *i-th* sample.
  Critical region is determined by the probability of that *H* is higher than certain value of *K* vinculated to a certain level of signification. H follows an asymptotic that nears to χ^2 with *c-1* degrees of freedom. If repeated observations are detected, then H is adjusted dividing by a correction factor, f:

$$f = 1 - \frac{\sum_j (\tau_j^3 - \tau_j)}{n^3 - n}$$

where τ_j represents the number of repetitions of each *j-th* streak. Then, the new test will be $H^* = H/f$.

Table 11. Mann-Whitney test

Bilateral test: Let two samples be of sizes n_1 y n_2. The statistics U are defined as follows:

$$U_Y = \sum_{i=1}^{n_1} \sum_{j=1}^{n_2} \psi ij$$

$$U_X = \sum_{i=1}^{n_1} \sum_{j=1}^{n_2} \psi ij$$

where ψ_{ij} is an indicator function of type:

$1 \text{ si } x_i < y_j$

$0 \text{ si } x_i > y_j$

for U_Y and changing inequities, for U_X. To determine the acceptation region, U_m statistic must be used:

$$U_m = \min[U_X; U_Y]$$

in such a way that U_m be among certain values, K_1 and K_2, for an specific value of *1-α*.

The Kruskal-Wallis test indicates clearly that –when applied on our three samples- these three samples are not equal in terms of medians. Rather, The Kruskal-Wallis test applied solely on GDL and COL indicates that are 'equal' in terms of medians. This result prove that GDL and COL are good groups of control because they are samples extracted from regions where there is little environmental education. When people is educated in terms of giving personal and social value to environment, then one can find good citizens.

Another test applied on the data generated by our three samples was the Mann-Whitney test (see Table 11).

When apply the Mann-Whitney test to our three samples, the results underline the ideas obtained with Kruskal-Wallis: GDL and COL seem samples from the same population. Rather, when OVD is used with GDL (or with COL),then the Mann-Whitney test shows that these pair of samples are not extracted from the same population.

CONCLUSION

As seen, the authors have started from a general framework of experimental analysis in order to measure, at least, the weight of three dimensions –let us say *attitudes*- in relation with an specific behaviour: separation of trash in origin. This problem had relation, not only with environment but with expectations, too. And the main aim of authors have consisted on clearing up the structure of individual behavior when face on this question.

The dimensions considered in this work were the following: i) attitude to risk (individual damage in terms of taxes), ii) attitude to environment (individual perspective of environmental social damage), and finally iii) attitude faced to cost-benefit analysis (individual potential revenues, in terms of public goods and services).

Risk-averse individuals, in all considered groups, pay less taxes because there is a conservative attitude to loose individual o personal incomes. Rather, in the case of risk-propense individuals, in all groups, pay more taxes as a consequence of their attitude to avoid their personal responsability to separate selectively their trash. Differences begin to appear with risk-neutral individuals in these three groups: both in GDL and COL groups can be observed how risk-neutral individuals pay more taxes. This is probably explained by a lack of sensitivity faced to trash selective separation.

Considering the environmental attitude, the experimental results show that Among the three cohorts, OVD stands out for its environmental sensitivity. Sociologically is a real fact. And experimentally can be observed the important influence of environmental education at this respect. The other two cohorts tend to behave without much sensitivity to that problem, but solely to their own personal income.

Finally, individuals in control cohorts tends not to considerate the social benefit. Only center their interest in the final net income, after all tax payoffs.

The Kruskal-Wallis test indicates clearly that –when applied on our three samples- these three samples are not equal in terms of medians. Rather, The Kruskal-Wallis test applied solely on GDL and COL indicates that are 'equal' in terms of medians. This result prove that GDL and COL are good groups of control because they are samples extracted from regions where there is little environmental education. When people is educated in terms of giving personal and social value to environment, then one can find good citizens. And when apply the Mann-Whitney test to our three samples, the results underline the ideas obtained with Kruskal-Wallis: GDL and COL seem samples from the same population. Rather, when OVD is used with GDL (or with COL),then the Mann-Whitney test shows that these pair of samples are not extracted from the same population.

More experiments will be needed to determined more clearly the interdependence between social environment and personal attitude. But, at least this modest contribution intends to be a first step to clearing up all these complex relations and interdependences.

REFERENCES

Allais, M. (1953): "Le Comportement de l'Homme Rationel Devant le Risque, Critique des Postulates et Axiomes de l'Ecole Americaine", *Econometrica*, 21, 503-546.

Allais, M. (1979): "The so-Called Allais Parados and Rational Decisions under Uncertainty", in *Expected Utility Hypothesis and the Allais Paradox*, M. Allais and O. Hagen, editors. Dordrecht: Reitel.

Andreoni, J. (1988): "Why Free Rider?: Strategies and Learning in Public Goods Experiments", *Journal of Public Economics*, 37, 291-304.

Arnau, J. (1954): "The Nature and History of Experimental Control", *American Journal of Psychology*, 67, 573-589.

Bostan, I.; Burciu, A.; Condrea, P.P. and Durac, G. (2009): "Involvement of Legal Responsibility for Severe Acts of Pollution and Noncompliance", *Environmental Engineering and Management Journal*, 8, 469-473.

Daughety, A. and Forsythe, R. (1985): "Regulation and the Formation of Expectations: A Laboratory Analysis", working paper, University of Iowa.

Davis, D. And Holt, Ch. (1993): *Experimental Economics*. Princeton University Press, United Kingdom.

Delucchi, M., Murphy, J. and McCubbin, J. (2002): "The Health and Visibility Cost of Air Pollution: A Comparison of Estimation Methods", *Journal of Environmental Management*, 64, 139-152.

Dufwenberg, M. (2000): "Measuring Beliefs in an Experimental Lost Wallet Game", *Games and Economic Behavior*, 30, 163-182.

Gardner, R., Ostrom, E. and Walker, J. (1990): "The Nature of Common Pool Resource Problems", *Rationality and Society*, 2, 335-358.

Gordon, S. (1954): "The Economic Theory of a Common Property Resource: The Fishery", *Journal of Political Economy*, 62, 124-142.

Groves, T. and Ledyard, J. (1977): "Optimal Allocation of Public Goods: A Solution to the 'Free Rider' Problem", *Econometrica*, 45, 783-809.

Harrison, G. and Hoffman, E., Rutstrom, E. and Spitzer, M. (1987): "Coasian Solutions to the Externality Problem in Experimental Markets", *Economic Journal*, 97, 388-342.

Kagel, J. And Roth, A. (1995): *Handbook of Experimental Economics*. Princeton University Press, United Kingdom.

Kahneman, D. and Tversky, A. (1973): "Prospect Theory: An Analysis of Decision Under Risk", *Econometrica*, 47, 263-291.

Kunreuther, H., Kleindorfer, P. and Knez, P. (1987): A Compensation Mechanism for Siting Noxious Facilities: Theory and Experimental Design", *Journal of Environmental Economics and Management*, 14, 371-383.

Landsberger M. and Meilijson I. (1982): "Incentive Generating State Dependent Penalty System. The Case of Income Tax Evasion", *Journal of Public Economics*, 19, 333-352.

Lindquist, E. (1953): *Design and Analysis of Experiments in Psychology and Education*. Hougthon Mifflin, Boston.

Magadán, M. and Rivas J. (1999): *Economía Experimental*. Minerva, Madrid.

Magadán, M. and Rivas, J. (2010): "Less Green Taxes and More Control over Pollutant Industries: A Theoretical Proposal", *Environmental Engineering and Management Journal*, 9, 1173-1177.

Mark, I. and Walker, J. (1988): "Communication and Free-Riding Behavior: The Voluntary Contributions Mechanisms", *Economic Inquiry*, 26, 585-608.

Ostrom, E. and Walker, J. (1991): "Communications in a Common: Cooperation without External Enforcement", in T. Palfrey, ed., *Contemporary Laboratory Research in Political Economy*. Ann Harbor: University of Michigan Press, EEUU.

Palfrey, T. and Porter, R. (1991): "Guidelines for Submission of Manuscripts of Experimental Economics", *Econometrica*, 59, 1197-1198.

Plott, Ch. (1983): "Externalities and Corrective Policies in Experimental Markets", *Economic Journal*, 93, 106-127.

Walker, J. and Gardner, R (1990): "Rent Dissipation and Probabilistic Destruction of Common Pool Resources: Experimental Evidence", working paper, Indiana University.

Walker, J., Gardner, R. and Ostrom, E. (1990): "Rent Dissipation in a Limited-Access Common-Pool Resource: Experimental Evidence", *Journal of Environmental Economics and Management*, 19, 203-211.

In: Research Studies on Tourism and Environment
Editors: J. Mondejar-Jimenez et al.
ISBN: 978-1-61209-946-0
© 2012 Nova Science Publishers, Inc.

Chapter 15

ADVERTISING, UNFAIR COMMERCIAL PRACTICES AND CONSUMER PROTECTION IN SPAIN[*]

María Ángeles Zurilla Cariñana
University of Castilla-La Mancha, Spain

ABSTRACT

Law 29/2009 of 30 December, amending the rules on unfair competition and advertising in order to improve the protection of consumers, transposes Community Directive 29/2005 to Spanish law and amends two very important Spanish laws: Law 3/1991 on Unfair Competition and General Law 34/1988 on Advertising.

It establishes a unitary legal framework on unfairness in misleading and aggressive practices, imposing the same standards regardless of whether the addressees are consumers or businesses. The traditional distinction between unfair practices and the regulation of unfair or misleading advertising is set aside. It is worth noting that the legislators have opted to maintain the General Advertising Law and that the concept of unlawful advertising in the framework thereof has been preserved, safeguarding actions and remedies allowing that practice to be combated. In this paper we will examine the most notable aspects of Law 29/2009, with special attention to those concerning the regulation of unlawful advertising and the mechanisms to protect consumers from it.

Keywords: Unlawful advertising, unfair competition, advertising practices, injunctive action

I. INTRODUCTION

The need for consumers to be protected stems from the recognition of the fact that a large number of people when making normal everyday transactions involving the purchase of

---

[*] This work places in the Project of Investigation financed by The Junta de Comunidades de Castilla La Mancha (Spain): "Regional Impact of the recent regulation on consumer, s law: the regime of arbitration and mediation of consumers, the TR of the LGDCU and the Board of Commercial Improper Practices".

goods and services are not in a position to secure appropriate quality and prices on their own. A typical consumer is one who is unable to make fair demands in relation to the products or services that he/she purchases and who lacks the necessary means to confront the companies with which he/she transacts.

It is precisely this progressive contractual inequality that is the origin of the rules aimed at protecting consumers in the framework of the market economy. The need for protection of the weaker contracting party takes the form of the notion of guaranteeing certain minimum requirements to private individuals, such as the quality and price of the products acquired or the services provided, and the possibility of demanding compliance both with the terms specified in the contract and with those appearing in advertising. In order to fully satisfy such demands our legal system has effective tools for protecting consumers, which will be analyzed in this paper. Special attention will be given to Law 29/2009 amending the rules on unfair competition and advertising in order to improve consumer protection. This law transposes Community directive 29/2005, and amends four highly important laws in the Spanish legal system: Law 3/1991 on Unfair Competition; the General Law on the Protection of Consumers and Users (LGDCU, Consolidated Text adopted by Legislative Royal Decree 1/2007); the General Advertising Law 34/1988; and Law 7/1996 Regulating the Retail Trade.

As stated in its Preamble, Law 29/2009 was conceived with the aim that consumer protection legislation should be consistently integrated with market regulations as a way of ensuring that such protection may be truly effective. To achieve this aim it provides a unitary legal regime on unfairness in misleading and aggressive practices, imposing the same standards regardless of whether the targets are consumers or businesses. Thus the traditional distinction between unfair practices and the regulation of unfair or misleading advertising is set aside. It is worth noting that the legislators have opted to maintain the General Advertising Law (hereinafter LGP) and that the concept of unlawful advertising in the framework of the LGP has been preserved, safeguarding actions and remedies allowing the practice to be combated, especially as regards advertising which undermines personal dignity or infringes the rights and values enshrined in the Constitution, in particular in relation to children, young people and women. In this paper we will analyze the most notable aspects of Law 29/2009, with special attention to those concerning the theme of our study, and consider the law's effects on the legislation reformed by it.

2. PROTECTION OF CONSUMERS AGAINST UNFAIR COMMERCIAL PRACTICES AND ADVERTISING

2.1. Introduction

The protection for consumers entailed by the principle of integrating advertising into consumer contracts is reinforced by other legal provisions adopting specific measures against unlawful advertising and misleading or unfair commercial practices: Law 34/1988 of 11 November of General Provisions on Advertising, and Law 3/1991 on Unfair Competition. As noted in the introduction, Law 29/2009 of 30 December amending the legal rules on unfair competition and advertising in order to improve the protection of consumers and users provides mechanisms for the coordination of those two laws, given the overlap between them that sometimes occurs. For these purposes it amends LGP so as to provide a single legal body of actions and remedies against all commercial practices detrimental to consumers' interests, without prejudice to the specific regulations on advertising. In line with this approach it repeals LGP Title IV, which is left with no content. Moreover, Law 29/2009 devotes a whole chapter to regulating unfair competitive practices which, though they also affect competitors, are regarded as liable to harm their targets only when these are consumers and users. Such is the case of misleading omissions or unfair practices, in whatever circumstances they occur.

2.2. UNLAWFUL ADVERTISING

LGP article 3 regulates *unlawful advertising*. In the wording given by Law 29/2009, the following is considered unlawful:

a) Advertising which offends personal dignity or infringes the rights and values enshrined in the Constitution, especially those referred to in Section 3, articles 14, 18 and 20. This provision is deemed to include advertisements representing women in a demeaning or discriminating way, particularly and directly using either their bodies or body parts as mere objects unrelated to the product to be promoted, or their image associated with stereotyped behaviours contrary to the bases of our legal system, helping to generate the violence referred to in the Organic Law of Measures for Integral Protection against Gender Violence of 28 December 2008.

b) Advertising aimed at minors encouraging them to buy goods or services by taking advantage of their inexperience or credulity, or in which they appear persuading their parents or guardians to make a purchase. Minors may not in any event be shown in hazardous situations. Advertisements must not be misleading as to a product's characteristics or safety or as to the abilities or aptitudes required in children to use it without harming themselves or third parties.

c) Subliminal advertising, deemed to be that using techniques that produce stimuli at levels verging on the thresholds of sensory perception or similar, which may affect the target public without being consciously perceived (LGP article 4, as amended).

d) That infringing the provisions regulating the advertising of certain products, goods, activities or services.

LGP article 5, as reworded by Law 29/2009, refers to special provisions relating to the advertising of health equipment or products, or products liable to generate risks for the health and safety of people or their property, along with that relating to games of chance, betting and gambling. Advertising of this kind may be subject to prior authorization arrangements. Such arrangements may also be established when so required for the protection of constitutionally enshrined values and rights. Authorizations must be granted in keeping with the principles of free competition, so that other competitors are not harmed. Reasons must be given for the refusal of authorization requests.

In any event, any regulations or provisions regulating a product or service that also include rules on the advertising thereof will also contain the requirement that such advertising must mention any risks arising from its normal use.

Article 5 restricts the advertising of narcotic or psychotropic substances and medicinal products for people or animals to the cases, forms and conditions specified in the special provisions regulating them. It also prohibits the advertising on television of alcoholic drinks with an alcohol content of more than 20 degrees. Also prohibited is the advertising of alcoholic drinks in places where their sale or consumption is forbidden.

e) Misleading advertising, unfair advertising and aggressive advertising, which will be treated as unfair competitive practices on the terms provided in the Law on Unfair Competition.

Article 18 of the Law on Unfair Competition, also as reworded, provides categorically that advertising regarded as unlawful by LGP will be treated as unfair.

The Supreme Court (TS) has on many occasions discerned the existence of misleading advertising. Its judgment of 20 March 2000 (RJ 200/2010) deemed that there was misleading advertising in a case of illustrated leaflets distributed to advertise wines including a box claiming that they were covered by the Bierzo designation of origin, though the advertiser was not registered in the records of the Bierzo Regulatory Council. The judgment highlights the unlawful usurpation and the misleading effect on consumers. The Supreme Court Judgment of 25 April 2006 (RJ 2006/2201) deemed that there was misleading advertising in a case of healthcare insurance for which an "ethical supplement" was offered along with prices called "franchises", which might have led the insureds to believe that they were part of the actual price, whereas they were the full price of the service plus the intermediary's commission. The judgment states that consumers' freedom must be protected, deeming that advertising leading the addressees to take transactional decisions under a misconception is misleading.

Misleading advertising has also been discerned on many occasions by the High Courts. As an example we may mention the judgments of the Catalan High Court's Social Division (SSTSJ) of 13 September 2001 (JUR 2001/1186) and of 4 May 2002 (JUR 2002/250330). The former concerned a case of distance selling promising a prize that proved to be non-existent. The latter also concerned a case of distance selling with a promised gift which, this time, proved to be substantially different from the one advertised.

It is also worth mentioning the judgment of the EC Court of Justice of 16 July 1998 (Case C-210/96) which deemed that the inclusion of equivocal statements on labelling may be regarded as misleading advertising.

The existence of misleading advertising was also discerned by the Álava Provincial Court judgment of 17 February 2000 (AC 2000/87), in the case of a hairdressing course for which the advertising led people to believe that the course being signed up to was officially approved by using terms similar to those used by approved centres; the Asturias Provincial Court judgment of 7 March 2000 (AC 2000/729) in the case of an investment savings product presented in the form of an insurance policy; the Balearic Islands Provincial Court judgment of 25 May 2006 (JUR 2006/236510), in the case of a hotel offering bookings over the internet whose characteristics did not correspond to those advertised on its website; the Cordoba Provincial Court judgment of 20 November 2006 (AC 2007/1027) in the case of advertising for a bank deposit by a financial institution offering double the normal interest rate; and the Oviedo Provincial Court judgment of 6 November 2007 (JUR 2008/56954) in another case of bookings for a hotel not corresponding to the category advertised on its website. More recently, the Madrid Provincial Court judgment of 24 January 2008 (AC 2008/516) deemed that there was unlawfully misleading advertising in a case in which tractors were being sold with a top speed of 50 km/h, whereas such vehicles can legally travel only up to a maximum of 50 km/h. The Court also took account of the added circumstance of the misconception that may be caused in consumers as to the maximum permitted speed for this type of vehicle.

Moreover, the Madrid Provincial Court judgment of 12 May 2008 (AC 2008/119) ruled that there was unfair advertising in leaflets for a course on "VAT in international trade" in which the advertiser claimed to be the only company in Spain engaged in VAT recovery (the Court deemed the advertising to be unlawful because it was misleading, unfair and contrary to the law). Moreover the Seville Provincial Court judgment of 9 February 2009 (JUR 2009/251631) ruled that there was unfair and misleading advertising in the case of an offering of debt consolidation and financing services because the advertiser had omitted the main characteristics of the loans and had not appropriately identified the credit institutions behind the offer. The Barcelona Provincial Court judgment of 2 July 2009 (AC 2009/464831) also deemed that there has been unfair advertising for a children's car seat in which the manufacturer claimed that its seats were the safest on the market, though this circumstance had not been really demonstrated.

To end this section we must refer to other legal provisions containing important stipulations on consumer protection against unlawful advertising.

The Law on the Protection of Consumers and Users in the procurement of goods with the offer of subsequent total or partial repayment of the price of 14 December 2007 provides for misleading advertising in its article 12. It requires that commercial communications for such goods should not cause confusion in consumers' minds with financial activities. To this end it provides that expressions characteristic of the financial sector should be avoided, such as saving by investment, returns, interest or equivalent. Given the alarm in society caused by the Afinsa and Forum Filatélico cases, the second paragraph of this article requires that all commercial communications should specify clearly and prominently that the goods on which the activity is based have no market value, and also mention, where appropriate, any capital appreciation offer or repayment guarantee.

For its part, Law 34/2002 of 11 July on *Information Society Services and e-Commerce*, after stating that the applicable provisions on advertising and data protection will apply to commercial communications by electronic means, refers to the need for electronic commercial communications to be clearly identified as such (article 19). Article 20 (amended by Law 56/2007 of 28 December on Measures to Promote the Information Society) provides requirements for commercial communications, promotional offers and competitions publicized by electronic means. If they are conveyed by email or other equivalent electronic media they must include the word "publicidad" (advertising) at the start of the message, or the abbreviation "publi". Promotional offers with discounts, gifts, prizes, games or competitions must clearly specify the terms for access and participation, which must be easily accessible. Also prohibited is the sending of commercial or promotional communications by email or equivalent electronic media that have not been expressly requested or authorized by the addressee (article 21). Article 30 allows actions for injunction to be brought against conducts infringing this law which are harmful to the collective or diffuse interests of consumers and users.

We must also refer, finally, to the *LGDCU (Consumer Protection Law) Consolidated Text* of November 2007, whose article 18.4 states that false or misleading advertising for goods or services will be prosecuted as fraud. It recognizes consumers' and users' associations as parties with standing to initiate and take part in legal proceedings to secure injunctions against such advertising.

2.3. Unfair Commercial Practices

Article 5 of the Unfair Competition Law, amended by Law 29/2009, gives a detailed list of the practices that may be regarded as *unfairly misleading*. Such practices are those which cause misconceptions in the addressee, or withhold vital information about the goods, activities or services. This information may concern the existence or nature of the goods or service, terms of purchase, full price or quoted price, reasons for the offer; advertisers' rights, after-sales service; treatment of complaints, extent of the commitments of the trader or professional, nature and characteristics of the trader or professional's rights, legal or conventional rights of consumers, or risks to which they may be exposed. The new article 7 of the Unfair Competition Law speaks of *misleading omissions*, stating that it is unfair to omit or conceal the information required for the addressee to take or be able to take a transactional decision with a due understanding of what is at stake, or if the information is unclear, unintelligible or ambiguous. Article 7 does not contain an exclusive list of the points whose omission constitutes a misleading omission.

There is such a list in the reworded article 20 of the LGDCU Consolidated Text. This article details the information required in an offering of goods or services. Such information must refer to the characteristics of the goods or service and must provide the consumer or user with adequate information to take a decision on a purchase. Hence it must contain at least the following details: name, trade name and full address of the trader responsible for the commercial offer, and if applicable of the trader on behalf of which it is acting; the main characteristics of the goods or service; full final price, including taxes, detailing, if applicable, the amount of any discounts applicable to the offer and any extra expenses passed on to the consumer or user; payment procedure, delivery and contract execution periods; and, where applicable, existence of the right to withdrawal.

As noted by Carrasco Perera, the result of this transfer is that article 7 of the Unfair Competition Law contains a list of potentially misleading omissions, while in respect of relations with consumers – but for the sole purposes of specifying administrative penalties – there is a closed list, in accordance with Directive 29/2005, which is transposed by Law 29/2009.

Article 20 of Law 29/2009 regulates *misleading practices that are confusing* to consumers, identifying as such those which create confusion with the trade marks, trade names or other distinguishing marks of a competitor, including comparative advertising. Article 21 refers to misleading practices regarding codes of conduct or other quality marks (those in which the trader falsely claims to be a signatory to a code of conduct, or displays a trust or quality mark without having obtained the necessary authorization). Article 22 deals with baiting and misleading promotional practices. By way of example we may mention the following: conducting clearance sales without any of the conditions provided therefor being applicable to the trader; or practices, including competitions or draws, in which an automatic prize is offered, without the prizes described or others of equivalent quality or value being awarded, etc. Article 23 refers to misleading practices concerning the nature and properties of the goods or services, their availability and the after-sales service. Article 24 states that pyramid selling practices are unfairly misleading. Article 25 refers to misleading practices that cause confusion. Article 26 deals with covert commercial practices. Finally article 27, under the generic title of "other unfairly misleading practices", lists a series of business practices such as presenting rights given to consumers in law as a distinctive feature of the trader or professional; passing on inaccurate information on market conditions or the possibility of finding the good or service, or falsely representing the trader as a consumer or user.

The new article 8 of the Unfair Competition Law regulates unfairly aggressive practices, specifying as such those which, by means of harassment or coercion, including the use of force or undue influence, significantly impair the consumer's freedom of choice or conduct with regard to the goods or service, and so may affect his/her transactional decisions. In determining whether a practice involves harassment, coercion or undue influence, account will be taken of its timing, location, nature or persistence; the use of threatening or abusive language; any onerous or disproportionate non-contractual barriers imposed by the trader; and any threat to take any action that cannot legally be taken.

Article 28 regulates *aggressive practices involving coercion* (deemed to be those which make the consumer or user believe that he/she cannot leave the premises of the trader or professional without forming a contract). Article 29 provides for *aggressive practices involving harassment* (paying personal visits to the consumer's home, and ignoring the consumer's requests to leave; making persistent and unwanted and solicitations by telephone, etc.). Article 30 refers to aggressive practices with regard to minors (including in an advertisement a direct exhortation to children to buy advertised products or persuade their parents or other adults to buy advertised goods or services for them). Article 31, under the heading *other aggressive practices*, states that such unfair practices include those requiring a consumer who wishes to claim on an insurance policy to produce documents which could not reasonably be considered relevant as to whether the claim was valid; informing a consumer that if he/she does not buy the product or service, the trader's or professional's job will be in jeopardy, etc.

Law 29/2009 also gives a new wording to article 10 of the Unfair Competition Law, which under the heading *comparative practices* permits public comparisons, including comparative advertising with explicit or implicit references to a competitor, if the following requirements are met: the goods and services compared must be intended for the same purposes and meet the same needs; the comparison is made objectively between one or more essential features, which may include the price; for products with designation of origin, the comparison may be made only with other products with the same designation; no goods or services may be presented as imitations or replicas of others bearing a protected trade mark or trade name. However, article 20 of the Unfair Competition Law, defines as unfair those commercial practices, including comparative advertising, which create confusion with any other goods or services, registered trade marks or trade names ... provided that they are likely to affect the transactional decisions of consumers and users (LGP article 6 bis, now repealed, introduced by Law 39/2002 of 29 October, dealt with comparative advertising, and its content was much like that of the current article 10 of the Unfair Competition Law).

Article 11, in its new wording, refers to *practices imitating business or professional initiatives*, which are considered lawful, unless such initiatives are protected by exclusive rights enshrined in law. Imitation will be regarded as unfair where it is liable to generate an association in the minds of consumers regarding the product or service, or it takes undue advantage of the reputation or efforts of others.

Law 29/2009 does not predetermine which authorities are competent to punish as consumer offences the unfair commercial practices specified by the Unfair Competition Law as acts of unfair competition. We agree with Carrasco Perera that such authorities will doubtless be the regional ones (LGDCU new article 47.3), as should be clear from the inclusion of such practices as specific consumer offences in LGDCU article 49. According to this article, the competent authorities on consumer affairs will moreover punish practices specified as offences against the rights of consumers and users by traders in sectors with specific regulations, as well as commercial practices unfair to consumers and users.

3. Proceedings Arising from Unlawful Advertising and Unfair Competitive Practices

LGP Title IV has been repealed by Law 29/2009 of 30 December amending the legal rules on unfair competition and advertising in order to improve the protection of consumers and users. Articles 25 to 32 of this Title were devoted to *Actions for injunction and rectification*.

The recent Law 29/2009 gives a new wording to LGP article 6, under the heading "actions against unlawful advertising". According to this article, such actions will be those of a general nature provided for proceedings arising from unfair competition in chapter IV of Law 3/1991 of 10 January on Unfair Competition.

Article 32 of this Law, also worded pursuant to Law 29/2009 of 30 December, states that against unfair competitive practices, *including unlawful advertising*, the following actions may be brought:

1) Action to have a practice declared unfair
2) Action for an injunction to have the unfair practice stopped or its future repetition prohibited. Proceedings for prohibition may also be brought, if the practice has not yet been implemented
3) Action to have the effects of the unfair practice removed
4) Action for rectification of misleading, incorrect or false information
5) Action for compensation for damages caused by the unfair practice, in the event of fraud or gross negligence by the trader
6) Action in respect of unjust enrichment, which will be applicable only where the unfair practice harms an entity covered by exclusive rights or others with similar economic content

In the case of the actions provided for in numbers 1 to 4, the court may if it sees fit, and at the respondent's expense, order the total or partial publication of the judgment, or, where the effects of the offence may persist over time, of a rectifying statement.

Article 33 regulates the *legal standing* required to bring action with regard to advertising. The first section states that the following have standing:

- Any individual or entity participating in the market whose economic interests are directly harmed or jeopardized by the unfair practice has standing to bring the actions provided for in article 32.1 (1-5).
- Standing to bring the actions provided for in article 32 (1-5) is possessed by any individual or entity that is affected and, generally speaking, anyone having a subjective right or a legitimate interest.
- Action for compensation for damages caused by the unfair practice may also be brought by anyone with standing under the provisions of article 11.2 of the Civil Procedure Law (LEC).
- Action in respect of unjust enrichment may be brought only by the holder of the infringed legal rights.

In accordance with the provisions of the first section of article 33, consumers will have standing to bring proceedings in respect of unfair competition in so far as they "are directly harmed". However, this procedural channel is confined to the exercise of actions for injunction, removal, rectification or compensation, which will be only ones in which consumers may have a particular interest. Otherwise contractual actions (contract fulfilment, contract invalidity, termination of contract, compensation for damages due to contractual default) will follow common procedure and be subject to common jurisdiction.

In the case of the actions referred to in article 32.1 (1-4), they may also be brought, pursuant to the second section of article 33, by associations, professional corporations or organizations representing economic interests, where their members' interests are affected. Such actions, when exercised in the defence of the diffuse interests of consumers or users, may also be brought, pursuant to the third section of article 33, by:

(a) The National Consumer Institute and corresponding bodies or entities in Spain's autonomous regions, and local authorities with powers concerning the protection of consumers' and users' rights
(b) Associations of consumers and users meeting the legal requirements provided in the LGDCU Consolidated Text
(c) Entities from other EU Member States established to protect the collective and diffuse interests of consumers and users, and authorized to do so by their inclusion in the official list published for the purpose in the EU Official Journal.

The Public Prosecution Service may also bring injunctive action to defend the general, collective and diffuse interests of consumers and users (article 33, section 4).

In addition, pursuant to LGP article 6, regarding unlawful advertising that uses the image of women in an offensive or demeaning way, the following parties will have legal standing to bring the actions referred in article 32.1 (1-4) of the Unfair Competition Law:

(a) The Government Coordination Office against Gender Violence
(b) The Women's Institute (IM) or its equivalent in economic matters
(c) Legally established associations whose sole purpose is the defence of women's interests, and which have no profit-making partners
(d) The Public Prosecution Service

The actions provided for in article 32 may be brought against any person who has performed the unfair practice, ordered it to be performed or cooperated in its performance. However, action in respect of unjust enrichment may be brought only against the beneficiary of the enrichment. Such actions are limited to one year after the time at which they could first have been brought and the person with legal standing learned of who carried out the unfair competitive practice, and will lapse in any event three years after the practice's termination.

The limitation on actions in defence of general, collective or diffuse interests is governed by the provisions of article 56 of the LGDCU Consolidated Text. This article provides that actions for injunction have no limitation period, without prejudice to the provisions of article 19.2 of the Law of 13 April 1998 on General Conditions of Contract in relation to the general conditions registered in the Register of General Conditions of Contract.

In: Research Studies on Tourism and Environment
Editors: J. Mondejar-Jimenez et al.

ISBN: 978-1-61209-946-0
© 2012 Nova Science Publishers, Inc.

Chapter 16

SOME ISSUES ABOUT THE INVIOLABILITY OF THE HOTEL ESTABLISHMENT IN SPANISH LAW

Silvia Valmaña Ochaíta
University of Castilla-La Mancha, Spain

ABSTRACT

The protection granted by the legal system to privacy in Spanish law is not in the majority of cases applicable to the hotels, hostels and other hotelier establishments. Casuistry is so large that it is necessary to distinguish the different nature of the object of protection. This protection would depend on the room concerned for the invasion of the privacy, the person who commits the intrusion and the moment of the entry. This chapter tries to systematize and clarify some essential points about the treatment given by the Spanish law, especially the criminal law, to these issues, distinguishing the application of the concept of domicile, and therefore the application of the crime of unlawful entry of dwelling-houses, from other unlawful entries in hotel establishments open to the public.

1. INTRODUCTION

The highest expression of the protection of the personal privacy reaches its heights in our days in the development which the legal ordering in general and the criminal law in particular, do of the entry in other people's dwelling against the will of the inhabitant. The protection of the address, the dwelling, or the establishment opened to the public supposes different forms to approach these entries, and they also have different answers based on the level from affectation of the legally protected interest, and the legal perspective from which the study of the action focuses.

Although this protection is formed of systematic way and widely in the modern criminal law, we cannot forget that the home has been the object of legal protection from the beginning of the Law system.

Some authors have seated the precedents of the figure of the unlawful entry in the *Lex Cornelia de injuriis* (Pellise, 1950, p. 616). For others it is in the Middle Age when a new and more ideal notion of unlawful entry begins, due to security reasons more than ideological one (Quintano, 1972, p. 951).

In the Spanish *Foral Law*[1] we have wide samples of this concept of protection of the dwelling that is formed like a possibility of the subjects to be against to the invasion of its last sphere of personal security, reduced to their own home. Of course, the protection was settled down *erga omnes* (Valmaña, 2009, p. XXX), which includes to the own servants of the king in the exercise of its functions. The *Fueros* de Najera and de Leon are good examples of this assert (Valmaña, 2009, p. XXX). In the Fueros de Logroño, from Leon, bestowed by Alfonso IX in 1188, and Cuenca, bestowed by Alfonso VIII towards 1189, there are also one exempts of responsibility for the inhabitants who killed their aggressors when acting against a unlawful entry (Alcalá-Zamora, 1969, p.281), and in the *Fuero* from Najera (1304) appears the comparison (De La Iglesia, 1996) between the inviolability of the particular homes and the palace of the king.

This protection reaches sometimes not only to the inhabitant, but also whom takes refuge in other people's dwelling, constituting this way a peculiar "right of asylum". A good example of this is contained in article 7 of the *Fuero* of Cuenca (Valmaña, A. 1978, p. 75) under the protection of the *law of inviolability of the houses*

Within the evolution that the protection of the figure presents in the Spanish historical Law, we can indicate that the Germanic tradition gives priority to the freedom of the individual over any other consideration, as are considered in the Spanish local *Fueros* (Orlandis, 1944, p. 107-161). This is the key at the time of considering the sanctity of the home and, therefore, its protection (*pax domus*), as opposed to the legal tradition which drinks in the sources of the Roman Law and that impregnates the *Partidas* of Alfonso X, nicknamed *The Wise* (Quintano, 1972, p. 951 and 952), that only refers to the house "in generic form" (Cerdá, 1967, p. 40)

The practical disappearance of the unlawful entry crime after *Las Partidas* is not momentary or accidental. It´s due to the primacy of the figure of the King, in front of which submit the rights of the individuals. The limits to the King´s authority and the appearance of an authentic domiciliary inviolability do not take place until the Spanish Codification. It appears in an indirect way in the Criminal Code of 1822, but it has not been until the Criminal Code of 1848 when appears a classic formulation of the crime of unlawful entry, that would remain with slight modifications until the Criminal Code of 1995 (LOPEZ-BARJA DE Quiroga, Rodriguez-Ramos, Ruiz, 1988).

Throughout the Spanish Criminal Codification, however, the reference that specifically is made to some establishments, that nowadays we would describe like hotel categories (coffee bars, taverns, inns) and other not specific (or, at least, euphemistic known as public houses), sets a specific regime about the violation of such places. However, the protection of the deprived room in the Spanish Law, with an exclusive and excluding use in these public establishments on the part of the client, is not express nor specific, and will be, as we will see ahead the more, indoctrinates and jurisprudence the ones in charge to elaborate the inclusion of the conducts that constitute the general category of the unlawful entry.

---

[1] In the Middle Age, the most important cities in Spain had a *Fuero*, given for the King to regulate the citizen relationships.

2. THE LEGALLY-PROTECTED INTEREST AND THE CONSTITUTIONAL RIGHT TO THE INVIOLABILITY OF THE DOMICILE

The question relative to which is the legally protected interest in the unlawful entry crime is far, still nowadays, of being pacific. At a first moment, it is the security preferential social interest in the unlawful entry crimes, the *pax domus* protected by the Germanic Law and the Spanish Fueros, referred in the previous pages. This consideration stays during the Spanish Codification, almost to the present time. Thus, until the current Criminal Code of 1995, that includes the crimes of unlawful entry grouped with the crimes of discovery and revelation of secrets under the heading of Title X, Book II of the Criminal Code "Crimes against the privacy, the right to the own image and the inviolability of the domicile", these crimes were located between the "Crimes against the freedom and the security", among a maze of different penal types as the unlawful imprisonment, the abduction of minors and the abandonment of family or children, the discovery and revelation of secrets, or the threats and coercion.

This conception of the legally protected interest is, without a doubt, inheriting of the medieval conception of the house as a sanctuary (Valmana, 2009, pp. 330-331), the unique and last redoubt of security in tumultuous time. Nevertheless, the security as legally protected interest goes gradually being replaced in the Spanish doctrinal conception by the privacy concept, even under the Criminal Code, Refunded Text of 1973. However, this formalist conception of the legally protected interest like protection of the security, coexisted with other proposals, welcomed with unequal fortune. Thus, the consideration of the freedom of the will as the legally protected interest in unlawful entry crimes, defendant in Spain by JASO (in Antón, Rodriguez-Muñoz, 1949) specially, was criticized widely by QUINTANO (Quintano, 1972, pp. 941-942), who understands that it is the real quality of the object, concretely the dwelling and its assimilations *ex-lege*, which characterizes objectively to the unlawful entry crime"; it is not enough the exclusion of others through any average expressed by the holder (signs like no entry, reserved the admission right, and so on) in order to consider as a dwelling house what it is not. In a similar way, the proposal of the personal freedom, in the aspect of *objectified* personal freedom (Quintano, 1972, p. 954), or of *located* personal freedom (Suarez-Montes, 1968, p.862-900), that grants a dynamic protection to a physical space not *per se*, but "in attention to the functions that fulfills" (Larrauri, 1984, pp.297-298), is majority forming as legally protected interest of these crimes, opening the way to the consideration (Larrauri, 1984, p. 298) of the privacy as the legally protected interest of this crime (JORGE, 1987, pp. 26 ff.).

All these considerations are perfectly acceptable after the approval of the Spanish Constitution of 1978 that, in its article 18, 2 °, consecrates the inviolability of the domicile as a fundamental right: *The domicile is inviolable. No entry or search will be done without consent of the holder or the judicial resolution, except in case of flagrant crime.*

In order to determine the legally protected interest in the unlawful entry, we must study necessarily the constitutional elaboration of the domicile concept, to determine if this one serves at the time of filling in the content that legally protected interest. The real thing is that, although the concepts of domicile and home are not necessarily equivalent, exists certain consensus in the majority constitutionalist doctrine, at the time of establishing the domicile concept, in looking for in the Criminal Law the valid criterion of interpretation for that one (Alonso, 1993, p. 78 ff.).

In this sense, the Constitutional Court has been pronounced in a reiterated and categorical way, when establishes that there is not a total identity between the constitutionally protected domicile with the concept of domicile settles down in the Civil Law.

On the other hand, for some authors not even legally protected interest, that underlies in the protection that the Constitution grants to the domicile, can be led back, without any consideration, to the privacy as the legally protected interest (Alonso, 1993, p. 82 ff.). Despite of that, the penal doctrine has been aligned in the position that leaves from the constitutional concept of domiciliary inviolability to set the legally protected interest in the unlawful entry (Lamarca, Alonso, Gordillo, Mestre, Rodriguez, 2008, p.193). Thus, the Constitutional Court affirms that even though the distinction that the Constitution makes between the right to the personal and familiar privacy (art.18,1) and inviolability of the domicile (art. 18,2), this does not exclude the consideration of the domicile as one of the spheres where that protection of the privacy becomes a reality.

The Sentence of the Constitutional Court[2], Plenary, number 10/2002 of 17 January (RCC \ 2002 \ 10), establishes the limits of the administrative performance in relation with the entry and search warrant and forms an elaborated constitutional concept of domicile that establishes *that the protection of the domiciliary inviolability has instrumental character respect to the protection of the personal and familiar privacy (SCC 22/1984, of 17 of February) although this instrumentality is not an obstacle to the autonomy that the Spanish Constitution recognizes both rights, being distanced to the unitary regulation of such rights that contains art. 8.1 of the European Agreement of human rights*

In other words, which is protected with the inviolability of the domicile is a determined space that, nevertheless, is not specifically defined by the Spanish Constitution, and it is not identified with the definition that it is used in the sphere of the Civil Law, *especially in art. 40 of the Civil Code, as a point of location of the person, or the place for the exercise by this person of his rights and obligations"*; in the other hand, the constitutional concept of domicile has *"greater amplitude than private or administrative legal concept" (SCC 22/1984, of 17 of February; 94/1999, of 31 of May), and does not admit reductionist conceptions... like this that compares it to the Criminal concept of habitual dwelling or room" (SCC 94/1999, of 31 of May).* (Constitutional Court Sentence (Plenary), number 10/2002, of 17 of January).

In this way, we can affirm that from the combination of the constitutional right to the inviolability of the domicile, with the legally protected interest *privacy*, and, still more, from the landmark of that one through this one would give the interpretation mainly defended by the Spanish doctrine (JORGE, 1987, p. 42). However, the existence of this relation cannot be interpreted "in terms of strict dependency" that could lead to affirm so colorful questions as the illegal entry in a domicile will not constitute violation of this one if the privacy of its occupants (FIGUEROA, 1998, p.101) has not been interfered with.

Nevertheless, some critics (SANCHEZ, 1998, p. 46 ff.) have indicated problems at the time of accepting this interpretation. The most obvious is the own existence in the Criminal Code a "specific penal protection" (SANZ, 2006, p. 29) of the constitutional right to the inviolability of the domicile, in the Title XXI, under the heading, *Crimes against the*

---

[2] SCC in advance.

Constitution, in article 534 of the Criminal Code, "within the frame of the most generic right to the privacy" (Toledo, 1987, p. 326).

In the last years a position has resurged with force that, unlike the majority one, pleads for the return to one more formal conception of the legally protected interest in the unlawful entry, and that is based on the *right of exclusion* ("ius exclusionis") of others, of a certain "*spatial scope*" (SANZ, 2006, p. 29; Oliver, 2008. p. 839), which somehow supposes a return to the inclusion of these crimes in the category of *crimes of contempt of the will* (SANCHEZ, 1998, pp. 50 ff.), in line with German doctrine (SANZ, 1989, p. 319 ff.)

3. THE CRIME OF UNLAWFUL ENTRY. THE CONCEPT OF DWELLING HOUSE IN THE TOURIST ESTABLISHMENTS

At sight of the little pacific doctrine at the time of analyzing the different possibilities that the configuration of the legally protected interest offers is unusual the practical unanimity from doctrine and jurisprudence at the time of determining the object of the crimes object of this study (articles 202-204 of Spanish Criminal Code, 1995).

Quintano (1972, p.955), indicates successfully that the dwelling-house concept is more a fact notion rather than a law notion; this is the reason why it is difficult to elaborate a concept of dwelling house without determining previously what requisite forms it:

First, it must be a "closed or open, or partly separated space of the outer world, in conditions such that make patent the will of the inhabitants to exclude the others" (Suárez, 1968, p. 869). From old times, the jurisprudence has admitted application of the figure of the unlawful entry of dwelling houses in the cases that a person introduced himself in a house inhabited through a half-closed door, or even totally opened one (Quintano, 1972, p. 958).

In this sense, so commented SCC of 17th of January of 2002 it establishes that "the essential characteristic that defines the domicile negatively, delimits the spaces that cannot be considered such domicile: in a hand, those in which it demonstrates, in an effective way, that they have been destined to any activity different from the private life, whatever was this activity a commercial, cultural, political activity, or of any other nature; on the other hand, those that, by their own characteristics, never could be considered apt to develop in them the private life, this is, the open spaces. So, it is necessary to set that, although all closed space does not constitute domicile, it is possible to be a dwelling house even if it´s circumstantially open; nevertheless, it is consubstantial to the notion of private life and, therefore, to the type of use that defines the domicile, the delimited character respect to the outside of the space in which it is developed. The own instrumental character of the constitutional protection of the domicile against to the protection of the personal and familiar privacy demands that, independently of the physical configuration of the space, their external signs reveal the clear will of their holder to exclude this space and the activity developed inside from the knowledge and interferences of others" (SCC, Plenary no. 10/2002, January, the 17th, 2002). As we see, in this Sentence are included some essential elements of the dwelling concept, which therefore, we must consider common to both constitutional and criminal concepts of domicile.

The nature of personal or real property of the dwelling-house is indifferent to be deserving of the protection that confers these criminal types (JORGE, 1987, pp. 48-49), as long as their use is present (Suárez, 1968, p. 870). In this sense, the idea of *spending the night* (Quintano, 1972, pp. 957 ff.) as the characteristic element of the dwelling is remarked by the doctrine not as much in the literal sense to demand that it is necessary in fact to spend the night in the dwelling-house, but is enough to grant such consideration that the fact to spend the night, is the destiny given to such physical space (JORGE, 1987, p. 51). This allows including the concept of dwelling-houses those temporarily uninhabited like temporary inhabited spaces (SANZ, 2006, pp. 46 ff.). In the same way, repeated jurisprudence establishes that, according with article 8,1 of the European Agreement of Rome of 1950 (RCL 1979, 2421), could be considered dwelling-house any place, whatever humble and precarious it was the construction, in where the person, the people or the family live, even if it's a temporary residence, from the "roulot" (sic), the tent or the shack, to greater of the palaces" (Sentence of the Supreme Court, Court Criminal no. 1448/2005, November, the 18th), as well as the staterooms of a boat (SSC Criminal Court no. 178/2006 of 16 February).

The main destination of the dwelling must be the "development of the activities of private life" (SUÁREZ, 1968, p. 870), in its strictest sense (SANZ, 2006, p. 46), compared to other authors who consider including in the private life concept both aspects of professional and family life (JORGE, 1987, p. 49). The Supreme Court has been delimiting negatively the concept of dwelling-house excluding of this premises and other dependencies that have not such consideration, including a garage (SSC, Criminal Court, no. 32/1995, January, the 19th; SSC no. 2057/1994, November, the 22nd), a "*zulo*"[3] (SSC, Criminal Court, no. 1151/1993, May, the 22nd), the cell of a prison (SSC, Criminal Court, no. 1165/1995, November, the 24th), a bar (SSC, Criminal Court, no. 1775/1993, July, the 9th), a video-store (SSC, Criminal Court, no. 327/1994, February, the 21st). or any other commercial premises (SSC, Criminal Court, June, the 11st of 1991; SSC of June, the 19th and October, the 5th of 1992; and SSC of September, the 16th of 1993), an industrial building (SSC, Criminal Court, no. 1225/1995, December, the 1st), a room rented for hours in a hostel (SSC, Criminal Court, no. 68/1995, January, the 28th), the locker in a barracks (SSC, Criminal Court, no. 55/1995, January, the 26th), a box room or another small space not suitable for housing (SSC, Criminal Court, July, the 3rd, 1993), the common elements of a building, specifically a lift (SSC, Criminal Court, no. 379 /1996, April the 30th), and not inhabited house (SSC, Criminal Court, no. 139/1995, January, the 31st).

A special consideration deserves the couchette of a train. The Sentence of the Spanish Supreme Court of December, the 28th, 1994, establishes that to have protection as a domicile, two elements are necessary: the absolute privacy of the activity developed inside and the ability to exclude others to enter in the private place.

Neither of those two essential notes are given in transitory occupation by a person of a berth in a couchette, collective and shared with other travelers, which even have no relationship of treatment or knowledge among themselves, just to be transported in a convenient way from one place to another by train. The transience of the occupation or the nature of mere passive and collective transport does not let you to develop an intimate activity in such a place, and the traveler has not the power to exclude others of the area occupied by, because it is compelled to share it with them, even changing in the course of the journey depending on the place of departure and destination of the travelers, or cannot prevent the entry of officials of the Railway Company, already for accommodation of other travelers,

---

[3] It is a hidden and closed place to conceal illegally items or kidnapped people.

already for control of banknotes, or for any other activity of their job, nor can finally, prevent visits for the compartment colleagues to those who come to see them, or even other train passengers.

Finally, the use of the dwelling-house must be legitimate to be subject to the criminal protection (Jorge, 1987, p. 50), no matter if the origin of this legitimacy is a legal relationship or a factual situation which the law recognized legal effects (Suarez, 1968, p. 871).

At this point, we are able to determine the scope of criminal protection of temporarily inhabited spaces in tourist accommodation. In the jurisprudence of the Supreme Court, SCC of 17 January 2002, we find the concept of rooms in hotels and guesthouses (SSC, Criminal Court, no. 1900 / 2 November 1994), as a domicile for the purpose of the crime of unlawful entry of dwelling-houses, and even in relation to the entry and search cases. Thus, the entry and search without a warrant in a hotel room infringe on right to the inviolability of domicile.

QUINTANO (1972, p.960) points the necessity of setting the difference of the places, for common use, such as reception, hallways, corridors or lounges or meeting rooms, ballrooms and so on, and the rented apartments, guests rooms, or private rooms for owners or employees; the first are essentially public, and the second could be object of the unlawful entry crime. In reference to the room for travelers, however, it´s assumed a tacit authorization to the establishment staff in order to the entry. Cause that, they are the keeper of the keys and other devices to the effect. About this question we will return later.

At the time of determining the quality of the room of a hotel as a domicile, the SSC no. 684/1993 of 17 of March indicates, in relation to the entry and search, that everything what serves as dwelling has to be had as domicile, even a vehicle destined to serve as accidental dwelling, i.e., also the named *"roulottes"* and, of course, the room of a hotel, a residence, a house of guests, because, in another case, the person that has not any kind of house as a home, would never have protected his privacy.

4. THE WILL OF THE INHABITANT OR HOLDER OF THE DWELLING-HOUSE AS AN ELEMENT OF THE "UNLAWFUL ENTRY" CRIME

The consideration of the room of hotel as a domicile, and the extension of the constitutional and criminal protection that enjoys this one are completed with the necessity of an express consent or, at least, an unquestionable one. In this sense the SSC no. 204/1995 of 15 February is pronounced when, in occasion of a drug trafficking, the Police got into a room of the Hostel *Dora*, with the announce of the owner, but without the express consent of the inhabitants of that room. Of course, there were no entry and search warrant (art. 558 Spanish Code of Criminal Procedure), neither flagrant crime (art. 18 of the Spanish Constitution). This Sentence denies the position established by the SSC no. 2043/1992 of 5 October, according to which, the tacit consent does legitimate a search conducted in a room of hotel without the corresponding warrant and must be admitted, when, having opportunity for it, the subjects are not against to this entry and search. Nevertheless, as the mentioned Sentence, of 15 of February of 1995, before indicates, to understand the consent granted, this one must be express, and reliable, with the purpose of avoiding the defenselessness of the subjects, that or do not dare to be against to a police requirement or do not know the rights that correspond to them. This express consent it must be only understood related to the entry and search in the room of the hotel establishment, or any other susceptible to be considered dwelling-house, to the effects of the crime of unlawful entry.

It is possible to affirm (Quintano, 1972, p.960) that such tacit consent exists to permit the holder of the hotel establishment, or hotel housekeeper, the access to the room occupied by the guest when this one is not in it, to carry out the usual or necessary cleaning or maintenance, and as long as an express exclusion of this entry has not been made.

The problem is declared when the guest refuses systematically to allow the entry into his room, preventing to the holder the workings of cleaning, conservation and inspection. Moreover, it could be possible, in the cases of non-payment, the subject opposition to the eviction alleging that such hotel room constitutes his dwelling.

The first element to consider is in the legitimacy requirement to appreciate a juridical dwelling concept. It is certain that in the cases of illegal squatting of the premises or uninhabited houses, we have even observed a gradual increase of the legal protection that prevents, if firm judicial resolution does not occur, the eviction of those who, by lack of legitimacy, would not have to be able to obtain such protection that the Constitution or the Criminal Code grants respectively to the domicile and the dwelling, unless in the first moment in which this place has not became a dwelling.

In the other hand, in the cases in which, being legitimate the residence in that place at the beginning, stops the legitimacy cause, or even in the cases in that such legitimacy of origin does not exist, a judicial, or at least administrative, resolution becomes necessary to come to the eviction. The first case could be, for example, the cases of benefit of a house linked to a labor relation (Sentence no. 538/2002 of 20 December, of the Social Court of Seville no. 4); the second one, the cases of "squatting" of buildings or abandoned premises, that, although it could be an usurpation crime in Spanish Law, an administrative file or a judicial order is needed to come to their eviction (Sentence of the Provincial Court of Madrid, Section 15th, no. 363/2003 of 10 September).

In this sense, the Sentence of the Constitutional Court, of 2 of November of 2004 (SCC, Court 1st, no. 189/2004), establishes that the rooms of the hotels deserve the domicile consideration, and therefore recognizes the right of the guest the inviolability of his domicile, comparing this situation to the case studied by such Sentence, about the eviction of a person from a military residence.

This Sentence analyzes the *modus operandi* of the Military Administration at the time of carrying out the eviction, indicating that, although this one took place in fulfillment of the Residence Policies, the Administration must respect the fundamental rights of the passive subjects of its execution proceedings, in such a way that when the entry or the search in the domicile is needed, it will be precise to carry out the requirements of the article 18th of the Spanish Constitution.

The own sentence, when establishes the parallelism between the room of a residence and the room of a hotel, is solving the question set out here, but only in part. It is clear that the resolution of the Constitutional Court raises the question from the point of view of the inviolability of the domicile, and therefore with the object of the protection of such a right. On the other hand, the unlawful entry of dwelling-house, as it is said previously, is only one of the possible ways to protect such legal interest. However, the content of both rights do not have to agree necessarily, and in fact they do not. Furthermore, the legitimacy of the title held by the inhabitant is a fundamental requirement for the existence of the crime of unlawful entry of dwelling-houses. Therefore, it is possible the existence of a crime of coercion or duress, if it is committed by individuals, or against the inviolability of the domicile or of prevarication, if it is committed by a government official, but nevertheless it could be a crime of unlawful entry of dwelling-house, by the absence of that requisite.

Finally, the tourist legislation, that at the present time has a regional character in Spain, establishes the possibility of the holder of the tourist establishment of having the accommodation unit, once exceeded the period by which it was contracted, but in different manners and degrees.

So, the article no. 15 of Decree 205/2001 (Castilla-La Mancha, Spain) of 20th of November, establishes, in order to set the duration of the contracted services, that the prolongation of them will do only with mutual agreement between the establishment and the client, and if agreement did not exist to prolong the stay, the establishment will be able to have the accommodation unit. The Decree 19/1997, February, the 11st (Valencia Community, Spain), in its article no. 7, expresses in the same sense. And the article 37 of the Law of Tourism of Catalonia (Law 13/2002, of 21st of June), in relation to the right of access to the tourist establishments affirms that the tourist establishments have the consideration of the public premises. The access and the permanence in the tourist establishments can be conditioned to the fulfillment of regulations of use and internal policy, as long as these regulations are not against the dispositions of this Law. The rules must be also announced of very visible form in the joining points to the establishment. The holders of the tourist companies ask for the aid of the agents of the authority to evict of a tourist establishment the people who fail to fulfill the usual rules of social coexistence and which they try to enter with purposes different from the peaceful enjoyment of the service that is lent or the activity that is developed.

On the other hand, the article 27 of Decree 75/2005 (Community of Murcia, Spain), of 24th of June only sets the obligation of the client to leave the unit of accommodation once passed the agreed to period of occupation, but without arbitrating a correlative right of the proprietor to evict the client. In the same sense, the 15 of Decree 47/2004 (Community of Andalusia), of 10 February, says that the right of the client settles down to remain more time of initially contracted in the hotel establishment "whenever agreement between the parts exists"[4].

The existence of all these norms in no case abolish the consideration of domicile given to the hotel room and similar units of accommodation, but they could constitute a negative element of the type, the legitimate exercise of a right, that, by another way, would exclude the consideration of these facts as a crime of unlawful entry of dwelling-house.

The necessity to make compatible the protection of the privacy in the tourist establishments, and the effectiveness in the reply in view of the possible conflicts that could produce, makes a specific regulation, at national level, desirable, to homogenizer and to make agile the evictions in cases like sawn.

---

[4] As well is in the Ministry of Commerce and Tourism Order, of 15th of September, 1978, art. 10; Decree 53/1994 (Catalonia), of 8th of February; Decree 193/1994 (Aragón), of 20th of September, art. 15.

REFERENCES

Alonso, A. (2008): *"Delitos contra la intimidad, el derecho a la propia imagen y la inviolabilidad del domicilio"*, in Lamarca, C. (coord..)/Alonso, A./Gordillo, I./Mestre, E./ Rodríguez, A.: "Derecho Penal. Parte Especial", COLEX, Madrid.

Alonso, A.L. (1993): *"El derecho a la inviolabilidad domiciliaria en la Constitución Española de 1978"*, COLEX, Madrid.

Cerdá, J. (1967): "Consideraciones sobre el hombre y sus derechos en las Partidas de Alfonso X el Sabio", Publicaciones de la Universidad de Murcia, Murcia.

De la Iglesia, V. (1996): "Contenido y significación del Fuero de Nájera",inhttp://www.vallenajerilla.com/berceo/rioja-abierta/fueros/fuero.htm

Figueroa, M.C. (1998): *"Aspectos de la protección del domicilio en el Derecho español"*, EDISOFER, Madrid.

Fuero de Logroño: www.geocities.com/urunuela27/fuero_de_logronio.htm

Fuero de Nájera: www.geocities.com/urunuela27/fuero_de_nagera.htm

Fuero de Oviedo: http://el.tesorodeoviedo.es/index.php?title=Fuero_de_Oviedo

Jaso, J., in Antón, J./Rodríguez Muñoz, J.A. (1949): *"Derecho Penal. Parte Especial"*, T. II, Universidad de Madrid, Madrid.

Jorge, A. (1987): *"El allanamiento de morada"*, TECNOS, Madrid.

Larrauri, E. (1984). *"Allanamiento de morada y derecho a la vivienda"*, in Cuadernos de Política Criminal, nº 23, p. 291-309.

López Barja de Quiroga, J./Rodríguez Ramos, L./Ruiz de Gordejuela, L. (1988): *"Códigos penales españoles (1822-1848-1850-1870-1928-1932-1944). Recopilación y concordancias"*, AKAL, Madrid.

Muñoz, T.: "Colección de Fueros Municipales y Cartas Pueblas", Tomo I, 1847, in: www.geocities.com/urunuela27/fuero_de_nagera.htm

Oliver, F.M. (2008): *"Allanamiento de morada"*, in Arnaldo, E (coord.) et alt.: *"Enciclopedia Jurídica"*, T. II, La Ley, Madrid.

Orlandis, J. (1944):"La paz de la casa en el Derecho español de la Alta Edad Media", Anuario de Historia del Derecho Español, Vol. XV, p. 107-161.

Penal Code, D. 3096/1973, de 14 de September (M. of Justice), by which the Penal Code is published, text refunded according to Law 44/1971, of 15 of November: http://www.boe.es/aeboe/consultas/bases_datos/doc.php?coleccion=iberlex&id=1973/01715.

Quintano, A.: (1972): *"Tratado de la Parte Especial del Derecho Penal"*, Tomo I, vol. II. EDERSA, Madrid.

Sánchez, M.B. (1998): *"Análisis del delito contra la inviolabilidad del domicilio del artículo 534 del Código penal"*, COMARES, Granada.

Sanz Morán, A. (1989): *"Algunas observaciones sobre el delito de allanamiento de morada"*, in Libro Homenaje al Profesor Rodríguez Devesa, Estudios de Derecho Penal y Criminología, T. II, Madrid, p. 313-344.

Sanz Morán, A. (2006):"El allanamiento de morada, domicilio de personas jurídicas y establecimientos abiertos al público", Tirant lo Blanch, Valencia.

Suárez Montes, R.F. (1968): *"El delito de allanamiento de morada"*, in RGLJ, nº 225, p. 862-900.

Toledo, E. O. de (1987): *"Revisión de algunos aspectos de los delitos contra la inviolabilidad del domicilio (art. 191 del Código penal)"*, Anuario de Derecho Penal y Ciencias Penales, T. XL, Fascículo II, p. 319-345.

Valmaña, A. (1978): "Fuero de Cuenca (Introducción, traducción y notas)", Editorial Tormo, Cuenca.

Valmaña, S. (2009): *"La protección penal de la habitación en los establecimientos turísticos. Concepto de domicilio y de morada"*, in Principales tendencias de investigación en turismo, Septem Ediciones, Oviedo, p. 325-348.

In: Research Studies on Tourism and Environment
Editors: J. Mondejar-Jimenez et al.
ISBN: 978-1-61209-946-0
© 2012 Nova Science Publishers, Inc.

Chapter 17

A HISTORICAL VIEW ABOUT THE LIABILITY BY SAFEKEEPING IN THE PARKING BUSINESS

Alicia Valmaña Ochaíta[*]
University of Castilla-La Mancha, Spain

ABSTRACT

The contract of parking has been regulated a few years ago in the Spanish Law, in particular in Act 40/2002, of 14th November, which tried to arrange a business activity more and more frequent from mid of the last century.

With the Act, was solved the existing scientific and jurisprudencial discussion about the juridical nature of the parking -deposit or rent-, and was provided a duty of vigilance and safekeeping connected to the obligation of restitution of the vehicle; however, the Act creates problems on its own scope and the extension of the liability.

In this chapter this question will be approached from the perspective of the Law in force and from the protection that the Roman Law gave to similar situations to the present ones, saved the distances, as it were the case of *stabularii*, owners of stables in which the cavalries could be left by a period of more or less long time, for reasons of terrestrial transport.

1. INTRODUCTION

The development of a society causes the birth of new commercial relations to which it is to be given an answer from the legal point of view; the time that passes between the birth, development and expansion of a new institution and its regulation is, normally, enough ample so, is frequent that, before new economic realities, the scientific doctrine and the jurisprudencial was the ones that take the first word at the time of describing this relation, once analyzed its legal nature[1].

---

[*] Contact author: alicia.valmana@uclm.es
[1] In relation to doctrinal and jurisprudencial status quaestionis, also with the controversy about the denomination of the contract, vide by all authors, Díaz Gómez (2004); Ballesteros de los Ríos (2000); Álvarez Lata (1997);

This is what it has happened in relation to the contract of parking in Spain where, firstly has been a long discussion, and secondly, its regulation in Act 40/2002 of 14th November, modified partially by Act 44/2006 of 29th December, about improvement of the protection of the consumers and users, which although have put in clear the applicable regime, do not enter the qualification of its legal nature.

Nevertheless the discussion around its legal nature is not trivial. We are before one of those economic realities, so frequent in our days, that participates in the characteristics of several typical contractual figures and that have given rise in the civil doctrine, fundamentally, to speak about atypical contracts, complex contracts, mixed contracts and about the different theories that can provide to them with a concrete regulation. For that reason, its normative regulation have not closed the discussion; it is logical: the Law is something more than strict legislative technique and the correct insertion of the new figures, contractuals in this case, in the legal system, demands a theoretical formulation of the institution taking care of its fundamental characteristics, especially in the cases, like which occupies to us, in which the used technique has not been the best.

I do not try in this chapter to make an analysis of historical antecedents of the parking contract; evidently, the life of this figure is very short and, in any case the jump from the Roman Law to the present Civil Law, immediately, without entering the intermediate Law, is probably an error; neither is to look for remote explanations, like archaeological rests, of figures that basically coincide with the premises of contract fact, in a kind of vindication of *nihil novum sub sole*; in this case, I only want to write a reflections about how certain concepts, yesterday and today, are so difficult to insert correctly: I am talking about to the liability by safekeeping.

2. SPANISH LAW IN FORCE

The regulation given by Act 40/2002 starts off with the idea that the fact of the "parking" can fit in different contractual figures, but the *"excessive quantity of the phenomenon"* of the parking of the motor vehicles is the one that deserves a specific legislative treatment in relation to other figures that will be governed by typical contracts. The *"excessive quantity"*, *"the strains"* of a type of parking, the public parking, and, *sensu contrario*, the *"smaller importance"* of other figures determines the exclusion of other "parkings" that, according to the Explanatory Preamble, *"can be dealt with the arrangement of other contractual figures"*. Although the Explanatory Preamble of the Act does not mention specifically to other reasons, it is evident that we are in front of contractual figures whose contents of the regulation of interests of the parties is determined, on the one hand, by the exercise of a mercantile activity of a business and, by another one, by the necessary protection to the consumer.

It is, in that measurement, that the "fact of the parking" appears with a double perspective of typical contract (Álvarez Lata, 2003, pp. 3 y 4), inasmuch as regulated in relation to the "public parking", and atypical contract, since the rest of parking does not deserve this specific regulation. Thus, the content of Act 40/2002 is the regulation of the contract of "public" parking that is conceptualized as is indicated in its art.1.1:

Martín Santisteban (2003); López Barba (2003); Herrada Romero (1994) and Espiau Espiau & Mullerat Balmaña (1996) (some of them, before the Act comes into effect).

Art.1.1. "This Act establishes the legal regime applicable to the parking in which a person yields, like mercantile activity, a space in a site or enclosure of which is holder, for the parking of motor vehicles, with the vigilance and safekeeping duties, during the time of occupation, in exchange for a price determined based on the real time of benefit of the service".

This dual vision of the "fact of the parking" has carried to exclude from the Act other cases very close to the "public parking", using the terminology of the Explanatory Preamble of the Act, as those which, not being gratuitous, does not mediate price, but rate (parking in zones ORA); those which are direct or indirectly gratuitous or, generically, all those which do not reunite the requirements of art. 1.

This writing is fruit of the legislative reform that underwent Act 40/2002 in 2006, due to the *"improvement of the protection of the consumers and users"* -as the title of the very Act 44/2006 says-; this norm, among other articles, modifies the 2nd, in its section b), in which, in its former redaction of the 2002, it excluded the parking that are made *in sites or dependent or accessory enclosures of other facilities, or that is gratuitous.*

The idea of the repayment as a result of the existence of a mercantile operation is, obviously, fundamental, and it seems that the reform of the 2006, besides to increase the protection of the consumer where there were *"diverse deficits of protection"*, went directed to incorporate supposed that the literal tenor of the 2002 text had excluded like, for example, the tied parking to commercial establishments, Hotels, Supermarkets or establishments when they lent it, as a service more of attention the client or, in other words, not directly onerous.

The redaction of art. 2, section b) non aid to the clarity, when says *"non-repaid parking, directly or indirectly"*, and excluding them from the scope of the Act. It seems that with the expression of *indirect non-repayment,* one talk about the parking non-repaid by the user nor by a third (Bercovitz Rodríguez-Cano, 2002 p. 12), but also to those parking in which nobody could think, in any case, that a payment or repayment exists, although was an indirect payment, so that, *sensu contrario*, would have been included within the scope of the Act those *"repaid parking direct or indirectly"*. In this way, it is possible understand that the parking of Hotels as of Supermarkets by which a certain and individualized price is not paid, but understood like an another service that the Hotel or the Supermarket render, they would be, really, parking repaid indirectly, due to the one that parks is a client effective, and non-potential, who paying the amount of the bill or the invoice by the lodging, pays, to a certain extent, the parking of his vehicle in the enclosure qualified for such effect by the commercial establishment or hotelkeeper (SAP of Málaga, 26th September 2000; with bibliography and jurisprudence, Represa Polo, 2004, p. 119ff).

Really, if the legislator wanted to underline the onerous character of the contractual relation, the discussion will have to be centered in what is considered gratuitous indirectly or, following the negative formulation of the article, non-onerous indirectly (art. 2.b); but, saving the onerous character of the relation, the ulterior issue will be to explain how these services rendered by Hotels and Supermarkets, without additional cost, can be inserted in art. 1; that is to say, if they can be excluded of the scope of the Act *ex* art. 2, c) fundamentally in relation to the money compensation of the user which must be *a price determined based on the real time of benefit of the service* (art.1.1) where there would be to integrate the indeterminate price by the benefit of the service in the determined price of the total invoice that, as well, it does not vary if the service is really lent or not.

But, solved "the deficit of protection" of the consumer with the modification of the 2006 when establishing payment by the real time of parking in the modality of rotating parking, the fundamental question in the matter of parking contract, is the regime of liability of the entrepreneurs of the parking by the vehicles introduced in them and their components and accessories in the measurement established by art. 3[2] (Carrasco Perera, 2002, p. 3) and, in this sense, the critics to the regime of responsibility consecrated in art. 5.1 in relation to art 3, which establish the obligations of the parties, and the 1, in its deficient reference to *"the vigilance and safekeeping duties during the time of occupation"*, go on in force.

The civil doctrine has showed its disagreement with this legal exposition fundamentally in two directions: in the first place, the inclusion of the reference to "the vigilance and safekeeping duties" of the entrepreneurs of the parking in art. 1 supposes a serious error when gathering the content of the contract within its definition (Álvarez Lata, 2003, p. 4.). If the parking regulated by Act 40/2002 is these which content is the cession of a space for the parking with the vigilance and safekeeping duties, it is enough, then, that the entrepreneurs of the enclosure imposes that its contract does not incorporate such to have, to conclude that the Act is not applied (Carrasco Perera, 2002, p. 3; in the same direction, Álvarez Lata, 2003, p. 4).

It has been, probably, the desire of emphasize the obligation of safekeeping in the parking as essential element of a contract regulated by the Act, with general character, the thing that has generated the effect opposite, and that opens the possibility that certain parking leave the scope of the Act (Carrasco Perera, 2002, p. 3). Whereupon we would be with, to the doubtful assumptions because of the difficult or impossible determination of the price based on the real time of parking, others could be united in which the entrepreneurs of the parking excluded *a priori* the obligation from safekeeping. This would take to us, without a doubt, to an Act, practically, empty of content but, mainly, to an Act that allows, with its non application, the thing that tried to avoid: that the entrepreneurs of parking can elude the obligation of safekeeping of the introduced vehicles, against which had decided an important number of judgments.

And thus, although it is pleaded for an interpretation of the Act that does not avoid "its own aims" and, consequently, maintains the obligation of safekeeping like a "essential obligation" of the entrepreneur of the parking, "inherent to the cession of the space" (Álvarez Lata, 2003, p. 4), the certain thing is that, although it is clear what the Act has tried, not as much what the Act has obtained. In fact, an interpretation of this type is necessary because if not, the Act would not be applied, since, in principle, few entrepreneurs of companies of

---

[2] Article 3: Obligations of the holder of the parking: [...] c) To restitute to the carrier of the voucher, in the state in which it has been given, the vehicle and the components and accessories which are functionally built-in -of way it fixes and inseparable- to that one, and are habitual and ordinary, by its nature or value, in the type of in question vehicle. In any case, the non fixed and removable accessories, like radiocasettes and mobile telephones, will have to be retired by the users, not reaching, in their defect, to the holder of the parking the liability by restitution.

The entrepreneurs of the parking that count with a special service for it, will be able to also accept and to take on liability by the restitution of other accessories different from indicated in the first paragraph of the section 1.c) of this article, as well as of the introduced effects, objects or equipment by the user in their vehicle, when:

a) Specifically have been declared by the user to the entrance of the parking and the person in charge of this one accepts his safekeeping.

b) The user observes the prevention and safety measures that are indicated to him, including the one of the parking of the vehicle or the deposit of the effects, in the zone or place that will be qualified to the effect for its vigilance.[...]

parking would want to welcome the vigilance and safekeeping obligation, being able not to do it. From my point of view, it is clear that the Act tried, for those parking contracts regulated in it, to establish the obligation of the vigilance and safekeeping as essential for the entrepreneurs of the enclosure, but only has arrived to be implicit in the obligation of result that is constituted in section c) of art.3, -something that already had been seated by the Jurisprudence-; on the contrary, the concept of the contract of parking regulated by Act 40/2002 and consequently, its scope and/or exclusion has been complicated.

We are, in the best one of the cases, like before the promulgation of the Act, where the work of interpretation of the Jurisprudence gave channel to the exigency of liability by safekeeping of the entrepreneurs of the parking, although, from the being in force of the Act, a new work of interpretation will be needed that avoids the opposite effects of its literal application - or, rather, his non application- that could be tried by the entrepreneurs of parking businesses (Álvarez Lata, 2003, p. 4).

Nevertheless, we are not either in the previous situation; if the obligation of vigilance and safekeeping is considered inherent to contracts of public parking regulated by Act 40/2002 and like so, informer of the obligation contained in art. 3.1.c) about the restitution of the vehicle, this one is included in a more limited form, in some aspects, that had been considered by the *minor*[3] jurisprudence in its judgments. In this respect, liability for effects introduced in the vehicle or by its components and fittings has been limited with the new Act, respect the criteria usually used by the *Audiencias Provinciales* (by all authors, Gómez Calle, 2004, p. 174), mainly in relation to the Act, that expressly excludes liability for damages and robberies of effects, and limits the liability for damages and robberies of components and accessories based on different criteria -functionally incorporated, permanently fixed and non removable from the vehicle, and ordinary and customary, by its nature or value in the type of vehicle concerned, *ex* art. 3.1. c)-, and, specifically excludes in paragraph 2, no fixed and removable accessories.

Thus, when the Explanatory Preamble of the Act proclaims that one of its purposes is to tackle *"the imprecise regulation of the liability of the entrepreneur of the parking business in order to the return of the vehicle and its accessories or other effects, in terms that come to pick up and resolve criteria and concerns raised by the Jurisprudence"*, this has been materialized in a limitation of liability of the entrepreneur to exclude the effects, components and accessories referred to, in relation to the criterion laid down by the courts quite widely, which does not follow. On the other hand, it has been enlarged, according to some authors, because it has been attributed to the entrepreneur of a parking business a liability by damage caused by force majeure -he is forced to "return" the vehicle in the state that he received it [(art.3c)], and responds if does not fulfill this obligation (art. 5.1), (Carrasco Perera, 2002, pág. 3)-.

It seems that the regime of liability for damages and robbery will presenting a high factual casuistry, and so, jurisprudential, to which will apply a norm or another, including the Act 40/2002, either because the particular case fits into its article 1, either by analog interpretation, either by article 1258 C.c., which requires the parties to the fulfillment of all that which by their nature derives in the good faith and practice, either by the LCU. However, after the establishment of the legal regime of limited liability by robbery or damage to components and accessories and effects introduced into the vehicle, for companies engaged in

---

[3] Jurisprudence from de Audiencias Provinciales (High Provincial Courts), not Supreme Court.

the parking business, will be difficult to return to the previous legal interpretations if we do not want to make the situation of the cases excluded by the law more burdensome than to the parking business contracts.

In short, if the jurisprudential construction of contract parking was passing through the idea of the existence of a "public use" which is the content of an "enterprise operation" whose facilities are an "organized service" (Elguero y Merino, (1993), pp. 9 and 17) with a system watched over of access and exit of vehicles and the obligation of vigilance and safekeeping of the vehicles in order to return them in the same state, with the Act, it has been started off same budgets but, limiting liability, not embracing, in this sense, *the inherent duty of vigilance and custody of article 1,* to certain components and accessories, and in no case, for the effects –on a contrary way, SAP of Madrid, 14th January 1991, and 14th June 1992 cited by Elguero y Merino, (1993), p. 18. Also, the SAP of Pontevedra 26th May 2002–.

This allows to think that the inclusion of *"duties of vigilance and custody during the time of occupation"* in the article 1 and not in art. 3, where it ought to be placed if it had been configured as obligation, is not just a misfortune of the legislator, but an expressly chosen way. The use of the phrase *"duty"* instead of *"obligation"*, in my opinion, is not gratuitous, in this case.

It another way, only generate liability the non-fulfillment of the duty of custody as far as was accompanied by the failure to return the vehicle considered uniformly, that is, only extending liability for damage to components and fixed accessories, non-removable, usual and ordinaries accessories by their nature and value –with the problem of the "the legal incorporation" and not only "functional" of certain elements such eyeglasses or reflective triangle, elements that prescriptive way must be checked into the vehicle (López Barba, 2003, p. 110)-; thus if the effects that were inside of the vehicle are theft, the entrepreneur is irresponsible by the value of them, but will be responsible by the damage produced breaking the glass, or forcing a lock, for example.

When the vigilance and safekeeping are left outside of the article that regulates obligations, the emphasis is on the object of restitution: vehicle considered unitary and exclusively; the content of the obligation of the entrepreneur of parking business is, to the Act, fulfill the duties of vigilance and custody enabling the return of the vehicle in the state in which it was given. Traditional qualification of the duty of safekeeping as an obligation of means, with the consequent implications in matter of proof (the possibility of the debtor to demonstrate that there was no non-fulfillment by proving that he used due diligence) loses sense to impose on the entrepreneur the obligation to return it well and in the same state that it was entrusted. This is certainly an obligation of result (Martín Santisteban, 2003, p. 96 and 97).

Thus means the specific reference to certain accessories, like radio cassettes, the most controversial, to the effect that they could have a functional relation with the vehicle, to exclude the less problematic ones, suitcases, portfolios, etc...

3. THE ROMAN LAW

The confluence of a duty of safekeeping together with an obligation of return is not new in our law, or under Roman law, where we find that the safekeeping is contained in certain contractual relations[4] and that is going to evolve under different perspectives: *periculum*, consideration, responsibility. Figure polyhedral, like so many others in Roman law.

In the Roman legal system, we find that in certain rents - *locatio conductio operis* – concurs the obligation to return, besides the obligation of *facere* (Rascón García 1976, p.97), with an obligation of safekeeping, old times ago; thus, cases of *locatio horrei* and also of the *fullo* and the *sarcinator*. The first are cases in which a person –*horrearius*- made available to other spaces in a room or store -*horreum*-, so they deposit their goods, assuming the safekeeping obligation of the local, consequently thereof, of the goods, who responds in the event of non fulfillment. However, the *horrearius* could exclude from its activity the safekeeping of certain goods, such as gold and jewels from the mechanism of the *propositum* which consisted of a public announcement (Torrent Ruiz (2005), *s.v. horrearius;* D. 19, 2, 60, 6[5]).

It is important to highlight in this figure how the fact of the existence of an obligation of safekeeping, that generates a special liability, falls within the compulsory core of the *horrearius*, not affecting to the cause of contract for the purposes of a hypothetical discussion on its legal nature[6]; in the case of the *horrearius*, there is no doubt that we are facing a *locatio conductio operis*, that includes the obligation of safekeeping as far as the return of the delivered goods is also content required, so non-restitution of goods would mean, not only the breach of this obligation, but the safekeeping of them. That is why the *horrearius* business ceases to be so, when rent the entire local: we are then before a *locatio conductio rei* where no reason would require such obligation of safekeeping (D., 19, 2, 60, 9).

---

[4] The contribution of the doctrine romanistic about safekeeping is indeed extraordinary; since at the beginning of 20th century classical custody has been studied from the perspective of liability and of the vigilance activity. The problem in Roman law is in the use of an expression custodiam praestare which has been understood, historically, as a form of contractual liability, objectively derived from the non fulfillment of obligations of return (reddere/tradere), independent of subjective considerations in the conduct of the debtor, in a view which exceeded the safekeeping as a criterion for the charge of subjective liability - diligentia custodiendo- based in Justinian texts. However, there is an intermediate line which considers the safekeeping as consideration, in which the debtor gets involved in obtaining a result through developing a specific behavior. In relation to the different positions offers an extensive reference to them, and to him, we send you, also to bibliographic effects, Serrano-Vicente (2006), pp. 61-109. This author thinks classic safekeeping, as the Roman jurists understood it, was a consideration, while the "custodiam praestare" is not assessable behavior according to general criteria that are directed to examine the behavior of the debtor. Safekeeping consideration is looked at by the judge from a technical point of view, so since the realization of that behavior achieves the purpose which it pursues, non-fulfill of the consideration is always attributable, vide Serrano-Vicente (2006) pp. 361-362. Especially interesting the opinion of Rascón who, on the basis of his previous work about Pignus and custody, (Rascón Garcia, 1976) examines the issue from a procedural perspective, from my humble opinion, in a very successful approach from which analyses certain figures as some locationes (rents), commodatum (loan of use) and the pignus (pledge) in which "the presumption of diligence that the actor had to destroy if he wanted to win the trial, became a presumption of negligence in front of which nothing could make the defendant" (Rascón García, 2000, pág. 275).

[5] References with D., are from the Digest of Justinian.

[6] Hypothetical discussion among deposit or rent, as occurred in the contract of parking that we have seen in the preceding pages; in Roman law should not be such, in part because the deposit be essentially free, partly because it was a real contract versus the consensual rent, but also, for not being the dealing purposes exactly the same. I think that this is essential, as we shall refer the idea that organized activity of the horrearius involves a different thought, and for other reasons, to the depositary in general.

There are also obligation of safekeeping in other cases, as the *fullo* and the *sarcinator*; dyers or launderers and tailors performed a major activity for some objects, clothes, which was the dye or washed in the first case, and the sewing the second, "generating a typical obligation of result" (Torrent Ruiz, 2005, *s.v.fullo*). To this activity is added the obligation of custody whose non fulfillment, mainly due to robbery and damage in delivered things, had reflected in the non-restitution of goods. But obligations were always individualized in the sense that, one is the principal and other accessory and, indeed, the paid *merces* it is by the *ars*, the activity carried out on the clothes and not by the safekeeping (D. 4, 9, 5, *pr*; the relationship of this text with Gaius, Inst., III, 205-207 has been analyzed by the doctrine that has worked about the liability for safekeeping). The key issue in these cases in which appears the obligation or duty of custody, is that they are associated with an economic and business activity in which the payment service is a key element[7].

But since the Republican age, we have a number of commercial activities that, due to the bad reputation of some of the service providers on the one hand, and on the other, to be related to the transport of people and goods - booming in Rome - just forming a unitary group receiving the same treatment in the Edict of the Praetor about liability by the non-return of certain goods and effects: thus was born the figure of the *receptum nautarum,* spread, more or less early (Salazar Revuelta, 2007, p. 63ff), to the *caupones* and the *stabularii*, which, basically, have in common the assumption by cited traders: *nauta*, understood as the ship-owner (D., 4, 9, 1, 2 and D., 4, 9, 1, 3, *in fine*; Robaye, 1987, pp. 78-80; Impallomeni, 1976, *s.v. nauta*), *caupo*, innkeeper and *stabularius*, owner of a stable, probably –but not necessary: rich citizens rested in houses of friends, while their horses rest in the stables Földi (1999)- annexed to an Inn, a special liability by safekeeping of goods and personal effects of travelers in maritime transport - *vectors*-, walkers - *viatores* - or who leave their animals in a stable, that is presented as a genuine strict liability. In this regard, the liability by negligence for the rent held by these entrepreneurs with their clients was amended, or the liability for malice, where the relationship was rated as deposit if had not paid price.

Really, the *receptum*[8] was a pact (Salazar Revuelta, 2006, p. 1091) held alongside the *locatio-conductio* -or deposit- whereby these entrepreneurs assumed the obligation to return "safe and sound" - *salvum fore* - all delivered or entered by customers in their establishments -"*Praetor says: I will give trial against the sailors, innkeepers and owners of stables if they did not restitute all that from anyone they had received so that is out of danger*",(D., 4, 9, 1, *pr.*)- and if did not it, they would be responsible for them in any case - *omnimodo* - even if the thing has perished without negligence -*sine culpa eius rei perit*- (D., 4, 9, 3, 1). This pact, with the time, went from having to be explicit to be a natural element of legal deal: if we look at the text of the Edict that Ulpian pick up, the expression *salvum fore receperint* refers to the explicit commitment of these entrepreneurs ensuring the integrity of the goods, objects or animals of customers; however, it is not dangerous to think that, soon, began to understand that the *recipere* had included, as a natural element of the established relationship, the safekeeping of the things that are the *receptum* object (Salazar Revuelta, 2007, p. 168, but this

---

[7] The justification for the consideration of safekeeping in the commodatum would be given for the benefit or advantage received the commodatarius by free use of the thing.

[8] Bibliography about Receptum nautarum, cauponum et stabulariorum is extensive and important so, without exhaustiveness spirit, and linked to the references cited, vide Huvelin (1929); Menayer (1960); Rodríguez (2004); Fercia (2002); Torrent Ruiz (2005) pp. s.v receptum nautarum, cauponum, stabulariorum); Metro (1966); Herrera (1993).

affirmation is not pacific). And, increasingly, was making firm the idea that the business activity exercised by these persons included the assumption of liability, to the extent of exist jobs within the ship specifically aimed at ensuring the custody of the goods -storekeepers y ναυθύλαχες; D., 4, 9, 1, 3, *in fine*-.

Among these three figures, which interests us is, without doubt, the figure of the *stabularius* inasmuch as it is near - saved the distances-the current parking contract where the assumption of fact is, basically, the same. About the other figures in this Edict about the *receptum*, today continue to maintain liability regime for hotel keeper (*receptum cauponum*) for the effects introduced in hotel although the C.c. in its articles 1753 and 1754 consider it a "necessary deposit" (1753 C.c.) and, consequently, liable as depositary for the hotel keeper for loss or damage of the effects except those caused by force majeure or armed robbery (1754 C.c) or, as it has been as manifest in any judgment, would be imposed by the implementation of the general criterion of good faith (Represa Polo, 2004, pág. 119 y ss). The regime of transport of goods and effects invoiced to carry by passengers themselves, has changed substantially regarding liability that usually remains limited -to negligent non fulfillments - by former statement about the value and paying a supplementary amount (Estal Sastre, 2002, p. 1 and 11-12).

Regarding the owner of a stable, the sources describe his activity as one that *permittat iumenta apud eum stabulari*, i.e. allow the horses to stay in the stable (D., 4, 9, 5, *pr*). From this perspective, the owner of a stable was obliged to provide a place for use in both as lasted the arrangement of the horses, for a price.

Under this liability by the *salvum fore recip*ere - receive something in order to be safe - was the *stabularius,* and was understood such, a one at the forefront of the business of the *stabulum* "or its factors". They were the only ones that could constitute a *receptum*: were excluded those that "occupied the lower offices" (D. 4, 9, 1, 5). His liability covered any damage and robbery of goods, things or animals of customers (D., 4, 9, 5, 1), also those that other customers had caused (D., 4, 9, 1, 8 and D., 4, 9, 2). The object protected under *receptum stabulariorum* are animals and their mounts, in implementation of the extension of protection that the praetor made over "one thing any or goods" whenever they had been received –including *"the things that closely match to the goods"*, D., 4, 9, 1, 6-.

No matter that brought things into the establishment were property or not of the client for the purposes of liability of the *stabularius* (D., 4, 9, 1, 7); inasmuch as when or how it should be understood that things has been introduced in establishments and, accordingly, since when it was assumed the liability for damage or robbery, the "receiving" –*recipere*- appears more abstract when in Ulpian, and picked up by Ulpian, Pomponio, (D., 4, 9, 3, *pr*.) infer the no need for an explicit act of consignment -*res assignatae sunt*- which must to be given, *in principle*, by the contest of both parties (D., 4, 9, 1, 8).

The liability *ex recepto* must have been originally absolute, responding *"in any case - omnimodo- who received, although the thing has perished or damage caused without negligence"* (D., 4, 9, 3, 1); however, according to the same passage of Ulpian, was introduced the exoneration of the same if *"has happened by irremediable accident"* and already Labeón, would have introduced a exception allowing the *nauta* to paralyze the action of the customer-plaintiff´s, when the loss had occurred by a case of *vis maior* –force majeur-, exception which would be extended to the *stabularius,* assimilating, again, the owners of stables´ regime to the ship-owners. Thus, becomes former liability *ex recepto* in a liability in which these entrepreneurs assume the *periculum* by damages in general or perish of the customers' things except in the case said -(regarding the relationship between liability and risk, Rodríguez, 2006, pp. 31 and 32)-.

As it can be seen in this brief tour of the regime of liability of the owner of stables in Roman law, from the liability set up in the Edict of the Praetor for those shippers, innkeepers and owners of stables, by the fact – *"if not return which has been received to be safe from any person"* - assumed by express agreement, absolute liability -*omnimodo*-without negligence, by the loss or damage in received things, it passed by the work of the classical Jurisprudence, to a reprocessing of all the regulation expanding in many cases, and, almost fully developing, the legal regime of the *receptum nautarum, cauponum et stabulariorum*. The path of reprocessing undertaken by the Jurisprudence was founded, basically, in the spiritualization of concepts and forms required for the establishment of a *receptum*: thus the *recipere*, used as a noun in *receptum*, alludes to the commitment, (López-Amor, 1994, p. 721ff; Torrent Ruiz, 2005, *s.v. recepta*, in relation to others *recepta*), and becomes a figure which, from express agreement, becomes a natural element of business activity that assume these Roman entrepreneurs –moreover, the *recipere* would be understood as the same business activity-, and the liability or risk for safekeeping, a very business risk over which is founded and justify the institution itself.

The liability assumed by the *stabularius* derives, as we have already mentioned, from his activity (D. 4, 9, 3, 2), for the exercise of particular profession or business, as the launderer or tailor - *fullo* et *sarcinator*-as reminds Gaius in D., 4, 9, 5, *pr.* to us (Serrano-Vicente, 2006, pp. 332-339; therefore, no matter who is the client or what position occupies (D., 4, 9, 4). Perhaps this is why in the *receptum* cannot be designed one express obligation of safekeeping -*custodiam praestare*- although, being obliged to return *salvum fore*, the normal behavior would be to deploy an supervision activity or custody mechanisms; the thing that assumes, is the *periculum* by damages or losses that would lead him in the position of non-fulfillment with restitution of animals and mounts "safe and sound".

This is the background behind, from my point of view, the creation of the figure by the Praetor and further development of the same by classical Jurists. They wondered why a new figure, when there were civil actions arising from contracts which the *nauta, caupo or stabularius* had concluded with its customers (D., 4, 9, 3.1). The reason provided by the very jurist: *"[...] because in locatio-conductio is responsible by negligence, and the only malice in deposit"*. But if this is the technical reason, aggravation of liability, the justification for the need of a change in the regulation of the regime of liability for these "entrepreneurs" is that *"most often it is needed to rely on them and things to instruct them safekeeping [...] and if this has not been set down, would be a reason to act in concert with thieves against those whom receive, when neither now, certainly, refrain from such scams"*(D., 4, 9, 1, 1; D., 4, 9, 3, 1. About the reliability of the justification of the pretoria intervention through the edict, De Robertis, 1952).

To punish these criminal activities and protect and promote the land and sea trade which such importance had taken to Rome and that needed some ways of transport of passengers and goods as much as secure as possible, justify the hardening of the regime of liability for these entrepreneurs, as it would have also justified, since antiquity, the existence of other ways of repression of damage and robbery in the area that we are analyzing, to the extent that alongside civil actions which we have shown relating to contracts -usually rents, and in some cases, deposits- and the *actio ex receptio*, are penal actions *in factum conceptae* as *actio furti vel damni adversus nautas, caupones et stabularios* (Giménez-Candela, 1990, pp. 129ff) where these *exercitores* could be sued by the robberies committed by their assistants.

The content differences between the *actio ex recepto* and criminal actions *furti vel damni adversus nautas, caupones et stabularios*[9] would have justified the coexistence of both at the same time, because the justification of both is not exactly equal. Both we find an idea of a liability in the exercise of their business (López-Amor, 1994, pp. 721 and 722; Cerami, di Porto, & Petrucci, 2004, p. 276): in some cases, those of the *receptum*, it assumed by the activity itself, and in the *actio*, since the employer must respond by acts of their employees in a regime of fault *in eligendo* (D., 4, 9, 7, *pr.*).

CONCLUSION

The analysis of the regime of the two figures, parking and *receptum stabulariorum*, protrude two ideas:

- The first is that the *receptum* was a voluntary agreement even when it was broaden its assumption, becoming a natural element of the contract; the scope of the contract parking as has been formulated in the Act 40/2002, art.1, makes that the entrepreneur of the establishment can say if included the duty of vigilance and safekeeping as a material element of his contracts or not.
- The second, that we have a liability that is specified in the return of certain goods "safe and sound". In this sense, in the *receptum,* we talk of a strict liability –except cases of force majeure- and the parking liability, even, in cases of force majeure.

These ideas lead us to ask ourselves some questions: what advantage can the voluntary assumption of a great liability bring to an entrepreneur? Why would a *stabularius* was going to formalize a *receptum* if he can respond for less, in the case of non fulfillment, if did it by *locatio*, or at least cases if it did by the *actio furti vel damni adversus nautas, caupones et stabularios*? Why, being voluntary, was reached to generalize so much? (in relation to these subjects, with bibliography, Serrano-Vicente 2006, pp,.326-329). Why the entrepreneur of a parking business may prefer that his contract is subject to Act 40/2002 which expresses the duty of vigilance and safekeeping, and probably the non fulfillment fall under consideration of strict liability?

Probably, because they are interested in. The first, perhaps because to assume the risk of losses meant not to be in the spotlight of the Praetor, and on the other hand, they get ahead and distinguished of other entrepreneurs of their gild, which not assumed commitment *salvum fore,* and remaining automatically marked as more suspects still; and the second, because the Act, although apparently seemed to punish the liability of the entrepreneur as had been recognized by jurisprudence, actually it was lowered significantly, or disappears there where

---

[9] In general, the liability of the receptum was broader than that of the actio furti vel damni adversus nautas, caupones et stabularios on relation to the persons that is responsible the owner of a stable and the place where damages or losses are produced; but the sentence would be lower in the receptum than in the actio. This, on the other hand, would have the possibility of the praedictio, that is, eliminate your responsibility, "which he is not responsible by the damage" - by expresses statement that "each passenger looks after their things" and "passengers have been agree with the notice" (D., 4, 9, 7, pr.).

there are the cases most conflictive; that is, in liability for robberies of components and accessories and effects introduced into the vehicle, except the possibility to assume safekeeping of these elements set with the requirements of the art.3. 2 shall entail, for sure, supplementary payment for this service (art. 3.3).

Finally, protection which received the client of the *stabularius* in Rome was extensive and intense, and from this point of view, we can, without fear of falling into exaggeration, talk about similar protections to the currently collected in subjects as the Law of the consumer, perhaps more than in the current contract parking.

REFERENCES

Álvarez Lata, N. (1997). Comentario a la sentencia del Tribunal Supremo de 22 de octubre de 1996 sobre la obligación de guarda y custodia en los aparcamientos de vehículos. *Anuario da Facultade de Dereito da Universidade da Coruña* (1), 559-570.

Álvarez Lata, N. (abril de 2003). El contrato de aparcamiento de vehículos. Notas a la Ley 40/2002 de 14 de noviembre. *InDret*, 1-16.

Ballesteros de los Ríos, M. (2000). *El contrato de aparcamiento.* Elcano: Aranzadi.

Bercovitz Rodríguez-Cano, R. (2002). Aparcamientos. *Aranzadi Civil* (15), 11-14.

Carrasco Perera, Á. (2002). Aparcamientos: una superfluidad normativa. *Actualidad jurídica Aranzadi* (557), 3.

Cerami, P., di Porto, A., & Petrucci, A. (2004). *Diritto comerciale romano. Profilo storico, Giappichelli, Turín, p.276.* Turín: Giappichelli.

De Robertis, F. (1952). *Receptum nautarum. Studio sulla responsabilità dell'armatore in diritto romano, con riferimento alla disciplina particolare concernente il caupo e lo stabularius.* Bari.

Díaz Gómez, M. J. (2004). Comentario breve a la Ley Reguladora del contrato de aparcamiento de vehículos. *Actualidad civil* (10), 1-18.

Elguero y Merino, J. M. (1993). *Garajes y aparcamientos ¿arrendamiento o depósito?* Madrid: Tecnos.

Espiau Espiau, S., & Mullerat Balmaña, R. (1996). Relaciones contractuales de hecho y contratos de adhesión: notas para el estudio del contrato de aparcamiento público. *Revista de derecho privado* (80), 787-814.

Estal Sastre, R. d. (diciembre de 2002). Estudio sobre el posible carácter abusivo de determinadas cláusulas de limitación de responsabilidad en los contratos celebrados por los consumidores con establecimientos prestadores del servicio de revelado fotográfico. http://www.uclm.es/centro/cesco/pdf/investigacion/2002/pdf/2.pdf, 1-16.

Fercia, R. (2002). *Criteri di responsabilità dell'exercitor. Modelli culturali di atribuzione di rischio e regime della nossalità nelle azioni penali in factum contra nautas, caupones et stabularios.* Turín: Giappichelli.

Földi, A. (1999). Caupones e stabularii nelle fonti del diritto romano. En *Mélanges Fritz Sturm* (Vol. I, págs. 119-137). Lieja.

Giménez-Candela, T. (1990). *Los llamados cuasidelitos.* Madrid: Trivium.

Gómez Calle, E. (2004). La extensión de la obligación de restituir a cargo del titular del aparcamiento. *Revista de Derecho Patrimonial* (12), 169-181.

Herrada Romero, R. I. (1994). Reflexiones sobre la caracterización jurídica del contrato de garaje. *Revista de Derecho Privado*, 827-873.

Herrera, R. (1993). Planteamientos generales sobre responsabilidad por custodia en el Derecho Romano, en la tradición romanística y en la Codificación. *Revista de Derecho privado* (LXXVII), 661-684.

Impallomeni, G. (1976). s.v. "Nauta". En *NN.DD, Novissimo Digesto* (Vol. 11, pág. 177).

López Barba, E. (2003). Consideraciones sobre la nueva Ley 40/2002, de 14 de noviembre, reguladora del contrato de aparcamiento de vehículos (RCL, 2002, 2649). *Revista de Derecho Patrimonial* (10), 101-115.

López-Amor, M. (1994). "Receptum argentarii". "Receptum nautarum, cauponum, stabulariorum". En J. c. Paricio, *Derecho romano de obligaciones. Homanaje al prof. J. L. Murga Gener* (págs. 721-728).

Martín Santisteban, S. (2003). Ley 40/2002, de 14 de noviembre, reguladora del contrato de aparcamiento de vehículos (RCL 2002, 2649). *Revista de Derecho Patrimonial* (10), 95-100.

Menayer, L. (1960). "Naulum" et "receptum salvam fore". Contribution à l'etude de la responsabilité contractuelle dans les transports maritimes en droit romain". *RHDF* (38), 385-411.

Metro, A. (1966). *L'obbligazione di custodire nel diritto romano*. Milán: Giuffrè.

Rascón García, C. (2000). Custodia seu presunción de culpa. En *Estudios de Derecho Romano en memoria de Benito Mª Reimundo Yanes* (Vol. II, págs. 263-276). Burgos: Universidad de Burgos.

Rascón García, C. (1976). *Pignus y Custodia en el Derecho Romano Clásico*. Oviedo: Servicio de Publicaciones de la Universidad de Oviedo.

Represa Polo, M. P. (2004). *Responsabilidad de los establecimientos hoteleros por los efectos introducidos por los clientes*. Madrid: Editoriales de Derecho reunidas.

Robaye, R. (1987). *L'obligation de garde. Essai sur la responsabilité contractuelle en droit romain*. Bruselas: Publications des Facultés universitaires Saint-Louis.

Rodríguez, A. M. (2004). *El "receptum argentarii" en el Derecho romano clásico. Una propuesta de análisis*. Madrid.

Rodríguez, R. (2006). Perfiles característicos de la denominada responsabilidad contractual en Derecho romano. Apunte histórico y de conjunto. En *Responsabilidad civil de profesionales y empresarios. Aspectos nacionales e internacionales* (págs. 1-52). Netbiblio.

Salazar Revuelta, M. (2006). Configuración jurídica del *receptum nautarum, cauponum et stabulariorum* y evolución de la responsabilidad recepticia en el Derecho Romano. *Anuario da Facultade de Dereito da Universidade da Coruña* (10), 1083-1100.

Salazar Revuelta, M. (2007). *La responsailidad objetiva en el transporte marítimo y terrestre en Roma. Estudio sobre el Receptum nautarum, cauponum et stabulariorum: entre la utilitas contrahentium y el desarrollo comercial*. Madrid: Dykinson.

Serrano-Vicente, M. (2006). *Custodiam praestare. La prestación de custodia en Derecho Romano*. Madrid: Tébar.

Torrent Ruiz, A. (2005). *Diccionario de Derecho Romano*. Madrid: Edisofer.

In: Research Studies on Tourism and Environment
Editors: J. Mondejar-Jimenez et al.
ISBN: 978-1-61209-946-0
© 2012 Nova Science Publishers, Inc.

Chapter 18

ECOTOURISM CAPABILITY IN SENSITIVE WETLAND CONSERVATION, CASE STUDY: CHEQAKHOR WETLAND, CENTRAL IRAN

Homa Irani Behbahani, Hassan Darabi and Zhale Shokouhi
Faculty of Environment, Tehran University, Iran

ABSTRACT

Wetlands environmental functions in semi-dry and dry regions are very important, sensitive and fragile at the same time. In order to compensate deficit, wetlands marginal rural communities exploit of wetland resources directly such as; wetland plants collection for feeding livestock, hunting of migratory birds and the like. These actions damage ecosystem seriously and put sensitive ecological systems at risk.

On the other hand, wetlands have considerable characteristics and landscape attractions that are introduced them as tourism attractions. Communities-based ecotourism provides alternative income possibility for rural households. Ecotourism provide training and participation of indigenous people chance. These opportunities would cause destructive agents change into the environmental protections and their interests will be depended on the protection of wetland resources. This chapter addresses the needs for conservation endangered sensitive ecosystems and ecotourism capability to protect wetland ecosystems. An example of community-based ecotourism in CheqaKhor Wetland, Chahar Mahal Bakhtiari province, central Iran as a sensitive wetland will be highlighted.

Keywords: Wetland, Ecotourism, Conservation, Local community

1. INTRODUCTION

Despite the importance of wetlands; people exploited of wetlands over the years based on short-term benefits. After clarifiing their specific roles and functions in ecosystems some activites were done to protect them in recent decades. wetlands habitats for different animal species, plants and other hydrological systems supply underground water supplies and fresh water(Marsh, 2005, 396) Wetlands are ecosystems that provide numerous goods and services with an economic value, not only to the local population living in its periphery but also to communities living outside the wetland areas. They are important sources for food, fresh water and building materials. they provide valuable services such as water treatment and erosion control. (Schuyt et. al,2004, 3)

An attempt was made at estimating the economic value of global wetlands, which showed $ 3.4 billion is a very conservative estimate of this economic value. Two main reasons for this conservative estimate are that:(1) many wetland functions were not valued in the economic valuation studies collected and used for this value. (2) only a fraction of world's global wetland area was used(63million hectares) this study also shows that Asian wetlands have the highest economic value at $ 1,8 billion per year. (Schuyt et. al, 2004, 4)

Economic value production based on wetland is vital for local community life. In different regions people earn economic values by various spectrums of activates but we take Add value productions in two different ways:

1. direct exploitation from wetland resources,
2. Indirect exploitation from wetland resources.

1.1. Direct Exploitation from Wetland Resources

In this way, local community faced with some shortages like population growth, insufficient understanding the functions and values of wetlands, unclear property rights and the most important lack of income that led to extensive direct use of wetlands resources like Plants, Fishery, hunting, cutting trees, converting wetland into agricultural lands, useing wetland water for irrigation and etc. All activities take place to compensate the lack of income. While the lack of property rights cause unwanted competition among local stockholders and ultimately will lead to destruction of the wetlands. The continuty of this process reduces the wetland production potential. At the same time income demanding is rising. Consequently, the imbalance between supply and demand will be happen taht leads to the fragility and the gradual destruction of wetlands.

1.2. Indirect Exploitation from Wetland Resources

In this way local Community is aware of different wetland values. Despite the need for income, people with the public participation try to plan and operate action to exploit the wetland resources indirectly. In this way, people know that their income depends on the protection of based wetland resources. For this reason, despite the lack of property rights, intensified competition didn't happen over exploitation of based resources, but also competition take place over attracting income from other activities that don't harm the ecosystem. Some activities like ecotourism are the best substitution to earn more add value from wetland productivity potentials.

In this framework local community will tries to find different ways to wetland conservation and protection.

2. WETLAND DEFINITION

There are Different definitions of wetlands. Any definition varies based on prupose. all wetland definition have some common characteristics like combination of hydrology, soils and vegetation. (kusler,et al, 1996 , 55, Lewis, 1995, 60, Tiner,1999 ,6-10, Mitsch , et al, 2007, 40-41) Based on Ramsar definition; wetlands are areas of marshes, fens, peatlands or water, wether natural or artificial, permanent or temporary, with water that is static or flowing, fresh, brakish or sallt, including areas of marine water that depth of which at low tide is not exceeded 6 m. wetlands may incorperate riparian and costal zone adjecent to wetlands or bodies of marine water deeper than 6m a low tide lying within the wetland. At same time Ramsar convention discuse the "*wise use*" concept.

Definitions of the key Ramsar Convention concepts of "wise use" and "ecological character" of wetlands were adopted by COP3 (1987) and COP7 (1999) respectively. Action 3.1.1 of the Ramsar Strategic Plan 2003-2008 requested the Convention's Scientific and Technical Review Panel (STRP) to "review the wise use concept, its application, and its consistency with the objectives of sustainable development". (Convention on wetlands ,2005, Agenda item 8.2) Wise use concept was further developed as sustainable utlization for the benfit of humankind in a way compatible with maintence of their natural properity is understood to be those physical , biological or chemical components, such as soil , water, plants, animal and nuriennts and the interactions between thems.(Heijnsbergen,1997, 89)

3. VALUES OF WETLANDS

Kluser and opheim argue that wet lands are important for community because thay have different functions. they explain wetland functions as follow:

Flood conveyance, barriers to wave and erosion, flood storage, erosion and sediment control, pollution prevention and control, fish and shellfish production, recreation(water based), water supply, aquifer recharge(kusler,et al, 1996 , 7) There exists a wide body of literature on Wetland functions, and one approach is provided by deGroot(1992), who classifies functions in to four categories (deGroot,1992):

(1) regulating functions As ecosystems regulate ecological processes that Contribute to a healthy environment;
(2) carrier functions, where ecosystems provide space for activities, like human settlement, cultivation and energy conversion;
(3) producting functions ecosystems provide resources for humans, like food, water, raw materials for building and clothing; and
(4) informing functions, where ecosystems contribute to mental Health by providing scientific, aesthetic and spiritual information.(Schuyt,2005,181, Schuyt, et. al. 2004.

4. ECOTOURISM AND CONSERVATION

Tourism is notorious for its potential to disrupt, disturb, or otherwise do damage to natural habitats and local communities. 448). Ecotourism is often positioned as an alternative to mass tourism, a sustainable nature- and culture-based tourism (Weaver, 2001; Fennell, 2003). A significant contribution to ecotourism's global following has been its potential to deliver benefits to communities remote from centers of commerce; benefits that do not involve wide spread social or environmental destruction. (Wearing, et al. 1999, 72) Principles and characteristics of ecotourism can be counted as:

- Consistent with a positive environmental ethic, fostering preferred behavior.
- There is no denigration the resource and erosion of resource integrity.
- Concentration on intrinsic rather than extrinsic values.
- Biocentric rather than homocentric in philosophy.
- Take advantage of the resources.
- Firsthand experience with the natural environment.
- An expectation of gratification measured in appreciation and education, not in thrill-seeking or physical achievement. These latter elements are consistent with adventure tourism, the other division of natural environment (wild land) tourism.
- High cognitive (informational) and effective (emotional) dimensions to the experience, requiring a high level of preparation from both leaders and participants. (Source: After Butler (1992), Acott et al. (1998).adopted From: Higham 2007, 5)

Base on this characters emphases are on positive impacts. It has been ascribed with the power to sustain rural livelihoods (Honey 1999), to catalyze new development (Weaver 1998), to renew cultural pride (Epler Wood 2002), to empower local peoples (Scheyvens1999), and protect biodiversity (Christ, Hillel, Matus and Sweeting 2003).(adopted from: Gordillo, et. al, 2008,449) increase social capital and social cohesion, local institute enforcement, increase negotiation power against exterior power, But it contain some negative impacts specially for local community:

- Relatively few communities have realized significant benefits of any kind.
- Contributions to problems of degradation.
- Habitat loss by enhancing the buying power for more labor, technology, and capital local residents use to expand resource use.
- Increased income, especially when poorly linked with conservation goals and

Backed by weak or no enforcement, "simply fosters more rapid resource extraction".

- Changing fundamental social and cultural patterns of resource use.
- Volunteer capability to boom-bust cycles and seasonal fluctuations of the tourism market.
- Social conflicts that are emerged from unequal earnings and increased gaps between rich and poor. (adopted from: Gordillo, et. al, 2008,450-2)

All Impacts show that: the interdependence of tourism, the social and physical environment are fundamental to the future of each, and seeking a way to accommodate the needs of all parties, without control being external to those who experience its effects most directly, is essential. Features of the natural and cultural environment and supportive host communities are the foundations of a successful industry. Neglecting the conservation and quality of life issues threatens the very basis of local populations and a viable and sustainable tourism industry. (Wearing, et al. 1999, 73)

In ecotourism domain different actors play their roles; while locale communities seek their benefits by earning more portions from income, job opportunities, less dependency to natural resources. Ecotourists interest to qualify landscape, healthy environment, protected improvement, damage correction, and environment restoration, tourism industry actors look for financial benefits, environment conditions that stabilize tourism market, long-term attractions and ensuring the investment return.

Base on different demand and wishes, planners have a heavy task for create balance between different actors needs and from other wise enhance positive impact with emphases on conservative and sensitive environment and decrease negative impacts to minimums. Notwithstanding of all this issue, it is notable that in Ecotourism, the option is to channel greater benefits directly to communities and act as incentives for conservation. However, it needs some requirement:

1. People training and warning.
2. Developing communities based on ecotourism.
3. Local community participation and empowerment.
4. Local awareness from conservation natural resources and it's benefits for communities.

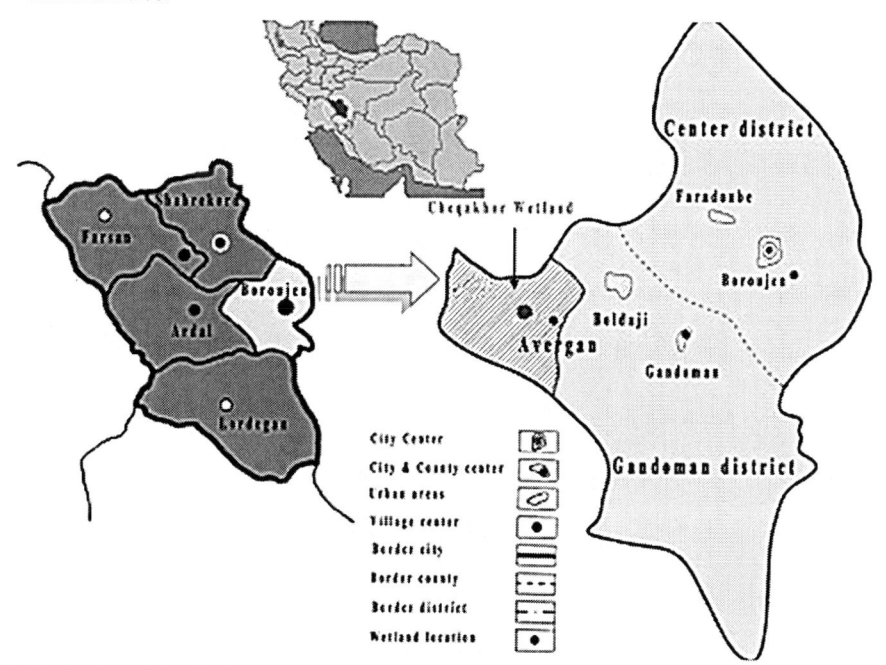

Figure 1. Wetland position in Iran.

5. CASE STUDY

5.1. Cheqakhor Wetland Attribution

The subject of chapter is a Cheqhakhor wetland. This wetland is the largest wetland ecosystem in the province of Chahar Mahal Bakhtiari in central part of Iran. It's located between 50 °, 52′ and 50 °, 57′ geographic north altitude and 31 ° 52′ and 31 ° 56 east latitude. It height from sea level is approximately 2270 meters. It is 65 km far from the center of the province (Shahr e kord). (Figure 1) Some village established in the east and north of wetland. These villages are: Sang chin, Avardegan, Dastgerd, Siback, Matvy, Seif Abad, Khani Abad, Saki Abad, Abad, khedr and Galogard.

Cheqhakhor wetlands located in a wide plain called Gandoman_Bldajy. It's surrounded by Bar Aftab Mountain in north, kelar in south and Shapour in west. Tectonic is an important and main factor shaping wetland. Hydrogeological Status in this region causes shaping springs in the mountain aquifer Basin of Klar Mountain. These springs water are permanent wetlands water supply resources. Cheqhakhor is an important ecological storage in region. Calm habitats, covering a variety of plants, diversity of natural habitats with the structure of water, grassland, woods and lamentable artificial habitats such as fields and gardens are surrounding the pond, Cheqhakhor to a diverse range of habitats for migrant birds. So that the resort and the habitat of 63 species of birds from 12 orders, 26 families and 38 different sexes and it hosts an annual average of 1600 next aquatic and aquatic birds. This wetland lies in eighth place among 75 countries placed pond based on criteria for determining conservation status of wetlands in the classification of Iran wetlands assessment was conducted in 2001. (Maunoury, Masoud 1383) Environmental Protection Agency because of the wetland ecological conditions and diversity of animal life and wetlands, and areas around it was recorded as a hunting prohibited area in 1999. Main source of water supply are spring and runoff water. Annual input water of wetland estimated about 70 million cubic meters.

In order to provide water for agricultural needs in this area, a dam was constructed in the eastern side of the wetland in 1989 and 23.9 million cubic meters of water exploit annually from. Water for land irrigation in wetland boarder take directly from this pond, and for lands and villages away from, exploit Aqbolaq River that source from wetland, and other springs or wells. Following table shows the monthly water balance in wetland output (input of Aqbolaq River) (Table No.1).

Table 1. Cheqhakhor wetland's monthly water balance

Area SKM	The average rainfall (mm)	Volume of rainfall(MCM)	Water exploit (MCM) Agriculture	Water exploit (MCM) Evaporation	Out put (MCM)	Cumulative output (MCM)
113	549	62	7	34	21	21

Source: Monavary, 2004.

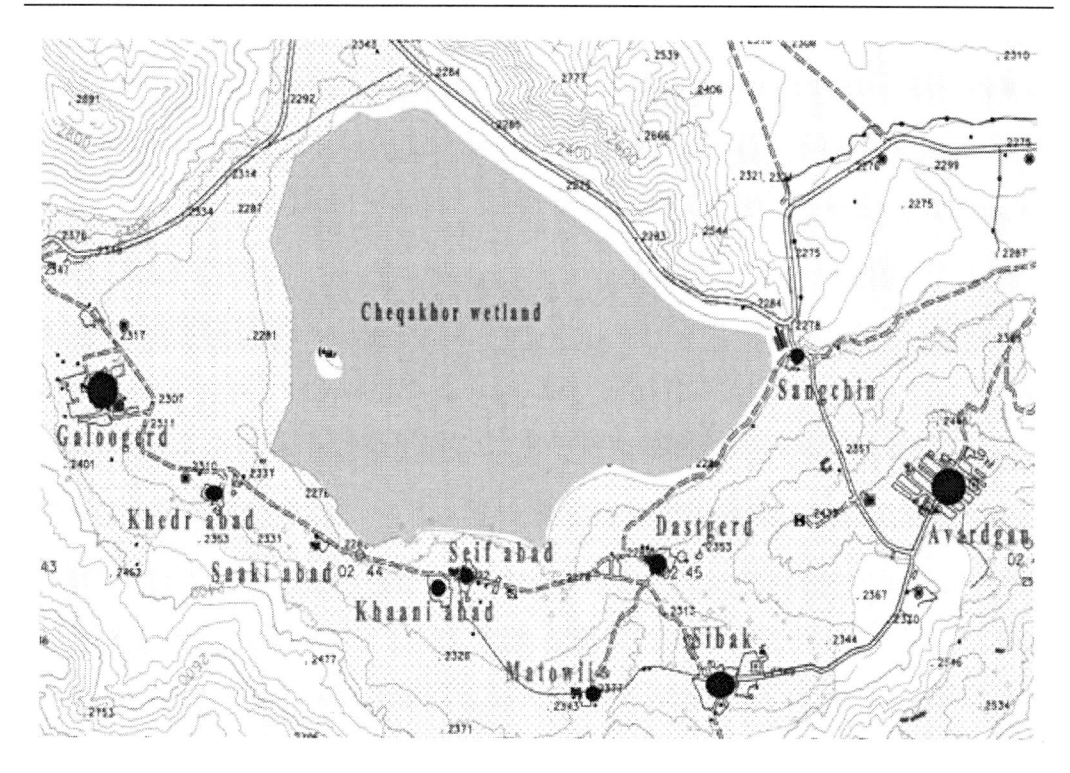

Figure 2. Cheqakhor wetland and villages around it.

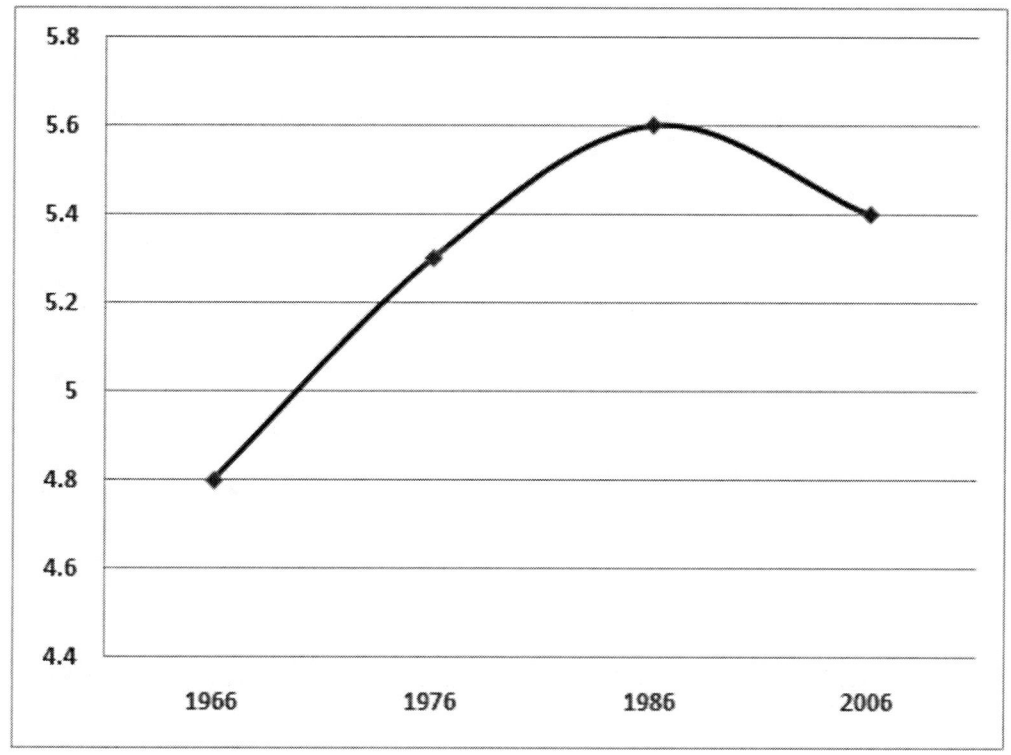

Source: SCI, 1966- 2006.

Figure 3. population growth of Villages from 1996 to 2006.

There are 10 villages around the pond at intervals of 250 to 2400 meters straight edge of the wetland: Sangchin, Avardgan, Dastgerd, Sibak, Matowii, Seif Abad, Khaani Abad, Saaki Abad, Khedr Abad, Galoogerd. Villages established as an arc in low lands of south slope wetlands. (Figure 2)

Nearest village to the pond _Sangchin_ is 250 meters far from in east by dam. Avardegn located at 2350 meters distance from the wetland in the southeast, 120 meters above the wetland water surface.

The study of Population for 40 years shows population growth about 5.3 percent in wetland margins villages. While growth of rural population in the country is about %4 and in the province is %4.5.

The growth rate about %5.3 means different subjects such as:

- Needs for more income;
- Needs for more job opportunity;
- Decrease of annual income of each person; all this condition means as more extensive exploitation from accessible resources.

Based on calculations performed in 2009, annual income for each person was 1026 $ in wetland margin villages. 97% of this income was from agricultural and gardening activities, 2.2% of the dairy and 0.8% from apicultural activities. According to official statistics, annual expense per person was equivalent to 1542 US dollars in 2009. Comparing costs with average household income per capita in the village pond margin represents lack of income to 516 US dollars yearly. The obvious result of this condition includes intensive exploitation of accessible resources. The table No. 2 represents the amount of annual income, expense and Lack of income for each village. (Us $)

Table 2. Income and expenditure of village in 2009(US $)

Villages	Population	annual income	Annual expenditure	Lack of annual income
Sangchin	40	41043.7	61698.4	20654.7
Dastgerd	578	593081.3	891542.1	298460.8
Seif abad	110	112870.1	169670.6	56800.5
Khaani abad	187	191879.2	288440.1	96560.9
Saaki abad	144	147757.3	222114.3	74357
Avardgan	2553	2619613.2	3937901.4	1318288.1
Sibak	2138	2193785	3297780.3	1103995.3
Khedr abad	126	129287.6	194350	65062.4
Galoogerd	1098	1126649.2	1693621.5	566972.3
Matowii	132	135444.2	203604.8	68160.6

Source: Field studies.

Table 3. Area of agricultural lands(hectares) from 1966 to 2003

Villages around wetland	Area of agricultural lands(hectares)	
	1966	2003
Sangchin	134	49
Dastgerd	203	181
Seif abad	94	113
Khaani abad	91	183
Saaki abad	-	30
Avardgan	377	782
Sibak	100	375
Khedr abad	108	132
Galoogerd	362	390
Matowii	24	67

Source: SCI, 1966, 2003.

Providing income deficit have five options:

- Increasing cultivation area;
- Increasing livestock and livestock activities;
- To increase the direct exploitation of wetland;
- Immigration;
- Increasing income from outside the village.

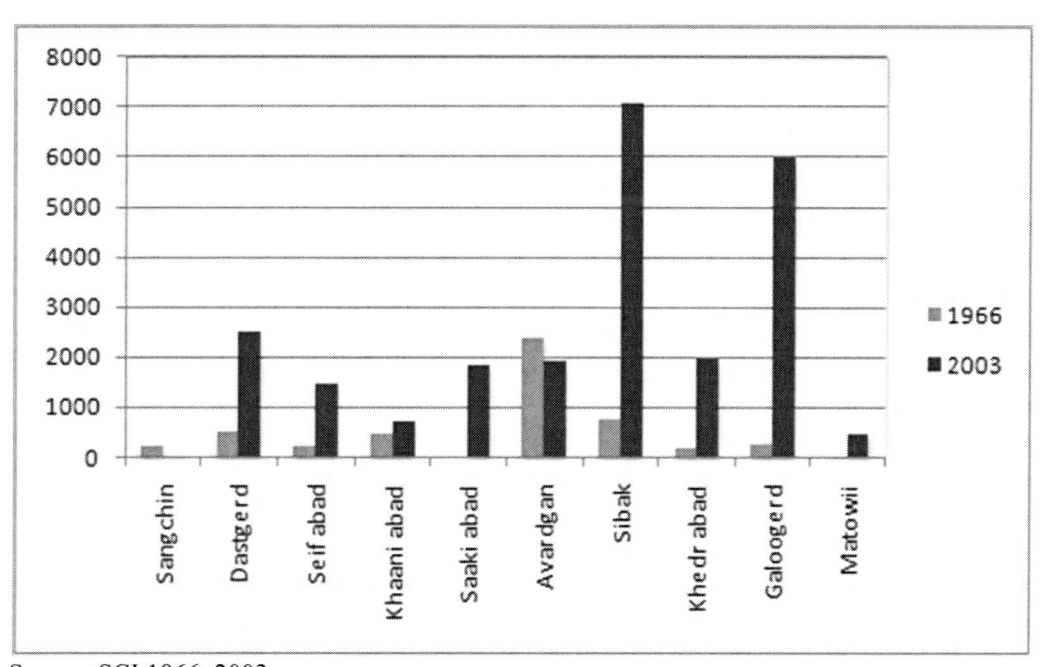

Source: SCI 1966, 2003.

Figure 4. Number of village's livestock.

5.2. Increasing Cultivation Area

Agriculture and horticulture are the main activities in the wetland surrounding villages. Main products are wheat, barley, forage plants, cereals, and sugar beet. The land area is about 2302 hectares in 2003. Gardens scattered between the villages Avardgan to sibak, sibak to Dastgerd and some small parts distributed in other areas. Products of garden are: grape, apple, yellow plum, peach, almond and walnut. Agriculture is main source of income for rural residents. Conversion of natural lands and pastures into the agricultural lands in this area is the most significant environmental change in the last few decades. Agricultural lands of the rural area has increased about 80% that is notable. The increase in cultivation area, need to water and direct utilization of the wetland water.

5.3. Increasing Livestock and Livestock Activities

To show increase of livestock rate, it has been used from the country's official statistical data. The data for two sections 1969 and 2003 are presented. Base on this statistics, 8 villages experienced a notable growth of number of livestock that show at means 80 percent increase.(Figure 4) Increase the number of livestock, the indiscriminate grazing capacity of rangelands and non-compliance causing loss in quality pastures and changed wetlands ranges into marginalized pastures. Pastures degree from first class and two changes into three degree. This means a grazing capacity of less than one animal unit per hectare in a one season. Increase the number of livestock, the indiscriminate grazing capacity of rangelands and non-compliance causing loss in quality pastures into wetlands have been marginalized.

The Pastures quality change from first class of quality to semi poor and poor. This means a grazing capacity of less than one animal unit per hectare in a season. On the other hand, tribal households also flow into this area. They are dependent on rangelands. The average number of livestock per household is headed by nomads about 77 sheep and goats. Nomadic livestock in some season of the year put more pressure on pastures and utilization of rangeland in this region will be intensified. In this case, increases livestock activities in the long term lead to total destroyed Pastures.

5.4. Direct use of Wetland Base Resources

Rural residents exploit directly from wetland base resources to compensate their lack of income. Directly exploitation of based wetland resources directly occurs on site in various forms:

1. Water withdrawal for irrigation farming;
2. Fishing for daily needs in addition recreational fishing;
3. Grazing animals in wetland margins and harvest wetland pant for animal feeding;
4. Use of native species such as Polygonum mite schrank Rumex Sp. and Polygonum hidropiper for feeding, especially in hot dry seasons;
5. Illegal birds hunting to provide income.

Wetland is severely damaged due to these actions and economical conditions. This process Continuity threats seriously of wetland ecosystem sustain capability.

Another two option (outside income and immigration) has been used by local people, that unfortunately there are any official information about it, But follow of income from outside always is unstable and it's amount are unaccountable. Because rural migrants are unskillful workforce, that occupied in marginal jobs like housing workers that decrease their saving capability. Nevertheless, with all this options any family faced with lack of income about 516$ per person and 2581 $ per family annually. Ecotourism is the best alternative to confronting against wetland sustainability threats. It can provide a certain source of income without damage to wetland base natural sources.

5.5. Ecotourism Capability

De Groot(1992), and Schuyt and Brander(2003) classified different function of wetlands as;

- Regulation functions;
- Carrier functions;
- Production functions;
- Information functions.

Based on this classification direct exploitation of wetland resources that occurs in this area's related to production function. In this method, competence will be happen over personal interest and any property right makes this competence more extensive and creates new environmental challenges. However, at the same time any wetland includes some function that can produce added value for local people without threatening the wetland base resources. This function categorized in carrier and information functions.

Ecotourism is the best alternative for produce new benefits for local people. Supremacy and ecotourism advantage lies on this golden key: When the tourist attraction is in place, people look to visit the area. Stronger attractions will attract more tourists. If tourism could be part of the lack of income to compensate, locale people try to preserve status and in later stages, income to be added. All of these efforts are highly dependent on the quality of wetland environments and conditions. It is quite clear that people understand the well established relationship between income and quality of the environment. Therefore, efforts will be directed to ensuring the ideal wetland ecosystems situation. Although ecotourism bugs may be appear in this area, but natural resources conversation is the primary purpose and the first priority in this region.

6. STATUS OF TOURISM IN WETLAND

Cheqhakhor Wetland Tourism base on origin can classified in to three categories:

1) Long range or Regional tourists:
   This group of tourists in clued people who come from different Provinces especially Khuzestan, Fars, Isfahan, Yazd and Tehran Provinces to spend their leisure time on wetland and margin villages.
2) Middle range inter- provincial tourists:
   This group of tourists includes people that their origin is in a province (Chahar Mahal and Bakhtiari), Neighborhood wetland with religious gravity (Hamza Ali Shrine Pilgrimage) has led to spend some Leisure time here, especially on holidays. Short-range or local tourists:

Some people of Borujen city (the nearest city to wetland) come to this wetland on their leisure time fishing and at least several times in a year.

Main Motivation of tourism in the region can be count as following:

1) enjoying the natural areas (landscape);
2) mountaineer and climbing, especially for youth (sports);
3) Hamze Ali Shrine Pilgrimage five km wetland wheat (Religious);
4) bird hunting and fishing by hook (hunting tourism);
5) short stay (two hours) for Relaxation specially for passing drivers (transit service);
6) Visiting nomads' life in summer.

At present the area lacks accommodations and facilities. On the other hand the tourist presence in the region has a high fluctuation. The presence of people is totally spontaneous and it is not planning. No institution is responsible for wetland Tourism Management. Thus, the presence of tourists in the region and their behavior does not follow any specific pattern, such as ecotourism. Tourist freedom and unmanagment Behavior of tourists in the region destroys the environment. The continuation of this process changes this behavior in to normal custom and behavioral patterns. In this condition ecotourism is an extraordinary alternative.

Ecotourism Within the new array of 'green' products and services, ecotourism claims to combine environmental responsibility with the generation of local economic benefits that will have both a development impact and serve as conservation incentives. Economic incentives are imperative for nature conservation, particularly in remote a dill-monitored (wunder, 2000, 465)

Achieving to ecotourism requires some actions in this area:

1. Choosing community based ecotourism as an ecotourism alternative;
2. Creating public institutions for Tourism Management;
3. Empowerment the indigenous people in a framework of local community-based ecotourism;
4. Teaching different groups of people involved in ecotourism;
5. Appropriate atmosphere for people's participation;
6. Building infrastructure needed by people participation;

7. Aware people and government responsible for the values of wetland ecosystems and the need to protect it;
8. Women and youth training for tourism services returns;
9. Preparing the grounds for the management impacts of tourism;
10. Create infrastructures for activities such as bird watching;
11. Institution and local legislation for visiting wetland;
12. Creating groups and nongovernmental conservation operations in the form of ecotourism tours.

With some actions, is expected to change direct use of wetlands resources into the indirect use of resources in wetland. Despite the problems and criticisms facing the ecotourism, it can play effective role in protecting the natural resources. Of course it is necessary to reduce the consequences of tourism in rural communities. Only some criticism from ecotourism cannot prevent the use of it and as an alternative for wetland conservation.

Conclusion

Ecotourism is the fastest growing sector of one of the world's largest industries. Nature tourism, a particularly dynamic sub-sector, is an important tool for generating employment and income in underdeveloped, biodiversity-rich Third World regions because it requires comparatively small investments. Conservationists also look at the nature tourism as a potential 'win-win' strategy of sustainable development, where tourist spending constitutes a much-needed instrument for capitalizing on biodiversity and natural sites. (Wunder, 2000, 466) Eco tourism is environmentally responsible for providing direct economical benefits for local residents and enhancing environmental conservation. All this condition could lead to sustainable development. Local people faced with income shortage and they try income compensation by increasing the number of livestock's, cultivated lands or utilizing directly from wetland base resources. All this activities help to wetland degradation. While wetlands have some interest attractions that without any planning effort attract tourist from different corner of the country. Degradation process that continues in wetland can be changed into environmental enhancement and conservation by community based ecotourism. Ecotourism activities are compatible with environment and it can help the ecosystem preservation by providing a source of income. Maintain income from tourism attractions are required environmental preservation by local people. This condition can be significant for the protection of sensitive wetlands ecosystem. to achieve sustainable ecotourism some actions are essential that the most important are as follows:

1. Raising awareness with decision-makers about the economic benefit of wetland conservation.
2. Training local population and community participation in planning and wetland management.
3. Predicting and prevent negative consequences of tourism in the area is also important to.

Without community based ecotourism, expected to accelerate the speed of destruction of the wetlands that ultimately will suffer the natural environment and the local community.

REFERENCES

31st Meeting of the Standing Committee Gland, (2005) Switzerland, 6-10 June, Agenda item 8.2.

Acott, T.G., La Trobe, H.L. and Howard, S.H. (1998), an evaluation of deep ecotourism and shallow ecotourism, *Journal of Sustainable Tourism* 6(3): 238–253.

Brander, L., Florax, R. and Vermaat, J. (2003), *the empirics of wetland valuation: a comprehensive summary and meta-analysis of the literature,* Institute for Environmental studies work paper, W30-30.

Butler, R.W. (1992), Ecotourism: it's changing face and evolving philosophy. *Paper presented to the IV World Congress on National Parks and Protected Areas,* Caracas, Venezuela.

Christ, C., Hillel, O., Matus, S., & Sweeting, J., (2003), Tourism and biodiversity: Mapping tourism's global footprint. Washington, DC: Conservation International.

Epler Wood Megan, (2002), *Ecotourism: Principles,* Practices and Policies for Sustainability, UNEP.

Fennell, David A. (2007), *Ecotourism,* Routledge.

Gordillo Javier, Stronza Amanda, (2008), *Community views of ecotourism, Annals of Tourism Research,* Vol.35, No.2, 448–468.

Groot, De R. S., (1992), *Functions of Nature: Evaluation of Nature in Environmental Planning, Management and Decision Making.* Wolters-Noordhoff, Groningen.

Heijnsbergen, P. van (1997), *International legal protection of wild fauna and flora,* IOS Press.

Higham, James E. S., (2007), *Critical issues in ecotourism: understanding a complex tourism phenomenon,* Butterworth-Heinemann.

Honey, M. (1999), Ecotourism and Sustainable Development: Who Owns Paradise? Island Press, Washington, *Journal of Travel Research* 28(3), 40–45.

Jones, Samantha, (2005), Community-based ecotourism, The Significance of Social Capital, *Annals of Tourism Research,* Vol.32, No.2, 303–324.

Kusler, Jon A. Opheim, (1996), *Teresa, Our national wetland heritage: a protection guide,* Environmental Law Institute.

Lewis, William M. Wetlands: (1995), *Characteristics and boundaries,* National Academies Press,

Marsh M. Willam, (2005), *Landscape Planning, Environmental Application,* John weily & Sons Inc.

Mitsch, William J., (2007), Gosselink, James G. *Wetlands, John* Wiley and Sons.

Monavary, Masoud, (2004), *Environmental Impact Assessment Cheqhakhor regional tourism master plan.*

Populations in Africa, *Ecological Economics* No.53, 177–190.

Scheyvens, Regina, (1999), *Ecotourism and the empowerment of local communities Tourism Management,* Volume 20, Issue 2, 245-249.

Schuyt, Kirsten, (2004), *The Economic Values of the World's Wetlands,* Luke Brander Institute for Environmental Studies Vrije Universiteit, Amsterdam, The Netherlands.

Schuyt, Kirsten, D. (2005), *Economic consequences of wetland degradation for local.*

Statistical center of Iran (SCI), *(1996), national agriculture census 1966,* Statistical center of Iran.

Statistical center of Iran (SCI), *(2006), national population census 2006,* Statistical center of Iran.

Statistical center of Iran (SCI), *(2003), national population census 2003,* Statistical center of Iran.

Tiner, Ralph W. (1999), *Wetland indicators: a guide to wetland identification, delineation, classification, and mapping,* CRC Press.

Wearing Stephen and Neil John, (1999), *Ecotourism: Impacts, Potentials and Possibilities,* Butterworth-Heinemann.

Weaver, D.B, (2001), *the Encyclopedia of Ecotourism,* Wallingford, Oxford. CABI.

Wunder, Sven, (2000), Ecotourism and economic incentives an empirical approach, *Ecological Economics,* 32, 465–479.

In: Research Studies on Tourism and Environment
Editors: J. Mondejar-Jimenez et al.

ISBN: 978-1-61209-946-0
© 2012 Nova Science Publishers, Inc.

Chapter 19

VINEYARD SITE QUALIFICATION AS A STRATEGY FOR VALUE CREATION IN WINE TOURISM: THE CASE OF "D.O. DE PAGO" PRODUCERS IN THE SPANISH REGION OF CASTILLA-LA MANCHA

Ricardo Martínez Cañas[*] and Pablo Ruiz Palomino
Universidad de Castilla-La Mancha, Spain

ABSTRACT

Wine regions in Spain are classified according to the quality wine they produce. The best quality of wine denomination comes from "*Denominación de Origen de Pago*", or "D.O. de Pago", that is the highest vineyard site legal qualification. This qualification is a very special guarantee of origin under which it is attempted to protect certain singular wines that are produced through a close relationship between vineyard and winery (like French *Chateaus*). D.O. de Pago wine has exceptional quality recognised both by specialists and by consumers. As approved plots, legislation allows the possibility of a guarantee of origin consisting of a name of a wine growing place that, because of its natural characteristics and the varieties of vines and growing systems, produces grapes from which distinctive wines are produced by means of specific preparation methods. In this paper we analyze the effect of this recognised quality trademark on wine tourism. Wine tourism in Pago's vineyards is a new and emergent industry that combines the synergic effect of commercial and tourism strategy from wine producers. To illustrate this phenomenon we analyze the case of wineries under that recognition that have emerged in the Spanish region of Castilla-La Mancha.

Keywords: Vineyard site, D.O. de Pago, Tourism strategy, Value creation, Spanish Wine

---
[*] Contact author: Ricardo.Martinez@uclm.es

1. INTRODUCTION

The wine producing sector is one of the best heritages of the Spanish region of Castilla-La Mancha, for both its economic and social relevance. It has become a strategic sector for the region that is the largest geographical area under vine cultivation in the world. Its 600,000 hectares under wine cultivation account for 50% of the Spanish vineyard surface area, representing 17,6% of Europe and 7,6% of the world's area under wine (Ministry of Agriculture of Castilla-La Mancha, 2007).

It is necessary to remark that Spanish wine is different from the perspective of the climate of the region where is produced, the grapes that are used, the elaboration process, technological advances employed, harvest tradition, etc. Quality wine produced in Spain is mainly labeled under a Spanish D.O.[1], such as the famous *Ribera del Duero* region in Castilla y León (MAPA, 2010), will have that region's D.O. stamp on the wine label or bottle seal. A D.O. is regulated by the Spanish Ministry of Agriculture, Fisheries and Food (MAPA, 2010) and follows strict guidelines to ensure the quality of the product. For wines in Spanish D.O. classification establishes the region of origin and ensures the quality of the wine produced within the D.O. boundary. The *Consejo Regulador*, a control council, enforces the rules and regulation. Each of the D.O. regions has its own *Consejo Regulador* that is also responsible for marketing the wines produced in their region.

Spain created the *Denominación de Origen* (D.O.) system in 1932 which was later revised in 1970. This denomination system shares many similarities with the hierarchical *Appellation d'Origine Contrôlée* (A.O.C.) system of France and Italy's *Denominazione diorigine Controllata* (D.O.C.) system (Wine Encyclopedia, 2009). As of 2007, there were about 67 DOs across Spain. In addition there is a *Denominación de Origen Calificada* (D.O.C.) status for DOs that have a consistent track record for quality. Also, additionally in the top of the quality recognition, there is the Denominación de Pago (D.O. de Pago) that is a legal denomination for individual single-estates with an international reputation of a wine of excellence. Each D.O. has a *Consejo Regulador*, which acts as a governing control body that enforces the DOs regulations and standards involving viticulture and winemaking practices. These regulations govern everything, ranging from the types of grapes that are permitted to be planted, the maximum yields that can be harvested, the minimum length of time that the wine must be aged and what type of information is required to appear on the wine label. Wineries that are seeking to have their wine sold under D.O., D.O.C. or D.O. de Pago status must submit their wines to the *Consejo Regulador* laboratory and tasting panel for testing and evaluation. Wines that have been granted D.O./D.O.C./D.O. de Pago status will feature the regional stamp of the *Consejo Regulador* on the label.

This legal recognition have made that during the last two decades wine producers of Castilla-La Mancha are trying to be more competitive and create a first class status with regard to grape varieties and growing methods, oenology technology and quality-price ratio of the wine. A few producers are generating a remarkable differentiation strategy (Porter, 1980) for defending and promoting the culture of D.O. Pago or single-state wines[2]. Their main resources are that their wines are produced on specific vineyards which reflect the

---

[1] *Denominación de Origen* (D.O.) in Spanish that can be literally translated as "label that certifies the quality of the origin".
[2] We are going to use D.O. Pago wines and single-state wines as synonyms.

unique personality of their owners choices, soils, subsoils and microclimates. They believe that this is an ages-old concept of quality wine that was only recently recovered in Spain and we believe that can be the key for their success in a niche of market of a so high competitive and global sector as the wine industry is.

In the region of Castilla-La Mancha there are nine Denominations of Origin: Almansa, La Mancha, Manchuela, Méntrida, Mondéjar, Ribera del Júcar, Uclés, Valdepeñas and Jumilla (part of which is located in the region of Murcia) and six Estate Wines for highly special denominations of origin that comprehends highly unique wines produced out of close relationship between the vineyard and the winery and with and exceptional quality recognized by both wine experts and consumers: Dominio de Valdepusa, Finca Élez, Pago Guijoso, Dehesa del Carrizal, Pago Florentino y Pago de Campo de la Guardia. The 9 denominations of origin and the 6 estate wines extend over a surface area near 280,000 hectares (Ministry of Agriculture of Castilla-La Mancha, 2007). In Castilla-La Mancha region those producers are only a small part of the industry that ranges between 20 and 25 million of hectolitres, out of which 50% is table wine, 22% must, 17% distilled wine, 6% wine with some Denomination of Origin, and 5% Vino de la Tierra de Castilla. That means that D.O. de Pago is less than 1% of the total production in the region which reflects that there is a tiny part of the winemakers that are interested in the search for excellence and the development of wines with the highest quality. So, winemakers are trying to facilitate the identification of a wine market demand other than that of the more common D.O. or table wines.

Recently the number of entrepreneurs/winemakers that want to create single-state wineries with high quality and with which it's key in the market a legal recognition. Their aims are the generation of a synergetic effect trying to link directly the excellence of their wines with other complementary activities like a wine tourism strategy for exploiting their unique and difficult to imitate resources. With activities like wine tourism we refer to tourism whose purpose includes the tasting, consumption or purchase of wine, often at or near the winery. Mainly, wine tourism can consist of visits to wineries, vineyards and restaurants known to offer unique vintages, as well as organized wine tours, wine festivals or other special events. From a business perspective that means that those winemakers are interested in developing a high value-add wine activities that differentiate them from the medium/low quality wines that are produced in the rest of the region.

To sum up the objectives of this chapter, we analyze the join effect of the D.O. de Pago quality trademark on wine tourism linking the synergic effect of commercial and tourism strategy in the case of winemakers from the region of Castilla-La Mancha. The rest of the paper structure is structured as follows. In the second epigraph we review the wine classification system in Spain to locate where the D.O. de Pago is. Next we analyze the wine tourism strategy as a complementary strategy for wine producers. Then we study the case of the six legal recognized single-state wine producers in Castilla-La Mancha. To finalize we claim some conclusions with regard to the present research.

2. WINE CLASSIFICATION SYSTEM IN SPAIN: THE D.O. DE PAGO DENOMINATION

After Spain's acceptance into the European Union, Spanish wine laws were brought in line to be more consistent to European systems (Law of Wine and Vineyard[3], 2003). The main development of this law was a five-tier classification system that is administered by each autonomous region[4]. The five-tier classifications, starting from the bottom, include (Wine Encyclopedia, 2009):

- *Vino de Mesa (VdM)* - These are wines that are the equivalent of most country's table wines and are made from unclassified vineyards or grapes that have been declassified through "*illegal*" blending. Similar to the Italian *Super Tuscans* from the late 20th century, some Spanish winemakers intentionally declassified their wines so that they have greater flexibility with blending and winemaking methods.
- *Vinos de la Tierra (VdlT)* - This level is similar to France's *vin de pays* system, normally corresponding to the larger Autonomous Community geographical regions and appear on the label broader geographical designations such as Andalucia, Castilla La Mancha and Levante.
- *Vino de Calidad Producido en Región Determinada (VCPRD)* - This level is similar to France's *Vin Délimité de Qualité Supérieure* (VDSQ) system and is considered a stepping stone towards D.O. status.
- *Denominación de Origen (D.O.)*- This level is for the mainstream quality-wine regions and are regulated by the *Consejo Regulador* who is also responsible for marketing the wines of that D.O. Almost two thirds of the total vineyard area in Spain is within the boundaries of a D.O. region.
- *Denominación de Origen Calificada (D.O.C.)*- This designation, which is similar to Italy's *Denominazione di Origine Controllata e Garantita* (DOCG) designation, is for regions with a track record of consistent quality and is meant to be a step above D.O. level. For instance, D.O. Rioja, the most well known wine trademark in Spain, was the first region that afforded this designation in 1991.

Additionally, and with more recognition of quality than the *Denominación de Origen Calificada*, there is a Denominación de Pago (D.O. de Pago) designation for individual single-estates with an international and national reputation of wine of excellence. As of 2009, there were 6 estates with this status in Castilla-La Mancha[5].

With a focus on this last denomination, the D.O. de Pago, which represent less than 1% of the wine industry in Castilla-La Mancha, we recognize the efforts of small wineries (worried about quality and the distinctive value of their wines) to have a legal recognition against the rest of the Industrial Wineries (that are the mainstream in the region) with very

---

[3] Ley de la Viña y el Vino (2003).
[4] Non-autonomous areas or wine regions whose boundaries overlap with other Autonomous Communities in Spain (such as *Cava, Rioja* and *Jumilla*) are administered by the *Instituto Nacional de Denominaciones de Origen* (INDO) based in Madrid.
[5] There are currently only 9 estates with the D.O. de Pago status in Spain: 6 in Castilla-La Mancha and 3 in the region of Navarra. The rest of Autonomous regions in Spain are not using this system of denomination.

large productions of name-brand wines made with grapes –often, bought-in– from different places. Clearly D.O. de Pago it's a commercial differentiation and a strategy for putting in value the quality of the product related to their place of production. Actually, only a very small group of producers following the legal criteria of single-estate wines are trying to emphasize the personality that defines their wines. That means that in the search of a excellence label they want to add a great value for their wines in the market. This is the main reason why at the beginning of this Century a few winemakers founded Grandes Pagos de Castilla (Great Growths of Castile) as a non-for-profit association responding to the requests of many colleagues in other parts of the country who wished to make the single-growth concept better known and to search excellence through the direct relationship between wine and its place of origin[6].

In the Spanish Law of Wine and Vineyard (2003) the D.O. Pago is a classification created for a single estate, which designates that Pago (vineyard-estate) as a fine producer of wine. The Pago is allowed to set its own rules for grapes and production. All grape growing, transformation, and bottling must take place on the estate.

So, to be a "state" or "state bottled" wine producers have to abide by specific rules:

1. The estate wine must designate an appellation for origin linked to a traditional way of cultivate the vineyard and with a very strict location criteria, often centering on specific plots of land, admitting only those wines produced in their immediate vicinity.
2. When the totality of a state-wine is located in a D.O.C., it can also be named as wine of D.O.C. Pago if certifies the requisites for both denominations.
3. Single-stated wines must be bottled by physic persons or commercial societies that must be the owners of the Pago or the owners of the wineries close to the state.
4. All grape must be harvested in the vineyard of the state and must be elaborated, bottled and aged in a separated place from other wines.
5. All state wine have to be under an integral quality management system applied during the whole process: from the grape to the commercial market.
6. Every state wine has to be managed by a regulator system, supervised by the Autonomous Region.

To our knowledge for the D.O. de Pago denomination we have to consider three key aspects that determine the quality of the wine:

- Wines and winery must be located in the same or near places;
- Grapes must come from vineyards owned or controlled by the winery;
- Wine must have been produced, from crush to bottle, in a continuous process without leaving the winery's premises approved by law.

---

[6] Actually, in this association there are about 20 wineries that want to differentiate the label of their products under the D.O. de Pago but some of them haven't fulfill the legal requirements to be considered as single-stated wineries so far. Some of them fulfill the basic requirements: have grapefruits in the same area than the winery, their vineyards are owned and controlled by the winery, and the wine is being produced, from crush to bottle, in a continuous process without leaving the winery's premises. Thus, some of them are also using the name of "Pago" in their denominations or in some of their commercial labels but not with a legal recognition. Actually, they are in the legal process to acquire the denomination and in few years it will be more D.O. de Pago in the region. For this chapter we only consider the winemakers that can use the D.O. de Pago with a legal recognition.

Those marked characteristics make the difference with the other denominations we introduced above. This can imply a lot in the market for both producers and consumers.

3. WINE TOURISM

Both wine and tourism industries have achieved high levels of growth within Spain in recent years, and are significant contributors to the GDP[7] as invisible exports (MAPA, 2010). Also, much has been written about wine and tourism industries, there is not a completely accepted definition of what is wine tourism (Getz, 1998). Most of definitions have a lot of common characteristics including (Charters and Ali-Knight, 2002):

- A lifestyle experience;
- Supply and demand;
- An educational component;
- Linkages to art;
- Wine and food;
- Incorporation with the tourism-destination image;
- A marketing opportunity which enhances the economic, social and cultural values of the region.

So, wine tourism experience can therefore be provided by means of high number of ways, the most notable being events and festivals, cultural heritage, dining, hospitality, education, tasting and cellar door sales, and winery tours (Charters and Ali-Knight, 2002).

Cambourne, Macionis, Hall, and Sharples (2000,p. 297) note that wine tourism is a concept that is still undergoing substantial development and there is, yet, a great deal to learn about how the two industries can make a positive contribution to each other and to their shared regions throughout the world. They also highlight the fact that wine can provide a major motivating factor for tourists to visit a destination as, more often than not, wine regions tend to be attractive places, and the vineyards themselves are aesthetically pleasing. The climate required to produce grapes is also congenial from a tourist's perspective.

Getz (1998) considers that wine tourism can be approached from three different perspectives:

- As a strategy to increase the touristic market and develop wine-related attractions and imagery.
- As a form of consumer behavior where wine lovers or those interested in wine regions travel to preferred destinations.
- As an opportunity for wineries to educate consumers and sell their product directly to them.

In this research wine tourism activities are those that includes the tasting, consumption or purchase of wine, often at or near the source where is generated (Carlsten and Charters,

---

[7] Gross Domestic Product.

2006). From this point of view, wine tourism can consist of visits to wineries, vineyards and restaurants known to offer unique vintages, as well as organized wine tours, wine festivals or other special events related to this industry.

In the recent years, many wine regions around the world have found out that wine tourism can be an interesting opportunity to promote tourism (Cambourne et al., 2000). Accordingly, growers associations and others in the hospitality industry in wine regions have spent significant amounts of money over the years to promote such tourism. This has been for both "*old world*" producers (such as Spain, Portugal, France or Italy), and for the so-called "*new world wine*" regions (such as Australia, Argentina, Chile, United States or South Africa), where wine tourism plays an important role in advertising their products. In countries like Argentina, the Mendoza Province is becoming one of the tourist destinations in the country as Argentine wine strides to gain international recognition. In Spain, in a similar way some well known regions like La Rioja are quite famous around the world for their wine tourism activities. Trying to imitate the famous and world recognition, other regions like Castilla-La Mancha are investing money and great efforts to promote wine tourism activities in order to generate regional wealth through exploiting their natural, cultural and culinary resources.

We think that for D.O. de Pago producers wine tourism can be approached as a mixture of the approach proposed by Getz (1988). From this view, wine tourism is a complementary strategy to reinforce the excellence of their wines and products and that can add high value for the commercial strategy of the firms. That means, in other words, winemakers are focusing their activities for wine lovers and experts interested in wine regions or in other related activities (like for instance hunting, food tasting and natural activities). And finally winemakers have an excellent opportunity to educate consumers interested in high quality wines and sell their products and services directly to them.

With the purpose to analyze their strategies, activities and services, we are going to make a exploratory study of the producers with the D.O. de Pago denomination and their wine tourism activities in the region of Castilla-La Mancha.

4. D.O. DE PAGO WINERIES IN CASTILLA-LA MANCHA: MAIN CHARACTERISTICS AND WINE TOURISM ACTIVITIES

Once we have made a direct link with D.O. de Pago and with wine tourism, we are going to analyze the case of the six single-state producers that officially are recognized by the Ministry of Agriculture of Castilla-La Mancha to use this denomination. From every producer we are going to synthesize the role of key characteristics of the owner, origin, philosophy, the microclimate, technology, land/state properties, planted varieties and finally the wine tourism activities developed to complement the value generation in their core activities.

4.1. Dominio de Valdepusa[8]

The Dominio de Valdepusa is located near Malpica de Tajo, in the province of Toledo 109 km from Madrid and 53 Km from Toledo. Within the historic location of the Casa de Vacas Estate which is owned by the family of its owner (Carlos Falcó, the Marquis of Griñón) since 1292. With an outstanding winemaking operation, it occupies an area of 50 hectares, of which currently 42 hectares are planted to wines, including 14 hectares of the original Cabernet Sauvignon vineyard planted in 1974. Since 1991, this was progressively expanded with new Syrah (1991) and Petit Verdot (1992) vineyards both of them, pioneering projects in Spain- as well as a new Graciano vineyard (2000), all under the Lyre and Smart-Dyson trellis systems[9].

A specific, single-estate *Denominación de Origen* appellation was awarded by the regional authorities to Dominio de Valdepusa on 19 July, 2002, which makes, together with Finca Elez, the first estate wine in Spain to receive this qualification. This was later ratified by the Spanish Ministry of Agriculture (21 February, 2003) and the European Union (14th April 2004). This state also has been amended by the Order of 20 Julay 2005 and then by the Order of 7 November 2008. Carlos Falcó is a well known entrepreneur in Spain as being one of the promoters of the Grandes Pagos de España[10] association. In 2006 he also was the promoter of the association *Divinum Vitae*[11] for promoting wine tourism in Castilla-La Mancha.

The main house and cellar of Dominio de Valdepusa are located in historical buildings from the 18th century. The underground barrel cellars are completely renovated in 1989 and also are equipped with modern air conditioning facilities and hygrometric control that permit aging under controlled temperature and humidity conditions.

Current production is centred exclusively on red wines produced from the estate's vineyards. The varietal collection includes the Dominio de Valdepusa Cabernet Sauvignon, Syrah, and Petit Verdot to which, since 1997, were added two blends, *Svmma Varietalis* and the top-notch *Emeritvs*, a blend of the best casks of each variety. All of them are regularly included among the best Spanish wines in guides and articles by national and international experts.

In 2009, as part of their wine tourism strategy, they are opening the winery for visiting the cellar, the warehouse and other historic buildings in an exclusive visit with selected and small groups. These visits consist in guided tours with technical staff and an explanation of the viticulture techniques that are pioneers in the world. With the visit they offer wine tasting (also of olive oil produced in the state) with a prestige professional and also with the speeches and commentaries of people from the Griñón Family. They are offering three different kinds of Lunch: homemade (products from the state), gastronomic (with products from the region) and with a high prestige chef for a special event or activity. Finally, Dominio de Valdepusa is offering, for direct sale in the state, special editions of some bottled wine with special wood boxes.

---

[8] http://www.pagosdefamilia.es/ (last access march 10th 2010)
[9] A system imported and adapted in Spain by Carlos Falcó from the California vineyards techniques.
[10] An association of wine states from all of Spain's regions united to defend and promote the culture of single-state wines, produced on specific vineyards and which reflects the unique personality of their soils, subsoils and climates.
[11] http://www.enoturismocastillalamancha.com/index.php?L=en (last access march 10th 2010)

4.2. Pago Finca Élez[12]

The vineyard of Finca Élez covers a surface area of 40 hectares that is located in the population of El Bonillo (Albacete), at the heart of the Sierra de Alcaraz, at 1,080 m above sea level. The owner is Manuel Manzaneque, an actor, director and theatre businessman (the winner of awards such as the National Cinema Prize and the National Theatre Prize). As the owner and main promoter of this state he felt a call to create an exclusive product of the highest quality leading him to found his own winery. To that end, after long years of analysing soils, microclimates, and types of vineyards, he placed his winery in the Finca Élez estate in southern Albacete province.

As special varieties in this estate the white grapes are Chardonnay and the reds are: Cabernet-Sauvignon, Merlot, Tempranillo and Syrah. The wine processing of the grapes harvested in these vineyards is performed in a winery located in the Manzaneque family Estate. With the same order from the 19 of July 2002 of the Regional Ministry of Agriculture amended by the Order of 25 November 2005, then by the Order of 10 January 2007 and finally the Order of 8 September 2008[13], confirms the status of the Finca Élez D.O. for some of the quality wines produced in the above vineyard.

Which differences this Finca Élez is that is quite near the Sierra de Alcaraz where the breeze blows constantly form the mountains, and large day-night temperature differences ensure that our grapes reach their full potential. The property is a real 'domaine' with varieties that were brought from Bordeaux and Burgundy by the same oenologist that formerly developed Valdepusa cellars. They also have Syrah vines originated in the Rhône[14] plus a Tempranillo vineyard that are planted immediately close to the winery so that the grapes don't need to be transported.

Another distinctive characteristic of Manuel Manzaneque is that it combines the highest technology with the care of a craftsman, from the vineyard to the bottle. This process ensures that the wines are different and can be qualified as single-estate wine what has made that their wines have received some of the most notable international awards. The Finca Élez red came in sixth, out of more than 100 Cabernets from all over the world, in the competition organised by the magazine *Weinwirtschaft* during the *Interwein* wine fair in Germany. For three consecutive years, the winery has won Spain medals in the prestigious 'Chardonnay of the world' competition in Burgundy, and a silver medal at the Vinalies Internationales 2001, in Paris. Manuel Manzaneque personal philosophy can be resumed in one paragraph, he thinks: *"that wine, just like tragedy or comedy in the theatre, is part culture and part pleasure. You must always give your best, one show after another, season after season, if you intend to win the public's favour, and that is our pledge day after day"*.

Actually in Finca Élez state, as part of their wine tourism strategy, they are offering visit to the cellar and warehouse of the state. Those visits include wine tasting courses. Further, as a special activity, they are offering a complete touristic weekend pack that includes wine and food tasting, accommodation in rural houses or in nearby hotels. This weekend activities are completed with the visit to places of interest in the Don Quixote de la Mancha Route and

---

[12] http://www.manuelmanzaneque.com/index_en.html (last access February 17th 2010)
[13] This D.O. de Pago has suffered some changes from their original order.
[14] It's a famous valley in France for their wines.

another visit to enjoy the landscape and natural environment of the Lagunas de Ruidera Natural Park.

4.3. Pago Guijoso[15]

The Pago Guijoso state is composed of twelve smallholdings of vineyards and located in the municipality of El Bonillo (Albacete), bordering onto the provinces of Albacete and Ciudad Real, at the start of the Guadiana river, and at an 1,000 m above sea level. This state is property of Sanchez-Multiterno Family and their grape varietals are: Chardonnay and Sauvignon Blanc for white wine, and Cabernet Sauvignon, Merlot, Syrah and Tempranillo for red wines.

The harvest is made by hand and very carefully produced. In their philosophy they believe that their wines are pleasure for the body and the spirit. Moreover, these are highly natural wines avoiding all herbicides, pesticides and so on. All these aspects are complimented with the fact that the winery is located in an extraordinary landscape with poor soil and 98 hectares of manicured vines, perfectly in tune with the surrounding natural environment. Their commitment is to ever-improving the quality in their wines through the application of cutting-edge oenological research as one of their principles. Quality is precisely one of Bodegas y Viñedos Sánchez Muliterno's guiding principles. This commitment has led them to apply for ISO 9001:2000 certification from BVQI, Spain's leading certification agency. They are amongst the few wineries in the world that have ensured comprehensive certification of the entire winemaking process, from the vineyard to the ageing cellar.

Their first wine was made by purely traditional methods in 1990. Having verified the estate's potential, the foundations of a bodega were built in 1993. However, official endorsement for D.O. Pago Guijoso only came in 1995, making it Spain's third single-estate D.O., or 'Pago', after D.O. Dominio de Valdepusa and D.O. de Pago Finca Élez. The Order 15 of November 2004 of the Regional Ministry of Agriculture sets forth the winemaking production standards that ensure the quality wines produced in the Guijoso vineyard.

As their tourism activities you can visit a natural environment of breathtaking beauty at the source of the River Guadiana, and close to the Ruidera National Park lakes. This whole area, including the nearby Montesinos Cave, was documented by Miguel de Cervantes as part of the route that was also covered by Don Quixote in the course of his adventures. This winery is also offering visit to wine cellars and accommodation in several rural houses inside the state. Furthermore there are wine and food tasting activities. The main difference of this state from others is that visitors can use the estate for hunting because the whole state has 3000 hectares.

4.4. Pago Dehesa del Carrizal[16]

Dehesa del Carrizal estate wines extends over a surface area of more than 22 hectares located in Retuerta del Bullaque in the province of Ciudad Real. The origin of this vineyard

---

[15] http://www.sanchez-muliterno.com (last access march 1st 2010)
[16] http://www.dehesadelcarrizal.com/ingles/index.html (last access march 15th 2010)

dates back to 1987, where the owner of the land, Marcial Gómez Sequeira, decided to plant 8 hectares with the first vineyard grapes of the red variety Cabernet Sauvignon. In the following years, more than new 14 hectares were planted with the white variety Chardonnay and reds such as Myrah, Merlot and Tempranillo.

The Doctor Marcial Gómez Sequeira, is a well known in his facet of businessman, game hunter and mainly as a wine enthusiast. During his hunting trips around the world he discovered a common component to many of the great red wines: the Cabernet Sauvignon variety. His curiosity led him to experiment with this grape on his estate in the Montes de Toledo. To his knowledge, no-one had planted vineyards in the Montes de Toledo region which was the domain of deer, mountain goats, and boar, so no one knew what the outcome of the project would be. He decided to take the grapes to the winery of his friend the Marquis of Griñón (Valdepusa), and so discover the true potential for quality of the grapes before making bigger investments. It only took five harvests to confirm the success of Cabernet Sauvignon in this region, and of Gómez Sequeira as a wine producer. He rapidly set about completing the project begun, almost as a game in the 1980's.

The Order 1 of February 2006 of the Regional Ministry of Agriculture confirms the Dehesa del Carrizal status for some quality wines produced in those vineyards and sets forth the regulations relating to standards of production.

As a wine tourism strategy, D.O. de Pago Dehesa del Carrizal is offering the estate shelters for visiting, also with the warehouse and other suitable facilities. Also Pago del Carrizal organizes events for social celebrations with individuals and companies. They are offering these activities for enjoyment in any season of the year. Those activities are complemented with tourist activities in the Cabañeros Nature Reserve.

4.5. Pago Florentino[17]

Pago Florentino, together with Pago Campo La Guardia, is the last of the D.O. de Pago approved in Castilla-La Mancha. The winery is located in the municipality of Malagón, in the province of Ciudad Real. The 58 hectare estate, known as *La Solana*, belongs to the Arzuaga group of wineries and contains the following red grape varieties: Tempranillo, Syrah and Petit Verdot. It was purchased in 1997 but the first wine was not made until 2002. This D.O. de Pago acquired its status with the order of 20th August 2009 of the Regional Ministry of Agriculture.

The planting density is between 1,300 and 3,000 vines/hectare, and the maximum authorized yield is 10,000 kg/ha for the three varieties, what means they produce few wine in the search of more quality. This is also related to La Mancha microclimate that is brutally hot in the summer and quite cold in the winter. The minimum alcohol content is 12.5°, and the wines must contain a minimum of 90% of the Tempranillo variety. The wine is aged for 8 months in a mixture of French and American oak what is part of Florentino Arzuaga, the owner of the winery, philosophy of winemaking. This way of doing things is related to all the D.O. de Pago what means that in addition to having a proven track record of consistent quality, the wines have to be both produced from estate-grown grapes and also have to be processed and aged in a winery (bodega) located on the estate.

---

[17] http://www.pagoflorentino.com (last access April 12th 2010)

As wine tourism strategy D.O. Pago Florentino has created courses for tasting wine and also for olive oil elaborated in *La Solana* state. As direct strategy for commercializing their products they have a shop for direct vending to consumers inside the state. Also, they organize traditional food degustation and business lunches only for special events.

4.6. Campo de la Guardia[18]

Campo de la Guardia D.O. de Pago is located in the municipality of La Guardia, in the province of Toledo. The 81 hectare estate belongs to *Bodegas Martúe* (with wineries in Spain and Portugal). Campo de la Guardia is the result of the effort of Fausto Gonzáez and Julian Rodríguez that planted vines back in 1990, as he realized the untouched potential in the area. They planted Syrah, Cabernet Sauvignon, merlot, Chardonnay, and of course, Tempranillo, Spain's noble grape. Located at 700 meters above the sea level, Campo de La Guardia is divided in 2 estates, *Campo Martuela* and *El Casar*. The sandy-calcareous soil, combined with cold winters and warm summers but with numerous hours of sun exposure, are favorable to fleshy and concentrated wines with own personality. In order to reach the best quality and to ferment the grape under controlled temperature, they harvest by night. During all year long efforts are all focused on a low production which culminates in only 800 grams per vine in certain grape varieties. This controlled low yield tends to obtain sensational wines which are totally faithful to the terroir[19]. This D.O. de Pago acquired its status with the order of 20th August 2009 of the Regional Ministry of Agriculture.

Bodegas Martúe philosophy is to produce a quality wine at accessible prices. They have realized that there were a huge hole in the market for good value top quality wines. The winery is designed in a *hacienda* style, like an old Colonial mansion, although it is quite new. The vineyards surrounding the estate are Chateaux-style with a special tasting touch. González family, the owners of Bodegas Martúe, established in the 80´s a new way of making wines in the traditional Spanish world of wine. Meticulous and avant-garde, Fausto González waited patiently for twenty years in order to reach an optimum quality before labelling under Martúe brand. Faithful to his goal, he always had in mind to make personal and singular as well as universal wines which would be able to spend a delicious and sensorial world perfectly understandable by everyone, either professional or amateur consumer.

The constant effort of Fausto Gonzalez and Julian Rodriguez has been the key of the increasing success of this family winery. Instead of increasing prices or the volume of production due to market demand, Bodegas Martúe decided to bet on repeating its philosophy in another projects (in other parts of Spain and Portugal) as encouraging as the first one. The winery is still finding its way in the market using a very optimistic commercial strategy to that become very well-known abroad.

---

[18] http://www.martue.com/ (last access April 12th 2010)
[19] Terroir is a French word used to describe a land with special characteristics.

Table 1. D.O. de Pago in Castilla-La Mancha

Title	Location	Pago date	Wine Varieties	Hectares	Nearby places of interest	Wine Tourism activities
Pagos de Familia Marqués de Griñón	Malpica de Tajo (Toledo)	2002	Red: Cabernet Sauvignon, Merlot, Syrah, and Petit Verdot;	50	50 Km near Toledo a World Heritage City	Wine tasting, visit wine cellars and historic buildings in the state. Lunch (homemade, gastronomic, chef's special).
Finca Élez D.O.	El Bonillo (Albacete)	2002	Red: Chardonnay, Cabernet Sauvignon, Merlot, Tempranillo and Syrah	40	Don Quijote de la Mancha Route and near Lagunas de Ruidera natural park	Visit to the warehose, wine tasting. Complete weekend pack (wine and food tasting, accommodation and visit to places of interest nearby)
Pago Guijoso	El Bonillo (Albacete)	2004	Red: Cabernet Sauvignon, Merlot, Tempranillo and Syrah; White: Chardonnay and Sauvignon Blanc.	98	Don Quijote de la Mancha Route and near Lagunas de Ruidera natural park	Visit wine cellars, accommodation in rural houses with wine and food tasting, hunting in the finca el Guijoso (3000 hectares).
Pago Dehesa del Carrizal	Retuerta del Bullaque (Ciudad Real)	2006	Red: Garnacha Tintorera and Syrah	42	Near Cabañeros' nature reserve	Visit estate shelters, use of warehouse for social celebrations with wine and food tasting
Pago Florentino	Malagón (Ciudad Real)	2009	Red: Tempranillo, Syrah and Petit Verdot	58	Near Tablas de Daimiel Natural Park	Wine and olive oil tasting. Shop inside the state. Food only for special events.
Pago la Guardia	La Guardia (Toledo)	2009	Red: Cabernet Sauvignon, Merlot, Malbec, Tempranillo, Syrah, and Petit Verdot; White: Chardonnay.	81	50 Km near Toledo a World Heritage City	Wine tasting and guided visit to the estate

As part of their wine tourism strategy they have a collaboration agreement with Cellar Tours[1] that is an American company that customized and tailored tours for VIP[2] wine tastings at top wineries with luxury tours. These tourism activities are complemented with wine tasting courses and guided visit through the state.

CONCLUSION

The region of Castilla-La Mancha has traditionally been known as a region producing vast quantities of dire wine, with a philosophy of quantity rather than quality. This has changed in recent years because a small group of entrepreneurs have increased their efforts to enhance the quality of the wine produced in this region. They know that grapes are already there, but if they want to make something special they have to differentiate their wines with quality. In a region that makes something like half of Spain's wine it's a high risk strategy.

With the Spanish legal recognition system of the quality of wine, created a few years ago, in this region only a few producers (less than 1% of the wineries) are trying to be more competitive and create a first class status with regard to being recognized by the D.O. Pago or single-state wines denomination. This denomination reflects that wines are produced on specific vineyards and show the unique personality of their owner choices, soils, subsoils and climates. D.O. de Pago winemakers think that this is an ages-old concept of quality wine that was only only recently recovered in Spain and can be the key for their success in niche of market of a high competitive and global sector like wine industry.

In this paper we have analyzed six D.O. de Pago wineries located in the Spanish region of Castilla-La Mancha in the search of the role of key characteristics of the owners, philosophies, differences in their microclimate, technologies used, land properties, and so on. We can conclude that as table 1 shows, every winery has peculiarities that allows the creation of high quality wine. This high quality wine, that is the core business of the winemakers, also can be complemented with wine tourism activities. These wine tourism activities mainly consist of visits to wineries, vineyards and restaurants known to offer unique vintages, as well as organized wine tours, wine festivals or other special events related to this industry.

REFERENCES

Carlsen, J. and CHARTERS, C. (2006). *Global Wine Tourism,* Edith Cowan University, Cabi Publishing, New York.

Charters, C. and Ali-Knight, J. (2002): *Who is the wine tourist?, Tourist Managament,* Vol. 23, pp. 311-319.

Cambourne, B., Macionis, N., Hall, M. and Sharples, L. (2000). In: Hall, M., Sharples, L., Cambourne, B., & Macionis, N. (Eds.), *Wine Tourism around the World: Development, management and Markets* (pp. 297–320). Oxford: Butterworth–Heinemann.

---

[1] Cellar Tours (http://www.cellartours.com) is a specialized travel agency for VIP's recommended by well-known magazines like Forbes Traveller, the Wall Street Journal and Travel & Leisure. This firm is a provider of private, chauffeured luxury food and wine tours around European countries like France, Italy, Spain and Portugal.

[2] Very Important Person.

Getz, D. (1998). Wine Tourism: Global Overview and Perspectives on its Development. Wine Tourism for Perfect Partners: *Proceedings of the First Australian Wine Tourism Conference.* Canberra: Bureau of Tourism Research.

Law of Wine and Vineyard (2003). Ley de la Viña y el Vino, Ministerio de Agricultura, Pesca y Alimentación. Available online at http://www.mapa.es/alimentacion/pags/vino/ley.pdf (last access march 15 2010).

Ministry of Agriculture of Castilla-La Mancha (2007). Analysis of the wine sector, Junta de Comunidades de Castilla-La Mancha, Spain.

MAPA, Ministry of Agriculture, Fishery and Food (2010). *Strategy for the Wine Industry in Spain 2007-2010.* Ministerio de Agricultura, Pesca y Alimentación.

Martin, E. (2009). *The Wine Encyclopedia,* Global Media Publishing: Delhi.

Porter, M. (1980). *Competitive Strategy: Techniques for Analyzing Industries and Competitors,* Free Press, New York.

Chapter 20

THE UNIFORM SYSTEM OF ACCOUNTS FOR THE LODGING INDUSTRY AND XBRL: DEVELOPMENT AND EXCHANGE OF HOMOGENEOUS INFORMATION

Adelaida Ciudad Gómez[1]

University of Extremadura, Spain

ABSTRACT

The internationalization of markets and users, new management tools using information based on a homogeneous mass, or the use of the Internet for exchange, has made as a whole that standardized information available internationally, not only in development but also in its publication and exchange, should become a requirement.

Therefore, this paper aims, first, to describe the possibilities for the lodging sector to enable it to develop, present and exchange information uniform, focusing both on financial and environmental information.

On the one hand, the *Uniform System of Accounts for the Lodging Industry (USALI)*, whose great contribution is to facilitate homogenization in the presentation of information worldwide, allowing comparison between hotels and build statistical databases from figures uniform and standardized, including providing segment information according to the requirements of *International Financial Reporting Standards* -IFRS 8 and *Statement of Financial Accounting Standards*-SFAS 131, and on the other, to allow that information to be transmitted electronically in a prompt, transparent, accurate, quality, implementation of the XBRL standard.

The second objective we aim to propose is those issues which should be addressed in a near future.

Keywords: IFRS, USALI, harmonization, XBRL, environment

---

[1] Contact author: adelaida@unex.es

1. INTRODUCTION

Globalization and internationalization is one of the most important factors in the need for uniform accounting information allowing the needs of internal and external users to be met. To achieve this goal, the *International Accounting Standards Board* (IASB) and the *Financial Accounting Standards Board* (FASB) agreed in 2002 to work together and to converge their respective codes and standards, as a result of which in 2008 the *Security and Exchange Commission* (SEC) began to accept financial statements prepared on the basis of the standards issued by the IASB.

Focusing on the tourism sector, expansion of the large hotel chains has brought internationalization of the users regarding accounting information, requiring use of accounting systems which allow uniformity of operating and financial reports. In response to that need, in 1926 the Hotel Association of *New York* created a reporting system called *Uniform System of Accounts for Hotels*[2] (USAH), which broadly unified accounting criteria in the lodging sector and allowed different hotels belonging to the same chain in different countries with different accounting regulations to be compared.

But the need for uniform, homogeneous information is not only a consequence of the internationalization of markets and users, but also of the ever increasing need to use useful decision-making tools such as *benchmarking*, which allows improvement opportunities and the reasons why one hotel has different levels of efficiency and efficacy to be seen, for which purpose it is indispensable that the information compared is guaranteed to be homogeneous; or *revenue* or *yield management*, which uses mass market information to predict dynamically the customer's behaviour at micromarket level, allowing the company to sell the right product to the right customer at the right time and at the right price, and therefore to maximize profits.

Finally, an increasing number of companies make use of Internet to publish their accounting information, although the amount and content of it varies, causing another problem, that the information cannot be quickly and simply downloaded for examination unless a common language is used for publication and transmission. To face this problem, basing on the initiative of the *American Institute of Certified Public Accountants* (AICPA), professionals have developed a language for financial reporting on the Web, called *eXtensible Business Reporting Language* (XBRL), which will allow reports to be generated digitally and processed automatically, making data exchange easier by using a method based on standards for the electronic exchange of economic and financial information.

Given the above mentioned, access to homogeneous information internationally has become a requirement, not only for its making but also for publication and exchange.

So the first aim of this paper is to describe the standards, codes and regulations arising from the process of convergence between the IASB and the FASB (IFRS & US GAAP) which allow companies in the lodging sector more, homogeneous information, on the one hand necessary for the use of management tools through implementation of the *Uniform System of Accounts for the Lodging Industry* (USALI), and on the other, opening up the possibilities in terms of publication provided by XBRL.

Secondly, we will think about the questions that should be considered in the near future to allow improvement to the USALI, and propose that its introduction should be

---

[2] Called *Uniform System of Accounts for the Lodging Industry* (USALI) in its 10th edition.

supplemented by the use of the XBRL-IASCF taxonomy, focusing especially on the internal information not standardized by this taxonomy.

To conclude, because the information supplied by companies should not only allow their financial and economic behaviour but also their social behaviour to be measured, we do not limit ourselves to financial and economic data, but also consider environmental information.

2. USALI, THE LODGING INDUSTRY'S ACCOUNTS REPORTING AND MANAGEMENT SYSTEM

The *Uniform System of Accounts for the Lodging Industry (USALI)*, a set of regulations published by the *American Hotel & Motel Association* (AHMA), is an accounting system applicable to the lodgings industry which over the years has become a reporting and management system used internationally by the hotel industry in general, though as HARRIS and BROWN (1998, p. 163) point out, it has become the industry standard especially for large hotel companies and international chains.

The USALI has been developing from the first American hospitality guide of 1926, and is now in its 10th edition.

The latest edition has attempted to provide answers to the big changes occurring in the lodgings industry and improve on the ninth edition, in use since 1996, eliminating alternative treatments, clarifying ambiguities and removing inconsistencies to meet the challenge of global accounting harmonization with the aim of providing information comparable throughout the world.

The contents of the USALI$_{10ed}$ are organized into four large sections:

- Financial statements aimed at external users, in accordance with the US GAAP.
- The standardized organization of financial statements aimed at internal users.
- Ratios and statistics.
- A dictionary of expenses, with over 1500 entries.

In the statements for internal users, the USALI's premise is that the hotel business is characterized by joint provision of a series of clearly differentiated services which individually take part in the hotel's profitability, and identifies each of these services with a centre of responsibility presenting information about each of them only taking directly attributable costs into account, also proposing that the *Summary Operating Statement*, the core of the management information, should be prepared not calculating the *Net Income* but the *Net Operating Income*, which provides more information to the owner or manager by concentrating on *operating cash flows*.

Therefore the USALI is a guide to carrying out modern management, being a conglomerate of principles and techniques customarily used by professionals in the sector, useful in assessment and control of the present and future effectiveness of operations from the financial information and management reports, made into a real, integrated business management system.

Table 1. Summary Operating Statement

Revenue
Rooms
Food and Beverage
Other Operated Departments
Rentals and Other Income
Total Revenue
Departmental Expenses
Rooms
Food and Beverage
Other Operated Departments
Total Departamental Expenses
Total Departmental Income-TDI
Undistributed Operating Expenses
Administrative and General
Sales and Marketing
Property Operations and Maintenance
Utilities
Total Undistributed Expenses
Gross Operating Profit- GOP
Management Fees-M
Income befote Fixed Charges
Fixed Charges
Rent-R
Property and Other Taxes-PT
Insurance-I
Fixed Charges-MRPTI
NET Operating Income-NOI
Less: Replacement Reserves
Adjusted Net Operating income-ANOI

Source: 10th edition USALI (2006, pp. 35).

Its operating principles, based on division of the hotel business into departments, and only attributing directly assignable costs to them, offer a model with simple, clear, precise presentation rules and can be used by any type of hotel establishment, regardless of category or size. In addition, the system produces a delegation of responsibilities which stimulates action in each department and motivates participation in the achievement of targets, facilitating monitoring and control of the actions carried out in every department and their efficiency, allowing them to be improved.

Its great contributions include facilitating homogenization of the presentation of information on a worldwide basis and allowing statistical databases to be built from homogeneous, standardized figures, as long as the system is introduced in hotels everywhere and applied homogeneously.

On this subject, CAMPA (2005, pp. 446 and 515) recognizes in his study that there is a high level of use of the USALI in Spain, although its implantation was not homogeneous, because customary hotel transactions were not treated homogeneously, the departments into which establishments were divided differ notably, and there were significant differences in how independent departments were.

Another of the USALI's notable utilities is that it facilitates presentation of segmented information, very important when it comes to decision making and internationally regulated by IFRS 8 *Operating segments*, recently published by the IASB, as a result of the process of convergence between the IFRS and the US GAAP.

In this sense, operating segments resemble centres of responsibility and so are related with the departmental information in the USALI system where, taking direct costs as reference, the coverage margins of the different operating departments are calculated, giving us the information on which decisions are based and helping to supervise manager's actions based on their achievement of set goals, allowing objective-based management.

It can be said that the USALI's operating departments fit the definition of operating segment and that the information supplied according to its criteria fits the segmented information prepared based on the company's operating segments, so it is very useful in the preparation and presentation of segmented company information.

As we have seen before, the USALI is useful for companies in the lodging industry, but it also has failings, including the lack of presentation rules for environmental accounting information.

In this respect, we think it is necessary that green or environmental accounting should begin to be included in this uniform accounting system, including rules related with the presentation of information about environmental aspects that might have an effect on business decision-making, because nowadays "*companies are beginning to consider environmental protection as an added value of their business image*" (RIPOLL and CRESPO, 1998, pp. 174), even more in the lodging industry where activity is dependent on the surrounding environment and its conservation.

For this reason, it would be desirable that coming reviews of the USALI should modify it to include environmental data which allow the level of use of environmental resources in the business and its associated costs to be assessed. For this purpose, information on both the environmental actions undertaken by the hotel and the environmental impact it causes should be provided, with identification of internal and external environmental costs, existing environmental income and cost reductions, and the Ratios and Statistics chapter should include indicators allowing environmental assessment and correct environmental management which improves companies' competitiveness, social image and results.

Whether or not to create an independent operating department, or to have environmental income and costs included as differentiated components within the company's different operating departments whenever they might be directly assignable should also be considered.

With regard to the standardization of non-financial reports from an environmental point of view and the choice of indicators providing useful, relevant information allowing the company's environmental behaviour to be assessed, we can use the suggestions of the several existing international organizations such as the *International Organization for Standardization* or the *Global Reporting Initiative* (GRI), or the Spanish *Asociación Española de Contabilidad y Administración de Empresas* (AECA).

In its *International Standard Organization* (ISO) 14031, the *International Organization for Standardization* provides guidelines on the design and assessment of an organization's environmental performance and on indicators.

As well as this standard on assessment of environmental performance, others to take into account on the subject of preparation of social reports and indicators are the *Global Reporting Initiative* (GRI), widely used internationally and identified by the European Union's Green Book as the model to imitate, because its *"guidelines on the preparation of reports related with sustainable development allow comparisons to be made between companies, and also include ambitious guidelines about the preparation of social reports"* (Commission of the European Communities, 2001, pp. 19).

In its *Sustainability Reporting Guidelines-G3/2006*, the GRI establishes that environmental indicators should cover performance related with input and output flows, and also performance related with biodiversity, compliance with environmental legislation and other relevant data.

But, as LIZCANO (2009) points out, although the GRI has become an international reference for the preparation of sustainability information in recent years, its guidelines have not resolved one of the main problems, how to compare the information in a simple, automatic, universal way. Because of this, complementarily to the GRI guidelines, la *Asociación Española de Contabilidad y Administración de Empresas* (AECA) has set itself the goal of standardizing information about corporate social responsibility (CSR) in Spain.

Because of all the foregoing, we believe that the following are the environmental factors related with the operation of hotel establishments that should be taken into account by companies in the lodging sector:

- Consumption of resources and use of pesticides.
- Emissions into the air and sewage.
- Hazardous waste matter and other waste.
- Conservation of biodiversity.
- Noise level.

This information should be broken down to the department level, allowing us to analyse and compare consumptions, emissions, etc. by department, taking into account the complexity of the lodging industry.

The indicators to be used to track environmental performance in companies in the lodging sector should be both absolute and relative, related with legislation and complaints, and about safety and hygiene.

3. XBRL, A STANDARD FOR ELECTRONIC INFORMATION EXCHANGE

The XBRL language came into being in 1998 with a proposal put to the AICPA by Charles Hoffman, with the intention of simplifying the automation of financial and accounting information using XML (Extensible Markup Language).

Following that study's guidelines, in July 1999 the AICPA and 12 companies[3], took part in a project called at the time XFRML (Extensible Financial Reporting Markup Language),

---

[3] Arthur Andersen LLP, Deloitte & Touche LLP, e-content company, Ernst & Young LLP, Edgar Online, Inc., FRx Software Corporation, Great Plains, KPMG LLP, Microsoft Corporation, PricewaterhouseCoopers LLP, and The Woodburn Group.

where the first XFRML prototype was presented; in April, 2000, its name was changed to XBRL (eXtensible Business Reporting Language), and in March, 2005, the SEC published its final provision 33-8529 in which it encourage companies to make voluntary presentation of financial information in the EDGAR system (*electronic data gathering, analysis and retrieval*) using the XBRL format.

This standard allows homogenization of electronically transmitted financial and accounting information, being internationally recognised and accepted and currently being promoted by the XBRL International consortium.

In terms of architecture, the economic and financial information coded to a specific taxonomy is converted into an XML document which can be found in the *Instance Document*. The taxonomy consists of a *schema* and *Linkbase*. The schema includes the definitions of elements (the basic units from which the XML documents are built) and how the computing application should treat them when it finds them, while the *Linkbase*[4] describes the relationships between elements and between the taxonomy and external information.

Related with the taxonomy in question are the DTS (*Discoverable Taxonomy Sets*) which define the set of taxonomies involved in the validation of a particular XBRL report, but there may also be extensions to the taxonomy which include elements not described in the base taxonomy but necessary, or modifications to the relationships between elements.

In addition, the various taxonomies currently in existence can be classed as universal, international, or national or sector-wide in scope, often representing extensions to the foregoing.

Those of global scope include the GCD (Global Common Document) taxonomy for the company's general information and the AR (Accountants Report) taxonomy for the audit report, produced by *XBRL International (XII)*, the not-for-profit consortium which develops the standard.

International taxonomies include those developed by the XBRL team of the IASC Foundation, which published the taxonomy of its *International Financial Reporting Standards 2010* (IFRS Taxonomy 2010) in April 2010, which conforms to both the IFRS and to the *IFRS for SMEs*, and is a taxonomy which can be used across an organization, as long as the information prepared by it has used the IFRS and the elements in its financial reports are common, while for those not meeting these requirements it will be necessary to prepare national, sector-wide and business extensions.

In this sense, it would be a good idea to develop an extension to the general taxonomy reflecting the particularities of the lodgings sector and the standardization of information supplied by companies applying the USALI, designing a taxonomy concentrated above all on the information of an internal nature, the Summary Operating Statement (EAR) and specific ratios, allowing reports to be prepared based on the XBRL taxonomy.

There is a risk, pointed out by BONSÓN, CORTIJO and ESCOBAR (2009, p. 46), of proliferation of a large number of XBRL taxonomies based on different accounting principles hampering the goal of achieving standardization, comparability and use of the information supplied by the XBRL.

Focusing on the environmental information presented in XBRL, there is the world-wide *Global Reporting Initiative* (GRI), which launched its first XBRL taxonomy for its G3 guidelines in 2006 (*G3 Guidelines XBRL Taxonomy*), while in Spain the most important

---

[4] The label linkbase, reference linkbase, presentation linkbase, calculation linkbase and the definition linkbase.

Muntaner, J. (2001). La igualdad de oportunidades en la escuela de la diversidad. http://www.ugr.es/ Consultado noviembre 2008.

Setián Santamaría, M.L., (2000) *Ocio, calidad de vida y discapacidad. Actas de las cuartas jornadas de la cátedra de ocio y minusvalías*. Bilbao: Universidad de Deusto.

Toledo Morales, P., (2001). *Accesibilidad informática y Discapacidad*. Sevilla: Mergablum.

initiative is being carried out by the AECA, which is leading a project for the "Development of a Corporate Social Responsibility XBRL taxonomy", with the support of the association XBRL Spain (a member of XBRL International).

In addition, in conjunction with the University of Huelva, AECA has produced a free software application for the creation of XBRL reports, and a repository has been created on the association's web site allowing publication of these reports.

CONCLUSION

Internationalization of markets and accounting information users, new management tools based on the mass use of homogeneous information, and the increasing use of Internet for the exchange of economic and financial information are leading to the creation of standards intended to allow homogenization of both the preparation of information and its international exchange and publication.

With regard to homogenization of the preparation and presentation of information globally, there is the process of convergence between the IASB and the FASB, while in the lodging sector, there is the *Uniform System of Accounts for the Lodging industry* (*USALI*), a financial reporting and management system internationally applicable in the hotel industry the main contribution of which is to facilitate homogenization of information presentation, allowing comparisons between hotels and the building of statistical databases from homogeneous, standardized figures.

An accounting system characterized by simple, clear, precise presentation rules, allowing it to be used by any kind of hotel establishment, regardless of its category or size.

Based on division of the hotel business into departments and calculation of departmental margins, allowing the actions carried out in each department and their operating efficiency to be monitored and controlled, leading to delegation of responsibilities stimulating action and participation towards goal achievement.

In short, the USALI allows information to be supplied which is useful for the decision-making process, although to be really useful it needs to be homogeneously and generally implemented by hotel companies.

It is also highly useful in facilitating preparation and presentation of the segmented business information required by both the IASB and the FASB, because the USALI's operating departments coincide with the definition of operating segment, so that information supplied using USALI criteria coincides with segmented information based on the company's operating segments, in compliance with the criteria of IFRS 8 and SFAS 131.

With regard to with international publication and exchange of financial and economic information, work has been being carried out recently on the standardization of communication of accounting information for decision making using XBRL, which will allow standardized publication and reception of financial and economic information, facilitating its dissemination, comparison and analysis.

These two instruments, the USALI and XBRL, are very useful for the homogeneous, standardized publication and presentation of information in the lodgings sector, and we consider it important that an extension to the general taxonomy be developed which reflects the peculiarities of the lodgings sector and standardization of the information supplied by companies using the USALI, designing a taxonomy concentrating especially on internal information, the *Summary Operating Statement* and specific ratios, allowing reports to be prepared based on the XBRL taxonomy, because if the homogeneous information provided by the USAL could be electronically exchanged by companies in the lodgings sector using the XBRL standard, we would be able to publish and receive information in a standardized way, facilitating its distribution and subsequent use for decision-making and management.

But because the information supplied by companies should allow assessment of not only its financial and economic behaviour, but also of its social behaviour from a triple point of view, economic, social and environmental, we believe that environmental accounting should begin to be included in future editions of the USALI, so correcting one of its deficiencies, by including rules for the presentation of information about environmental matters which might affect business decision-making.

In this respect, measurements should be made and reported of both the environmental actions carried out by the hotel and of the environmental impact it generates, identifying internal and external environmental costs, environmental revenues, and the indicators to be included in the *Ratios and Statistics* chapter to allow environmental assessment of the company.

Based on the international environmental standards set by organizations such as the ISO o GRI, or in Spain by the AECA, we propose these environmental factors to be taken into account by companies related with operation of hotel establishments:

- Consumption of resources and use of pesticides.
- Emissions into the air and sewage.
- Hazardous waste matter and other waste.
- Biodiversity conservation.
- Noise level.

This information should be broken down to the departmental level, allowing comparison and examination by department, taking into account that the lodgings industry is highly complicated. The indicators to be used to monitor environmental performance should be both absolute and relative, related with legislation and complaints, and safety and hygiene.

Finally, we should point out that both the codes and standards arising from the process of convergence between the **IASB** and **FASB** and XBRL and the USALI are related standards, so implementing them jointly should allow companies more value creation, shifting from an annual, quarterly or monthly presentation of heterogeneous accounting information to a model of presentation of homogeneous information available in real time in electronic format, so allowing fast information searching and use, and comparability, to the benefit of its use in companies' decision making.

In this direction, one possible line of research, considering the importance of the hotel sector in Spain, would be a study showing that joint application of the **IFRSs, the USALI y** and XBRL provides hotel establishments with a competitive management advantage which allows hotel companies greater value creation.

REFERENCES

American Hotel & Lodging Educational Institute (2006): *"Uniform System of Accounts for the Lodging Industry"* Hotel Association Of New York City New York City. (Tenth Revised Edition). Lasing, Michigan.

Asociación Española De Contabilidad Y Administración De Empresas, Aeca (2003): *"XBRL: Un Estándar para el Intercambio Electrónico de Información Económica y Financiera"*. Documento núm. 2 de la Comisión de Nuevas Tecnologías y Contabilidad. Madrid.

Asociación Española De Contabilidad Y Administración De Empresas, Aeca (2009): *"La Taxonomía XBRL de Responsabilidad Social Corporativa"*. Documento núm. 6 de la Comisión de Responsabilidad Social Corporativa y Documento n°. 7 de la Comisión de Nuevas Tecnologías y Contabilidad. Madrid.

Asociación Española De Contabilidad Y Administración De Empresas, Aeca (2007): *"Gobierno y Responsabilidad Social de la Empresa"*. Documento n°. 4 de la Comisión de Responsabilidad Social Corporativa. Madrid.

Bonsón, E.; Cortijo, V. Y Escobar, T. (2009*): "Towards the global adoption of XBRL using International Financial Reporting Standards (IFRS)"*. International Journal of Accounting Information Systems 10, pp. 46-60.

Campa Planas, F. (2005): "*La Contabilidad de gestión en la industria hotelera: estudio sobre su implantación en las cadenas hoteleras en España*". Tesis Doctoral. Universitat Rovira I Virgili. Available in: http://www.tdx.cat/TDX-0530107-120008.

Chan W., Wilco y LAM C. JOSÉ (2002): *"Prediction of pollutant emission though electricity consumption by the hotel industry in Hong Kong"*. International Journal of Hospitality Management, Volume 21, pp 381-391

Ciudad Gómez, A. (2009): *"Uniform System of Accounts for the Lodging Industry: presente y future"*. XV Congreso AECA Decisión en época de crisis: transparencia y responsabilidad. Valladolid (España) Available in: http://www.aeca.es/pub/on_line/comunicaciones_xvcongresoaeca/general.htm

Ciudad Gómez, A. (2010): "*Principales cambios en la decima edición del sistema de reporting de la industria hotelera*" Partida Doble, n°. 221, pp. 40 - 57.

Comisión de las Comunidades Europeas (2001): "*Libro Verde. Fomentar un marco europeo para la responsabilidad social de las empresas*". COM (2001) 366 final, julio, Bruselas.

Damitio, J. W. Y Schmidgall, R. S. (1998): *"The New Lodging Scoreboards: The Uniform System of Accounts for the Lodging Industry. Focus: Pre-Opening Expenses"* The Bottomline, pp. 18-20. Available in: www.hftp.org/members/bottomline/backissues/1998/oct_nov/usali.htm

Debreceny, R. S., Chandra, A., Cheh, J. J., Guithues-Amrhein, D., Hannon, N. J., Hutchison, P. D., et al. (2005). "Financial Reporting in XBRL on the SEC's EDGAR System: A Critique and Evaluation". *Journal of Information Systems*, 19 (2), pp. 191-210.

Debreceny, R. S.; Fendel, C. Y Piechocki, M. (2007): "New Dimensions of Business Reporting and XBRL". DUV, Wiesbaden.

Educational Institute of the American Hotel &Motel Association (1996): *"Uniform System Of Accounts For The Lodging Industry"*. Hotel Association Of New York City New York City. (Ninth Revised Edition)

Financial Accounting Standards Board (FASB) (1997): *Statement of financial accounting standards* n° 131: Disclosures about segments of an enterprise and related information. Stamford, CT: FASB. pp. 1- 48.

Global Reporting Initiative -GRI (2006). *"Sustainability Reporting Guidelines (Version 3.0)"*. Amsterdam, The Netherlands. Available in: http://www.globalreporting.org/Reporting Framework/ReportingFrameworkDownloads/

Harris, P. Y Brown, B. (1998): *"Research and development in hospitality accounting and financial management"*. International Journal of Hospitality Management 17, pp. 161-181.

Heskett, J.L. (1986): *"Service Management: An Evaluation and the Future"*, International Journal of Service Industry Management, Vol. 65, N°. 2, pp. 118-126.

Iasb (2006): International Financial Reporting Standard- IFRS N°. 8 "Operating Segments".

International Accounting Standards Committee Foundation (2009): *"International Financial Reporting Standards (IFRSs) Taxonomy Guide"*. London, United Kingdom.

ISO 14031 (1999): "Environmental Management -Environmental Performance Evaluation Guidelines" Switzerland: International Standard.

Lizcano, J. L., García, I. Y Fernández, A. (2009a): *"Normalización de la información corporativa sobre responsabilidad social. Estudio empírico sobre la elaboración de un Cuadro Central de Indicadores (CCI)"*. XV Congreso AECA Decisión en época de crisis: transparencia y responsabilidad. Valladolid (España), Septiembre. Available in: http://www.aeca.es/pub/on_line/comunicaciones_xvcongresoaeca/general.htm

Lizcano, J. L., García, I. Y Fernández, A. (2009b): *"Normalización de la información sobre responsabilidad social."*. Jornada AECA Normalización de la información sobre Responsabilidad Social Corporativa y el estándar XBRL. Madrid (España), Noviembre. Available in: http://www.aeca.es/comisiones/rsc/rsc.htm

Ortega Rosell, F. J. Y Ciudad Gómez, A. (2001): *"Contabilidad y gestión de la moderna hospitalidad: Alojamientos. El USAL estadounidense, un marco conceptual sectorizado de la información financiera para los usuarios externos e internos"*. XII Jornadas Luso-Espanholas de Gestão Científica. Covilhã (Portugal), pp. 273-283.

Ripoll Feliu, J. V. y Crespo Soler, C. (1998): "Costes derivados de la gestión medioambiental". Técnica contable, n. 591, marzo, pp. 169-180.

Sasser, W.E.; Olsen, R. P. Y Wyckoff, D. D. (1978): *"Management of Service Operations"*. Ed. Allyn and Bacon, Inc., Boston.

Schmidgall R (1997): *"Hospitality industry: Managerial accounting"*, American Hotel & Motel Association. Michigan.

Schmidgall, R. S. Y Damitio, J. W. (1999): *"Hospitality Industry Financial Accounting"*. American Hotel & Motel Association. (2ª Edición). Michigan.

Chapter 21

TOURISM: ACCESSIBLE DESTINATION?

Isabel Mª. Ferrandiz Vindel, Jose Luis González Geraldo and Ana Mª. Bordallo Jaén
University of Castilla-La Mancha, Spain

ABSTRACT

The link between Accessibility and Tourism has been one of the greatest challenges in recent years, and the reality has proved its possibilities. Today, one of the market sectors with more relevance is the one which takes care of people with reduced mobility (PRM) and/or disabilities. Its relevance is justified, mainly, by two reasons: the vast number of people who can be numbered within this group as well as their growing participation in the tourism market.

Accessibility in tourism should not only be understood as requirement that a specific service or destination must accomplish in order to foster access to people with reduced mobility or disabilities. On the contrary, accessibility in tourism has turned out to be one of the intrinsic factors of touristic quality because there cannot be real quality in tourism if the destination is not reachable for everyone. None should be excluded from tourism for any reason or circumstance.

In this chapter, we would like to increase the awareness of the institutions responsible for tourism, as well as the general public, about the relevance of tourism with quality; a tourism which everyone can join and enjoy. Having said that, we will approach this issue from the perspective of those collectives which traditionally speaking had problems with tourism accessibility.

1. INTRODUCTION

Our planet, which is ruled by the principle of life, is structured and developed from the concept of diversity. If we take a quick glance at nature, we can easily understand this amazing process of constant balance renovation seeking adaptation. If we think about human diversity, we find a reality where the need for adaptation is also always there. But, in this case, with some slight, but significant, differences from those of nature.

Following Muntaner (2001), we should understand diversity as a value to promote and foster because, in diversity, there is mutual respect, collaboration and knowledge and where all members of a community are worthy of consideration and esteem. Therefore, social diversity and plurality have to be the starting point of an inclusive social system that guarantees equal rights to everyone, without exceptions, without parallel or segregated social paths.

Within the universe of human diversity, individuals with different capabilities are a clear example of the search for adaptability and the hindrances which make this adaptability complicated and difficult. Most probably, this has to do with the resistance that human beings experience when we understand, assimilate and accept the difference (Ferrándiz, 2003)

Because of the tourism sector this reality is not alien to us, on the contrary, every day there are more people with different capabilities included in the touristic dimension and seek that kind of quality in tourism that we already commented.

We are all aware of the enormous gaps that exist in most societies between reality and what we hope for. Those gaps could be in terms of infrastructures, or adapted activities. Individuals with different capabilities are, more than often, unable to exercise their rights just because of the absence of public and private policies that do not invest in this aspect of human diversity (MCT, 2007)

Tourism has become a social phenomenon of extraordinary relevance, mobilizing millions of people around the world, especially in Europe, being not only a factor of wealth and unprecedented economic progress, but also a critical element in improving knowledge, communication and the grade of relationship and respect between citizens of different countries (Marcos Pérez y González Velasco, 2003)

Tourism is a social good of first order which should be within reach of all citizens, any group or collective shouldn't be excluded regardless of personal, social, economic or other circumstances.

The initiatives launched over recent years pursued a common goal: the full integration of people with reduced mobility, and/or disabilities, to the access and enjoyment of touristic services. Understanding these touristic services from a full perspective which includes: hotels, restaurants, touristic resources, etc. (see Setián Santamaría, 2000)

If we go back in time, to the 27th of September, Manila (Philippines) 1980, we will find the so-called *Manila Declaration*, made by the World Tourism Organization (UNWTO). In that declaration, we could find, for the very first time, the term accessibility attached to the term tourism. The Manila Declaration recognized tourism as a fundamental right and a keystone of human development and recommended the regulation of touristic services targeting the most important challenges about accessibility in tourism. These recommendations were mirrored in the document *"Para un turismo accesible a los minusválidos en los años 90"* (Accessible Tourism for disabled people in the 90's), approved at its General Assembly in Buenos Aires in 1990.

At the same time, the concept of Accessible Tourism was gaining significant relevance, in 1989 a group of British experts in tourism and disability published the report "Tourism for All". This report outlines the progress that had been made since 1981, which was the international year of disabled people and has the aim of promoting Tourism for All within the tourism industry sector regardless of age or possible disabilities. In addition, the report also defines Tourism for All as that form of tourism which plans, designs and develops leisure touristic activities in a way in which can be enjoyed by everyone regardless of their physical, social or cultural circumstances.

Later, during the celebration of the European year of tourism (1990), the Council of Ministers of the European Union adopted the Community Action Plan to promote tourism. The General Councils of the member states with competences in social affairs and tourism agreed to create, in 1994, a group of experts with the main aim of coordinating the activities related with "Tourism for All", eliminate hindrances, develop tourism for people with disabilities and the exchange of information about this issue at a national level in each state.

This Community Action Plan was reflected in the Communication on Equal Opportunities for Persons with Disabilities which was presented by the European Commission in Brussels in 1996.

More recently, the European Union (EU) continues recommending policy adaptations and guidelines to the member states which are aimed at improving accessibility. For example the Communication of the EU Presidency in Bruges, held on the 2nd of July, 2001, which stated the need for "Tourism for All" or the resolution adopted on the 15th of February in 2001 by the Committee of the Ministers in the Council of Europe in which universal based design policies were recommended.

At the 8th Regional Conference of Southern Europe and the Mediterranean, held in Strasbourg between the 3rd and 4th of December in 2008, it was agreed that, for 2009, the Day of Solidarity would be devoted to Accessible Heritage. This proposal aimed at rising citizen awareness and the institutions in charge of the historical sites and monuments and ensuring a barrier-free accessibility for people; residents and visitors, with any type of permanent or temporary disability or reduced mobility so that they can also enjoy the Heritage, not only as infrastructure is concerned, but also in terms of information and treatment they receive.

As we can observe, all these initiatives have a common goal: the full integration of any kind of people, with or without disability or mobility problems, the enjoyment of tourism through the full accessibility to any kind of touristic services in the whole sense of the term (hotels, restaurants, tourist resources, etc.) Tourism for All seeks to give a plausible answer to all recommendations and declarations which put emphasis in the possibility of creating a tourism aimed at all collectives of the population equally speaking, without any kind of discrimination (Cruz Ayuso, 2004)

Accessible Tourism development is one of the main bets for the future of the sector. According to United Nations (UN) figures, in 2050, 21% of the population will exceed the age of 60 (what is about 2,000 million people). The aging population is one of the most significant reasons of the increasing relevance of accessibility in tourism, but not the only one. Another relevant reason is that almost the entire population, at some point in our lives, suffers a temporary disability caused by illness or accidents. In addition, we should bear in mind that the traveler rarely travels alone but, on the contrary, usually does it with family members or friends and we could find children or pregnant women amongst them, for example. In this sense tourism is a right for all and we must ensure its accessibility to all citizens (Figini y Arch, 2007)

Tourism is a broad term, able to cover a natural or geographical area as well as the existing equipment and facilities used in a specific place (Marcos Pérez, 2002). This offer relates to what has traditionally been called the touristic product and would be compound by:

- Tourist Accommodation;
- Land, sea or air transportation;
- Leisure and recreation centers;
- Restoration;
- Travel agencies;
- Touristic commerce and crafts;
- Touristic resources.

Tourism activities should be organized in a way that ensures the freedom and voluntary participation of the people involved in them, in conditions of comfort and dignity. We, the authors, are aware that despite the many efforts made, a high percentage of the touristic offer is not accessible to a segment of the population in which could be included a number of people with reduced mobility and/or disabilities. There are many barriers that affect the planning of a trip (inaccessible public environment, non adapted transportation, accommodation without facilities for disabled people, etc.) but there are multiple possible solutions for those problems that would enable these people to integrate into tourism (see COCEMFE, 2005)

Tourism: Accessible destination?

Researching the destination	<ul><li>Internet: Web-site
	   – Accessibility of the urban areas
	   – Accessibility of the tourism resources</li></ul>

Booking	<ul><li>Internet: Web-site</li><li>Travel Agency:
	   – Environment
	   – Staff disability training</li></ul>

Transportation	<ul><li>Moving from the starting point</li><li>Access to transport terminals</li><li>Getting into the bus/car, etc.</li><li>Getting out of the bus/car, etc.</li><li>Access to the tourist facilities</li></ul>

Destination	<ul><li>Accommodation:
	   – Determine accessible hotels
	   – Select accessible hotels</li></ul><ul><li>Moving around the environment</li><li>Access to the Tourism resources (natural, cultural, etc.)</li><li>Leisure activities (theatres, cinemas, etc.)</li></ul><ul><li>Human resources.
	   – Selection of specialized guiding companion
	   – Selection of specialized companion at the final destination
	   – Specialized support staff (monitors) in the case of groups</li></ul>

Going back home	<ul><li>Going back to the starting point</li><li>Access to the station/airport, etc.</li><li>Getting into the bus/car, etc.</li><li>Getting out of the bus/car, etc.</li><li>From the station to your home</li></ul>

(Adapted and translated from Marcos Pérez y González Velasco, 2003, pp. 32 - 54)

There are three important aspects that should be considered when organizing Accessible Tourism travel. They are:

- Access to information;
- Physical access to places;
- Access to content and procedures.

If we take those three aspects into account, it would be easier to provide, in some cases, possible solutions to the accessibility offer in tourism to those people with reduced mobility and disabilities.

Accessibility in the touristic offer must be a requirement if we want the participation of everyone in the proposed activities. Providing accessible services means accepting that we are living in a plural and diverse society and that the needs are not equal for all of us. In the planning of touristic or leisure and free time activities for people with reduced mobility or disability we must consider each and every one of the following aspects to ensure that the destination is truly accessible (AENOR, 2005)

CONCLUSION

Since the Universal Declaration of Human Rights (adopted by the UN in 1975) we assume that every human being has the right to rest and leisure, including reasonable limitation of working hours and paid holidays. From this reflection, we can say that every individual, just because his unrepeatable and unique difference, is special and what is more important; all of us have the right to carry out activities in our free time (Izuquiza, 2000)

In the tourism sector there is a real discipline of study which is called Accessible Tourism and Tourism for All which is creating an increasing awareness in all enterprises, organizations and institutions in the tourism sector to improve accessibility for people with disabilities and/or reduced mobility. And what is even more significant, not only because of the lucrative earnings they could obtain but also because of the social and equity reasons already explained (Millán Calentí, 2005)

The accessibility of tourist services is one more condition to accept that we live in a diverse and plural society in which the needs of the people who live in that society are not the same for all.

Tourism for All has been designed, from its very beginnings, as tourism that assures the use and enjoyment of tourism for those people with physical, mental or sensory disabilities. It is common to read and hear about "Accessible Tourism" referring to the complex set of activities, occurring during free time and oriented towards tourism and recreation, that enable full integration of the people with reduced mobility and/or communication difficulties, obtaining the social and individual satisfaction of the visitor and, therefore, a better quality of living (Art. 1 of the National Law 25.643 of Accessible Tourism)

This unquestionable reality has led the European association movement of people with disabilities to proclaim inspiring criteria in tourism and disability that can be summarized in a kind of Decalogue in which is synthesized what kind of social aspirations they have in relation to tourism (see MCT, 2007)

This new paradigm of disability from a social conception presents a more complex point of view about equity in the rights of any person for tourism and leisure activities and the role that the Government should take to guarantee those rights regardless of whether he or she has or not a disability. For this reason, the concept of Accessible Tourism is expanding within the concept of Social Tourism on a broader perspective of potential beneficiaries because it fights against inequalities and the exclusion of those who have a different culture, have fewer resources or live in disadvantaged regions. In sum, the combination of these two concepts, Accessible Tourism and Social Tourism, make possible the achievement of a true Tourism for All (Cruz Ayuso, 2004)

Our social and economic reality, in regards to accessibility and tourism, shows the need of opening new areas of professional expertise to develop both areas of knowledge comprehensively collecting the different techniques and competences that will make this union one of the most potential actions for occupation and development.

Professionals specialized in Accessible Tourism are a keystone for the future of tourism. They should conduct activities in enterprises, public and private organizations and both in the field of accessibility as in tourism, and in both at the same time.

REFERENCES

AENOR (2005) *Sistemas de Gestión de accesibilidad global: guía de interpretación de la norma une 170001-2*. AENOR.

CERMI. (2003). *Plan Estatal de Accesibilidad*. Madrid: CERMI.

COCEMFE (2005) Guía accesible del viajero con movilidad y/o comunicación reducidas. Ed. Polibea.

Cruz Ayuso, C., de la. (2004). *Los retos del ocio y la discapacidad en el siglo XXI*. Bilbao: Universidad de Deusto.

Ferrándiz Vindel, I.M. (2003) "Ocio y Tiempo Libre en adultos con deficiencia mental" en Buisán, C., Freixa, M., Panchón, C., Paula, I., y Pastor, C., (2003) en *Educación y diversidades: Formación, Acción e Investigación. I Congreso Internacional. XX Jornadas de Universidades y Educación Especial*. Barcelona: Universitat de Barcelona.

Figini, L., y Arch, M. (2007) *Espacio libre de barreras*. Nobuko.

Izuzquiza, D., (2000) *El ocio en las personas con Síndrome de Down*. Tesis Doctoral. Madrid: Universidad Complutense

Juncá Ubierna, J.A., (2004). *Accesibilidad Universal: diseño sin discriminación*. IMSERSO-CAJAMADRID.

Ley Nacional 25.643/2002 de 11 de septiembre, de Turismo Accesible Publicación (BOE núm.1125 de 12 de septiembre de 2002).

Marcos Pérez, D. (2002). *Accesibilidad hotelera en el siglo XXI*. Madrid: Centro de Publicaciones y Documentación.

Marcos Pérez, D., y González Velasco, D.J., (2003) *Turismo Accesible*. Madrid: CERMI

MCT (2007). *Decálogo de Buenas Prácticas en Accesibilidad Turística*. Madrid: Ministerio de Ciencia y Tecnología.

Millán Calenti, J. C. (coord.) (2005) *Mayores, accesibilidad y nuevas tecnologías de la información y comunicación*. A Coruña: Universidade da Coruña.

In: Research Studies on Tourism and Environment
Editors: J. Mondejar-Jimenez et al.
ISBN: 978-1-61209-946-0
© 2012 Nova Science Publishers, Inc.

Chapter 22

THE TEACHING OF SPANISH AS A SUSTAINABLE RESOURCE FOR TOURISM

Pilar Taboada-de-Zúñiga Romero[*]
University of Santiago de Compostela, Spain

ABSTRACT

The promotion and development of Cultural Tourism has become one of the primary objectives of tourism policy for many countries, attempting to bring about less invasive tourism, with less marked seasonal trends, targeting a public with higher cultural awareness and consequently more respectful of the environment.

In a globalized world the learning of languages gains more importance every day, not only amongst young people as part of their education, but also amongst adults. This desire to learn other languages has brought about the concept of 'linguistic tourism', consisting in the provision of organised trips, where the primary objective is the learning or perfecting of a language, usually in another country, combined with tourism activities.

Linguistic tourism has a long tradition in countries like the United Kingdom (UK) and France. In Spain, however, it is a more recent phenomenon. Nevertheless, there is significant willingness on the part of those bodies responsible for the promotion and commercialisation of cultural tourism to develop this sector, whilst maintaining quality.

This study presents two well defined sections. In the first section, we analyse the great potential for growth in the learning of Spanish as a foreign language, and how this contributes to the development of linguistic tourism in Spain. In the second, the study focuses both on the particular case of Santiago de Compostela as well as using evidence from Salamanca to make a comparison of two similar cities differently developed in terms of linguistic tourism. The demand for Linguistic Tourism in Santiago de Compostela is demonstrated through a survey.

Keywords: cultural tourism, linguistic tourism, language tourism, multiplying effect, Spanish, Santiago de Compostela, Salamanca

---

[*] Contact author: pilar.taboadadezuniga@usc.es

1. INTRODUCTION

The 21st century will be remembered as a period of globalisation, multiculturalism and multilingualism, bringing together people of different native languages. From this arises a renewed significance for those languages that can function as common ground between speakers who do not share a common tongue.

One of the consequences of globalisation is the necessity to learn other languages in order to be able to communicate. The necessity of a 'vehicle language' has been a constant throughout the history of humanity. Numerous languages have taken the position of the *lingua franca*: Latin in ancient times, later, during the Enlightenment, French became the international language and now in the twentieth century, after the period of the Great Wars, English has become the principle language of international communication and consequently, a significant part of the global population have acquired it as their second language.

Currently 6,909 languages are spoken worldwide (e.g. Lewis, 2009) but just a small group of eight languages are each spoken by more than 100 million people. This group is made up of the following languages: Chinese, Spanish, English, Arabic, Hindi, Bengali, Portuguese and Russian. But what is the current demolinguistic trend?

- A downwards trend in Chinese, not in terms of numbers of speakers, but in its use as a language for international communication,
- A gradual and constant increase in Arabic,
- A slower increase in the Hindu/Urdu speaking world.

The linguist, David Graddol, presents a pessimistic demolinguistic projection. He predicts a rather drastic reduction in the number of languages in the world in years to come. By 2100, his prediction translates into a reduction by half in the number of languages spoken in the world. He estimates that only some 1,000 languages will survive, however, in this context, the Spanish language has some promising prospects. The author shows us the great weight of Spanish as much in its geographic as its demographic spread, this upward trend translating into an equal or greater increase in Spanish than English speakers which in the future could position the Spanish language as the second most spoken language in the world (e.g. Graddol, 2006).

All languages are important, for they are the linguistic heritage of humanity; however, some are more influential than others. We have chosen *the Spanish language* as the object of this study because according to all indicators and expert opinions, in the twenty-first century, it is the language with the greatest potential to expand. It presents a range of characteristics:

- It occupies, depending on author, second (e.g. Lewis, 2009) or fourth (UNESCO) place in the most spoken languages of the world;
- It is the official language of 21 countries;
- It is present as an official language of numerous international organisations such as the UN, UNESCO, and UNWTO etc;
- It enjoys a good position in new technologies as a language of communication, specifically third place on the internet (World Users by language, 2010);

- It is estimated that around 14 million people throughout the world study it as a foreign language, which makes it the second most studied language after English.

But Spanish is not only influential for its great demographic extension in the last 100 years, but also for its potential to further increase. This potential we have based on four pillars:

1. *Spanish as the second most studied language in Europe.* There has been an increase in the level of demand from students choosing to study Spanish as their second foreign language in Europe, provoking a fall in the demand of German and French. In Europe, currently 3,455,634 students study Spanish as a second foreign language (Ministry of Education and Science of Spain, 2006). The countries where Spanish is most taught are: France, Germany, Italy, Switzerland and the UK.
2. *The great commitment of Brazil.* The signing of the Mercosur treaty in 1991 was a milestone that marked an upward trend in the study of Spanish as a foreign language. One of the consequences of the treaty was the introduction of obligatory teaching of Spanish in secondary schools, subsequently provoking a great demand in teachers of Spanish as a foreign language. There are currently some 5 million students studying Spanish in comparison to 1 million in 2006. (Anuario Instituto Cervantes 2006 and 2009)
3. *The accession of Spanish to the second language in the United States (US).* Spanish is the second most spoken language in the US, although this has not led to it being declared an official language. It is estimated that by the year 2050 the US will become the country with the highest number of Spanish speakers. This incomparable development can be put down to the many Spanish speakers reclaiming their mother tongue – in fact the labour market favours workers with a bilingual education (e.g. López Morales, 2008) – and the demand for the teaching of Spanish as a foreign language, overtaking other languages such as French, proof of which are the 6 million students of the ELE (*Español como Lengua Extranjera* – Spanish as a Foreign Language) that there are today in the country (Ministry of Education and Science of Spain, 2006).
4. *The rise of the internet.* In modern times, new technologies are increasingly present, not only in the world of work, but also in the field of leisure. For this reason, it is necessary to briefly reflect upon the role of Spanish in new channels of communication. Of the ten languages most used on the internet, a small group of three (English, Chinese, and Spanish) have the majority of users (64%). Spanish, with 131 million users, is in third position, although at a great distance from English (499 million users) (Internet World Stats, 2010). Another interesting piece of data about the use of Spanish in new technologies is the incomparable rise in registered ".es" domain names (e.g. Taboada-de-Zúñiga, b 2010)

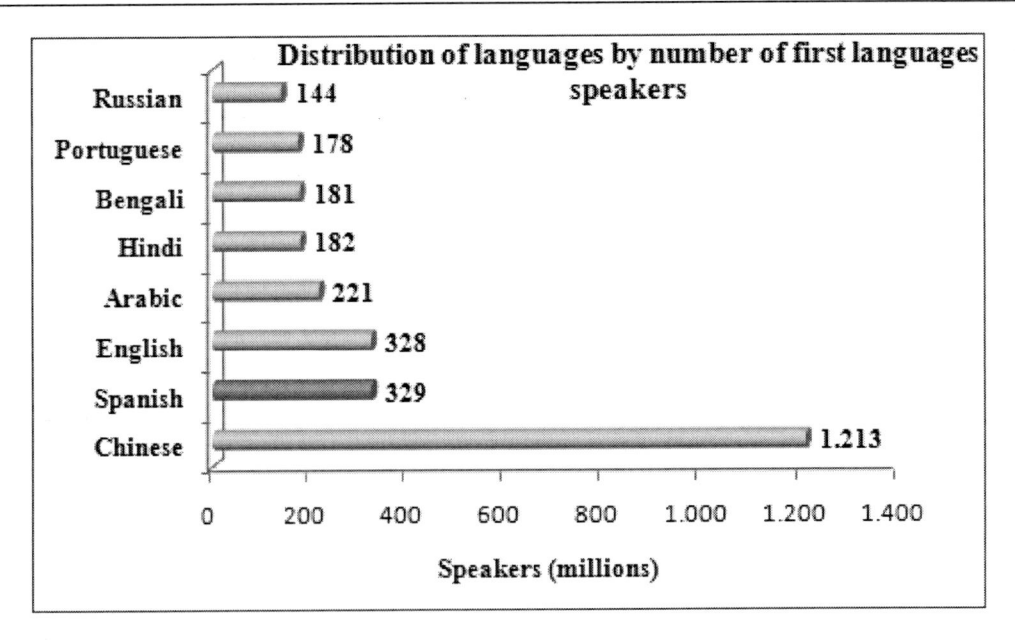

Distribution of languages by number of first languages speakers

Language	Speakers (millions)
Russian	144
Portuguese	178
Bengali	181
Hindi	182
Arabic	221
English	328
Spanish	329
Chinese	1.213

Graph 1. Distribution of languages by number of first languages speakers. Source Ethnologue Languages of the world, M. Paul Lewis. Authors own.

These four pillars help us to affirm that every day more countries decide to include the study of Spanish within their academic education; consequently many of these students contemplate the possibility of learning or perfecting their language skills in Spanish speaking countries, giving rise to the phenomenon of linguistic tourism that we are going to explore in this article.

2. OBJECTIVES AND METHODOLOGY

The objective of this article is to present the teaching of languages, and specifically the Spanish language, as a new resource for tourism, giving rise to the term 'linguistic tourism', which combines the teaching of a language with economic exchange in the sphere of tourism. Through this article we will try to demonstrate that this sector of cultural tourism possesses characteristics that make linguistic tourism a very attractive sector: it is less aggressive than other types of tourism, it has a less pronounced seasonal concentration, it promotes the local culture and produces an important economic impact in the destinations where it is carried out. But for this to be possible, coordination between the management and planning bodies of the tourist destination and its stakeholders is fundamental.

The method employed for the first part of the study has been secondary sources, published articles and books about philological, economic and tourism research. For the second part, it was necessary to carry out field work, for there is no published data about language tourism in Santiago de Compostela. The field work consisted in qualitative as much as quantitative analysis. Within the qualitative analysis we used in-depth interviews with stakeholders, such as: the directors, professors, and administration of the centres where Spanish is taught to foreigners, who provided very valuable information about the state of the teaching of Spanish in the city. At the same time workshops with small groups of students of

Spanish for foreigners were carried out (from now and into the future with ELE students) to find out more about the reality of the situation.

Within the quantitative analysis it was considered fundamental to carry out a survey of language tourists that took a Spanish course for foreigners in the city, with the aim of analysing their preferences and attitudes, and in this way it is possible to put forward new proposals for improvement.

In order to carry out the survey we used a semi-structured questionnaire consisting of twenty-five mainly closed questions. The fundamental objective of the survey was to analyse the following aspects of linguistic tourism in Santiago de Compostela:

- Socio-demographic profile
- Motivation and source of information on destination
- Planning the trip: Method of reservation, transport and accommodation
- Activities carried out during their stay and evaluation of the destination
- The expenditure of linguistic tourists and the multiplier effect

The questionnaire presented a range of characteristics:

Characteristics of the Survey
• A single model of questionnaire
• Chosen language: Spanish
• Length of the questionnaire: 25 questions
• Approximate time required to carry out the survey: 20 minutes
• The questionnaires were completed anonymously
• A pilot test was carried out during March 2009 with a group of ELE students with the aim of testing their level of understanding and accuracy
• The total number of people surveyed was 186 with a sample error of (+/-) 6%
• The surveys were carried out in the ELE centres of Santiago de Compostela: Centro de Lenguas Modernas (CLM) (Centre for Modern Languages), Cursos Internacionales S.L(International Courses Ltd.)., Academia Iria Flavia (Iria Flavia Academy) during May 2009
• The survey presents a certain bias towards the CLM students, due to the fact that at this time of year, there are more CLM students.
• SPSS was used to analyse the data (Statistical Package for the Social Sciences).

In terms of the data used from the ELE students of Salamanca, we worked with reports kindly loaned to us from the Tourism Observatory of Salamanca *(Observatorio Turístico de Salamanca)* and the office of "The Spanish City" *("Ciudad del Español")*.

3. LINGUISTIC TOURISM. DEFINITION AND CHARACTERISTICS

Linguistic tourism is a sector of cultural tourism. A cultural tourist is characterized by their primary motivation being the desire to know, understand, contemplate, participate and live in other cultures, languages and countries, but especially those which have ample historical, artistic and gastronomic heritage.

For Richards (1996), cultural tourism can be defined from two points of view: the conceptual and the technical. The conceptual definition is "the temporary transfer of people to a cultural attraction far from their usual place of residence with the intention of satisfying their cultural necessities" and the technical is "all the transfers of people to a specific cultural attraction, such as heritage sites, artistic and cultural exhibitions, art and theatre, outside of their usual place of residence". (e.g. Richards, 1996)

Cultural tourism prevents ceasing of local-cultural values in the face of globalization. The relation between culture and tourism is to a large extent a symbiosis. Both increase incomes and cultural resources and provide sustainability. The people participating in cultural tourism are well educated, wealthy and being interested in travel, they are generally more acceptable, upper-level "tourists". From that point, cultural tourism forms a type of culture that is new, improving and attractive (e.g. Cengiz, 2006)

Linguistic tourism is considered a sector of cultural tourism because many authors include the study of languages as a resource of cultural tourism (e.g. Swarbrooke 1996), (e.g. Smith 2003), (e.g. Coltman 1989), (e.g. Kennett 2002). Although more recent studies are beginning to describe linguistic tourism and also label it in different ways. Over the course of this study of linguistic tourism we have encountered a wide range of terms used to refer to it, such as: *Linguistic tourism (e.g. Baralo, 2007), Linguistic stays (e.g. Davó, 2002), Language tourism (Plan Impulso al Turismo Cultural e Idiomático, 2002), Linguitourism [Turilingüismo] (e.g. Herranz, 2008)*. Although they have differing nuances, they all come down to the same basic meaning: tourism which consists of the provision of organised trips, where the primary objective is to learn or perfect a language in combination with tourism activities. Normally the trip is in a different country to that of the tourist's origin and commonly for a period of up to a year. It has a series of characteristics:

- Offered in all destinations, primarily urban, but not exclusively, and often those rich in artistic and cultural heritage.
- The perception of the destination is an important factor when the tourist is deciding on their destination.
- Important characteristics of a language tourist are: longer stays than other kinds of tourists with higher than average expenditure. To emphasise here is the economic impact derived from their stay, both for the resident population as well as for general employment within the geographical area, producing a series of impacts: direct, indirect, and induced.
- It demonstrates less pronounced seasonal variation than other types of tourism, although peaks are visible in spring and summer. In the most consolidated destinations this seasonality is less marked than in those destinations still in the early stages of development, depending on the different marketing campaigns and tourism policies to decrease seasonal concentration on behalf of the organisations responsible for the promotion and commercialisation of the sector.
- The age of the language tourist spans a much wider range.
- Due to the length of the stay and greater knowledge of the destination, users tend to develop a bond with it, and a large number of these language tourists return

accompanied by family or friends. This tendency to revisit, in most cases, does not mean that the tourist takes a language course again in that destination, but pays a normal tourist visit. For this reason we state that in the majority of cases it is simply *loyalty to the destination.*
- A feature which makes it very attractive and at the same time differentiates it from other types of tourism is that which we have called the *multiplier effect*. A new phenomenon, in which the tourist, during their stay at the chosen destination, is visited by other tourists: family members, friends, or work or university colleagues. This feature makes linguistic tourism a more interesting sector to develop as it could generate greater economic impact for the chosen destination (e.g. Taboada-de Zúñiga, P., a 2010).

In conclusion, linguistic tourism is less invasive than other types of tourism due to less seasonal concentration, it promotes local culture, it is respectful of the environment and it is sustainable so long as it is developed through well planned tourism policy.

4. THE TEACHING OF SPANISH AS A TOURISM RESOURCE

The relevance of the tourism sector as a driver for development and growth in the Spanish economy has been an unquestionable fact for many years. Tourism is of vital importance in Spain as it represents 11% of GDP. Spain is a leader in 'sun, sand and sea' tourism (3rd most visited by tourists with 52.2 million tourists in 2009, after France and US) (Instituto de Estudios Turísticos, 2009) and with a very consolidated position in the market.

One of the features of the Spanish model of tourism is its seasonal concentration in July, August and September when the sector is more dynamic. However, studies carried out by the IET (*Instituto de Estudios Turísticos de España*) show us that, although slowly and thanks to a wide range of tourism policies, this tendency towards seasonal concentration is becoming less pronounced, though not equally around the different regions or autonomous communities of Spain. For this reason, from a central government level, in order to boost Spain's tourist sector, in 2001 a series of measures were initiated, such as:

- The diversification of tourism provision,
- the search for tourist products with lower seasonal concentration,
- geographic deconcentration, and
- higher standards of quality.

In the context of putting these measures into place, Spain brought together many of its artistic and heritage resources, numerous historical and artistic monuments declared World Heritage Sites, currently 44 (UNESCO, 2010) as well as other less tangible resources such as the teaching of Spanish as a foreign language. They found the teaching of Spanish in Spain to have little structure, lacking accreditation of the teaching centres and certification of the teaching of Spanish as a foreign language unlike the case in countries such as the UK or France who have many years of experience in this area. For this reason on 26 July 2001, the 'Plan for the Promotion of Cultural and Linguistic Tourism' was announced.

The plan reflected the strengths and weaknesses of the cultural linguistic programme offered in Spain and proposed six action points using forty measures. In particular, the teaching of Spanish as a tourism resource was given special attention due to its specific nature both in terms of supply and demand. The concept of travelling to Spain with the motive of learning Spanish was presented as a by-product of cultural tourism, and further more, as a very attractive idea for Spanish tourism in general, both for direct and indirect economic repercussions as well as the contribution it would make to Spain's overall image as a destination.

The Plan developed various proposals and within the Marketing Plan they incorporated measures relating to promotion and commercialisation of the teaching of Spanish as a tourism resource. These actions bore fruit as the idea of Linguistic Tourism as a product became established, resulting in a tangible increase in language tourists.

As a consequence of this, the policy of promotion increased, not only on the part of Turespaña, The Cervantes Institute, the Ministry of Culture and the ICEX, but also as the Autonomous Communities and local bodies came together to roll out this new sector, recognising its growing potential. The work of the autonomous associations of Spanish schools for foreigners and the national association FEDELE should not be forgotten, as they played, and continue to play, a very important role both in the commercialisation of this product and in achieving quality in the services offered.

Currently, due to the world economic crisis, the Spanish government, conscious of the important role that tourism plays in Spain as a creator of wealth and employment, and recognising that it wasn't only the crisis that was affecting the demand for tourism, but also new competitors that were emerging within the world tourism market, making it more and more difficult to maintain the level of demand for tourism in Spain. On 25th July 2009 the *Plan de Promoción de Turismo Cultural (2009-2010)* was approved, which aimed to promote cultural tourist products overseas and to foster the destination *"España Cultural"* (Cultural Spain). In this plan, they established four action areas:

- Museums,
- Theatre, music and dance festivals,
- The promotion of themed cultural itineraries,
- The promotion of language tourism. Within the area of language tourism they contemplate a series of measures, where the Cervantes Institute's mission will be to encourage visits from foreign students to Spanish teaching centres located in Spain.

Spain is unlike the United Kingdom, France or Germany, which have numerous bodies at state, regional and local level which promote and commercialise language tourism.

5. THE SUPPLY AND DEMAND OF LINGUISTIC TOURISM IN SPAIN

5.1. The Supply

If we briefly analyse the supply and demand of Language Tourism in Spain, we can see that the provision of services has not been built up in a structured way, with a planned approach, but that the increase in supply is simply based on market forces. The most important increase has developed over the past 6 or 7 years, in a market lacking monitoring and regulation to assure quality criteria in the teaching centres.

In regards to the supply, it is very diverse and its current make-up is based, according to data form Turespaña, on 627 centres offering Spanish courses for foreigners. These consist of 53 universities, both public and private, 56 EOI and 518 private centres.

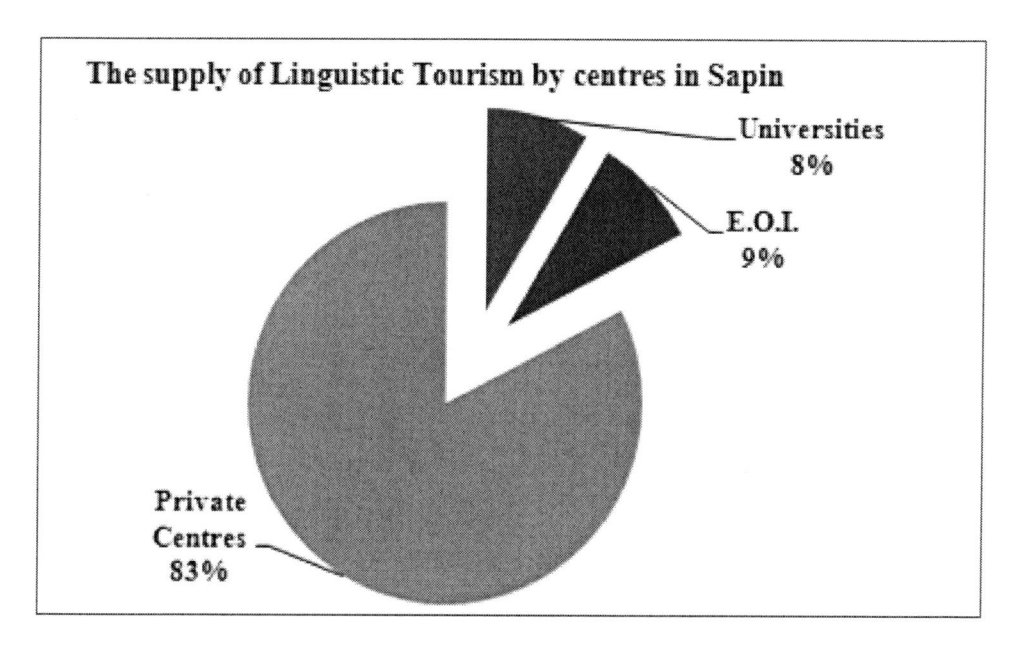

Graph 2. The supply of Linguistic Tourism by centres in Spain Source Turespaña. Authors own.

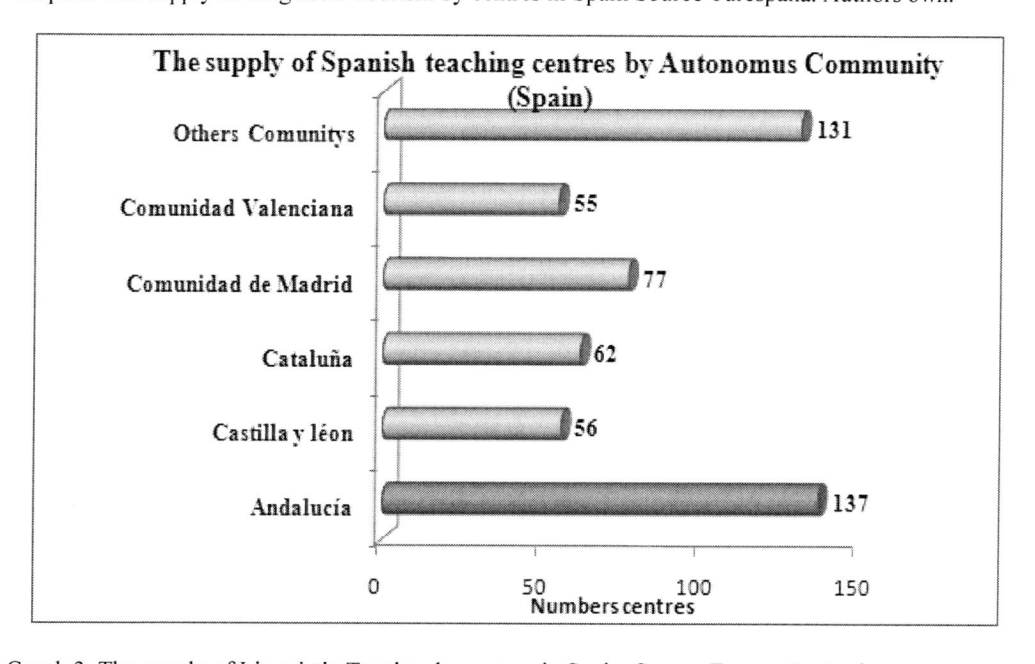

Graph 3. The supply of Linguistic Tourism by centres in Spain. Source Turespaña. Authors own.

There exists a strong geographic concentration in the supply of Spanish teaching centres for foreigners, Andalucía being the Autonomous Community (23,4%) with the highest index (23.4%), followed by C. Madrid (15.6%), Cataluña (14.8%), Castilla y León (11.3%) and C. Valenciana (10.4%). These figures represent 76% of the supply of Spanish teaching in only five of the Autonomous Communities. A reflection of this concentration is manifested in just a few cities: Málaga, Granada, Seville, Salamanca, Madrid, Barcelona and Valencia.

5.2. The Demand

According to studies form the Cervantes Institute, currently there are around 14 million students of Spanish as a foreign language, making Spanish the second most studied language in the world, after English. Studies of its potential show us that the greatest concentration of students is centered on the American continent. In North America there is a concentration of around 6,100,000 ELE students, where the majority are students from the United States with 6,000,000 students (Anuario Cervantes 2009). In South America, Brazil is the country which represents the highest increase, in recent years the rate of demand from ELE students has grown spectacularly passing from 1,000,000 to 5,000,000 students (Anuario Instituto Cervantes). However, despite such high numbers of potential tourists, it does not correspond to the actual level of demand of language tourists in Spain.

Cultural, historical and gastronomic factors are the preferred activities for potential language tourists, greatly outweighing those activities related to 'sun, sand and sea'. Further more, as expected, the prestige of the centres of study, questions of an economic nature and the provision of cultural activities (e.g. Montero et al., 2009).

In respect of the actual demand, from 2001 to 2007 there has been an increase of 9%, with Spain receiving some 237,000 language students in 2007. One of the characteristics of the market is an imbalance in terms of teaching centres. Private centres clearly show a greater concentration with 83% of the market, followed at some distance by universities with 9.26% and the EOI centres holding an insignificant 0.05 %.(Turespaña, 2008)

In terms of the distribution of the demand throughout the regions or communities of Spain, the main characteristic is the great geographic concentration in five autonomous communities: Andalucía, Castilla and León, C. de Madrid, Cataluña and C. Valenciana, amounting to 88.5% of the demand for language tourism in the country. Andalucia being the indisputable leader in language tourism in Spain, as much for its number of centres (147) as for its number of students (62,500) (Turespaña report, 2008).

6. SANTIAGO DE COMPOSTELA AND SALAMANCA: TWO LANGUAGE DESTINATIONS

One of the principle characteristics of language tourism is that, although it is offered in all destinations, it is the urban areas that present a greater demand and, above all, those that possess a wealth of heritage and culture. But one factor is key for the language tourist at the moment of choice of a destination: the position of the image of the destination.

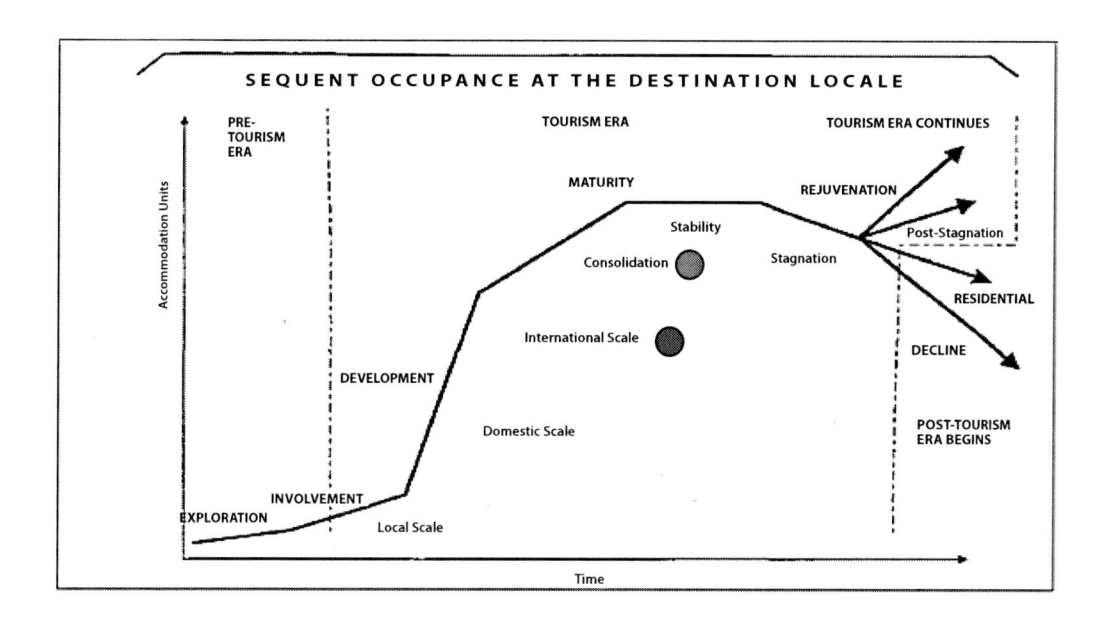

Graph 4. A revised version of Butler's (1980) "classical" sequence of stage for a resort town. Santiago de Compostela and Salamanca. Authors own.

We have chosen as the object of our study two Spanish cities with great artistic and cultural heritage: Santiago de Compostela and Salamanca, both cities being World Heritage Sites (1985 and 1988) and also European Cities of Culture (both 2002). They boast internationally prestigious, centuries old universities (1495 and 1218) and have long traditions of the teaching of Spanish (1940 and 1929). Due to this they have a well positioned image within the cultural tourism market.

However, positioning within language tourism behaves differently. If we take as our basis a revised version of Butler's (1980) "classical" sequence of stages for a resort town (Butler, R.W., 2006). Santiago de Compostela, despite its long tradition of the teaching of Spanish, would be located in the "development" stage (International Scale), however, Salamanca is a destination whose position would be located, according to the Butler model, in the "Consolidation" stage, in the field of language tourism.

6.1. Study of the Supply of Language in Santiago de Compostela

Provision in Santiago de Compostela, in spite of its great quality and tradition, is scarce and presents the following different types: Public provision and private provision

- The public provision is made up of the University (The Centre of Modern Languages (C.L.M.) and International Courses (C.I.)) and The Official School of Languages.

- The private provision is scarce with just one centre, The Iria Flavia Academy (the only centre in Galicia accredited by the Cervantes Institute)

The Centre of Modern Languages (CLM). This is a *service* of USC dedicated to teaching foreign languages, and has a very prominent place in the teaching of *Spanish for foreigners*, being geared especially towards students of Socrates, Erasmus and those whose universities have special agreements with USC. The channels of promotion are through the USC agreements with others universities. As a defining characteristic relative to the rest of the supply, its courses are exclusively language-based and offered as a complement to academic training for foreign students. For this reason they do not incorporate tourism activities within their programme.

International Courses Ltd. This is the oldest institution in the field of those providing Spanish as a foreign language in Santiago, born in the 1940s under the auspice of the University of Santiago de Compostela. In the present day it is a limited company in which the following members participate: USC (50%), Town Hall of Santiago de Compostela (20%), the General Secretariat of Tourism and Turgalicia (20%), the Association of Hotels (5%), Acotes (5%).

Provision at C.I. is founded in two bases: the teaching of Spanish as a foreign language and the teaching of Galician. A characteristic of International Courses is the combination of Spanish teaching with tourist activities. For this reason the main characteristic of C.I. is its business policy to differentiate and innovate its product, evident in the specific, high quality courses cited previously that the C.I. runs, being the pioneer in the language tourism sector within the city, and holding the greatest share of the market.

The Official School of Languages. The supply of Spanish for foreigners is very recent and consequently of little significance.

The Iria Flavia Academy. This is the representative of private provision, as well as the only academy in the city that specialises in the supply of Spanish courses for foreigners. It has a long professional tradition and is the only centre in Galicia accredited by the Cervantes Institute. The main characteristic of its marketing policy is that it has specialised in an unexploited niche in the local and national market, by working with the youngest people. This specialisation makes them unique amongst their competitors. In the same way as C.I. they combine language teaching with cultural and tourist activities, as well as specific language courses.

In summary, we can state that the principle characteristic of the provision of teaching in Santiago de Compostela is different to that of the rest of Spain in that the supply is concentrated in public provision because private provision is scarce, although, both represent excellent quality and experience.

6.2. Study of the Supply of Language in Salamanca (Spain)

On the contrary, Salamanca has a more varied supply: two universities - *USAL* and *La Pontificia* – and twenty private schools. Private provision is made up of recently opened schools; the great majorities have less than twenty years of experience. In terms of the physical capacity, 68% of the schools are small and supply less than 200 places each. One exception is the *Colegio Delibes* with 400 places and the *USAL* that has no limit of capacity, that is to say, it supplies for whatever demand there is. (Ciudad del Español y Observatorio Turístico de Salamanca).

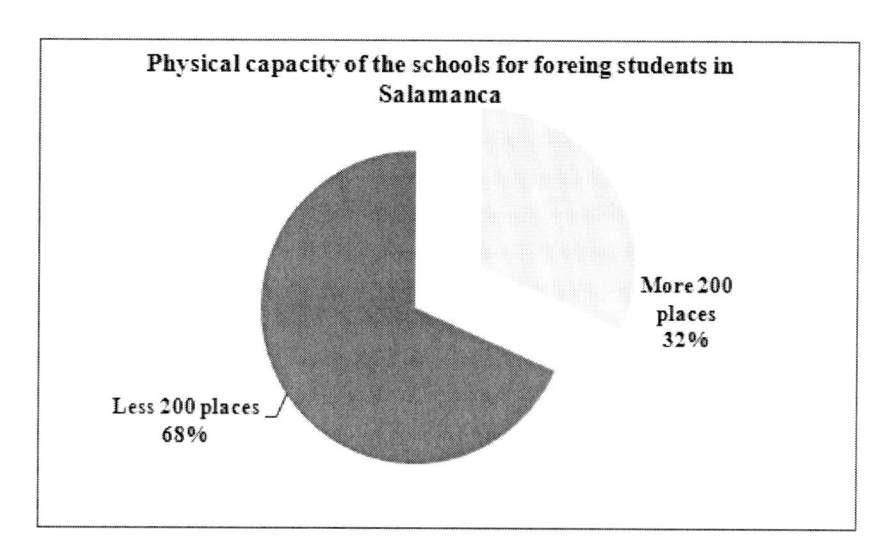

It is important to highlight that quality is very evident in the teaching centres in Salamanca and out of the twenty private centres, fourteen are accredited by the Cervantes Institute.

The concentration of the centres differs in the two destinations. In Santiago de Compostela the centres where they give the ELE courses are spread throughout the city, which is of benefit in terms of the minimal effect of the over capacity of language tourists on the resident population. However, Salamanca has a greater spacial concentration of its ELE centres, which are mainly concentrated in the historic centre of the city.

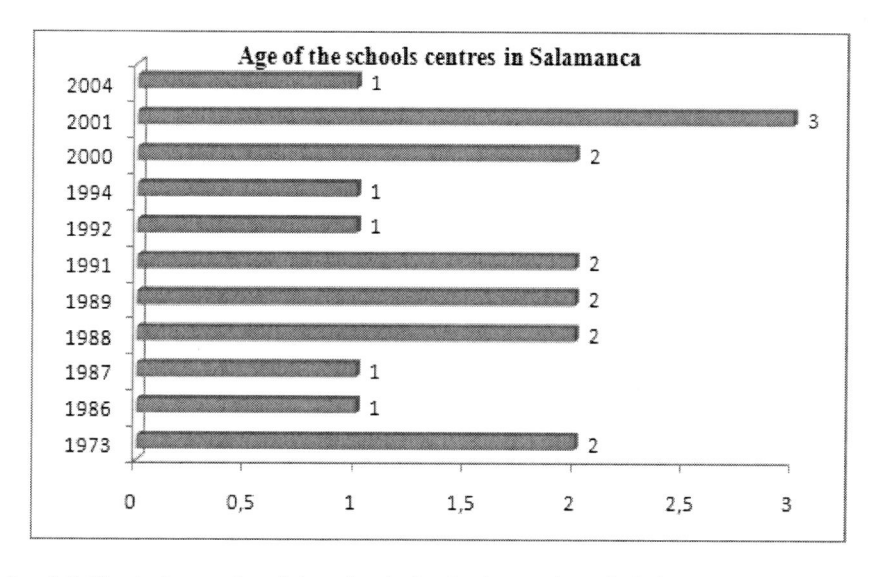

Graphs 5 and 6. Physical capacity of the schools for foreign students in Salamanca and Age of the schools centres in Salamanca. Source Tourism Observatory of Salamanca and "The Spanish City". Authors own.

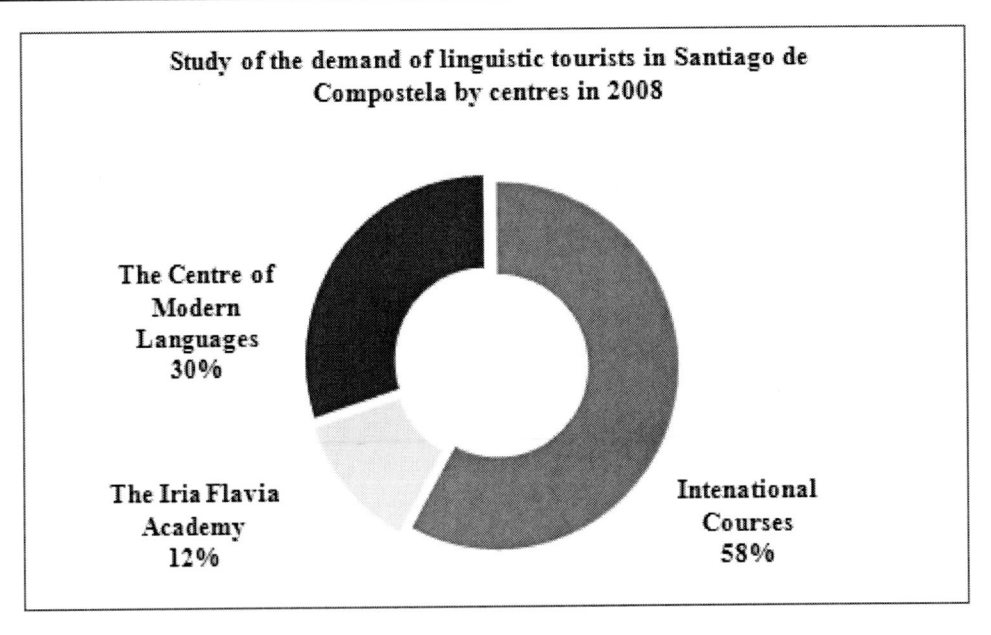

Study of the demand of linguistic tourists in Santiago de Compostela by centres in 2008

- The Centre of Modern Languages 30%
- The Iria Flavia Academy 12%
- Intenational Courses 58%

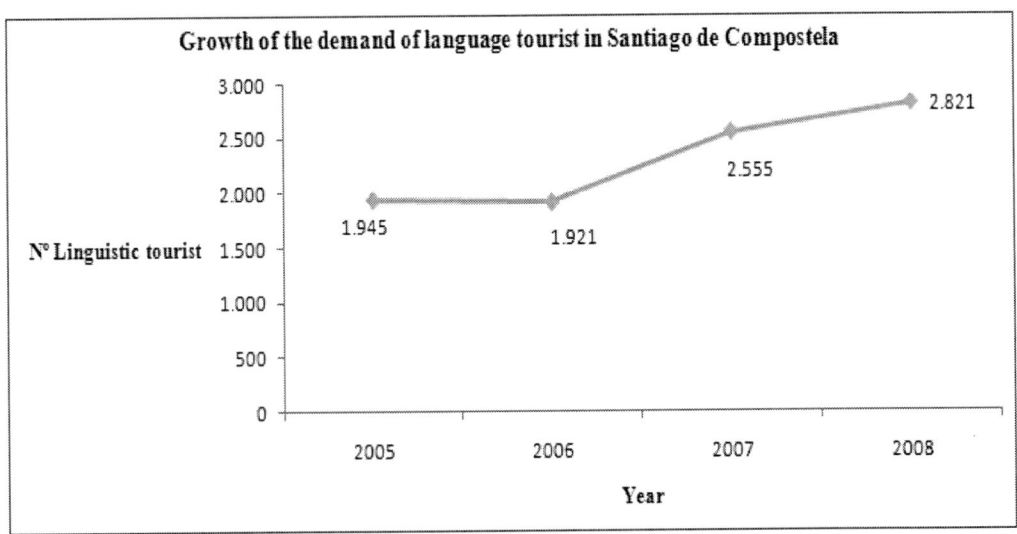

Growth of the demand of language tourist in Santiago de Compostela

Year	Nº Linguistic tourist
2005	1.945
2006	1.921
2007	2.555
2008	2.821

Graphs 7 and 8. Study of the demand of linguistic tourists in Santiago de Compostela by centres in 2008 and growth of the demand of language tourists in Santiago de Compostela. Authors own.

6.3. A Study of the Demand of Language Tourism in Santiago de Compostela (Spain)

A particular feature characterises the demand for linguistic tourism in Santiago de Compostela: the high number of pupils within public provision (88%), against a low level of private provision (12%), as is indicated in the adjoining graph. This situation clearly runs counter to the demands on the part of ELE students in the rest of Spain, which are centred primarily on private provision. The growth of demand for ELE courses in Santiago de Compostela presents a positive evolution with strong potential. It is striking, however, that this market is so underdeveloped when compared with others possessing similar characteristics such as Salamanca, for example, whose demand is very high with 25,619 students in 2007 (Observatory of Salamanca and "The Spanish City "report, 2008). The adjoining graph shows the growth in the demand of language tourists in Santiago de Compostela, the most important factor being its continued increase.

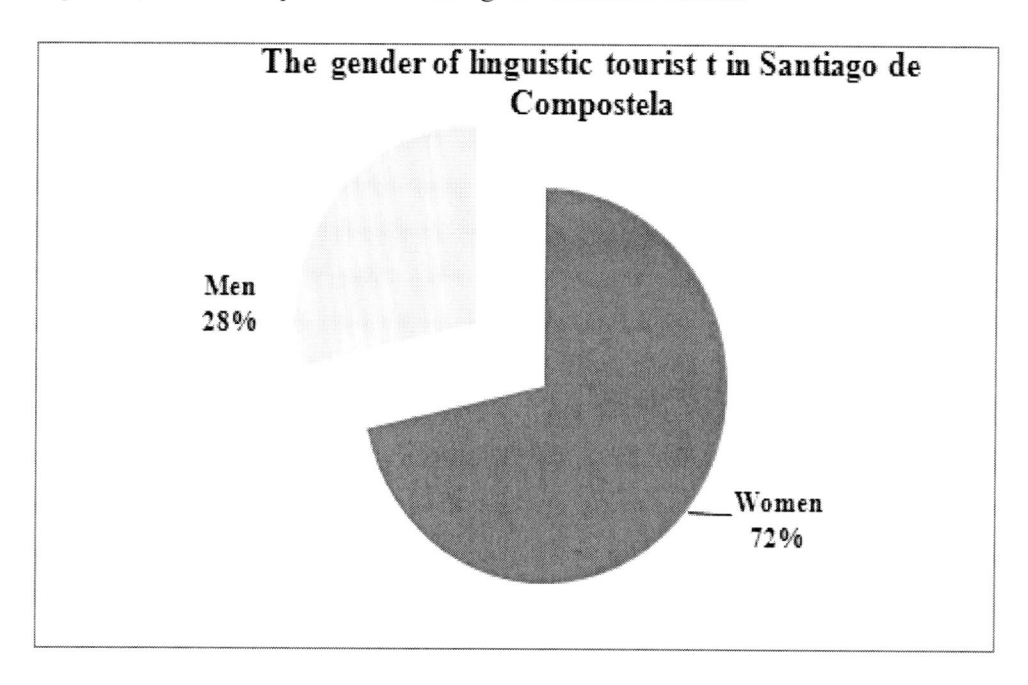

The gender of linguistic tourist t in Santiago de Compostela

- Men 28%
- Women 72%

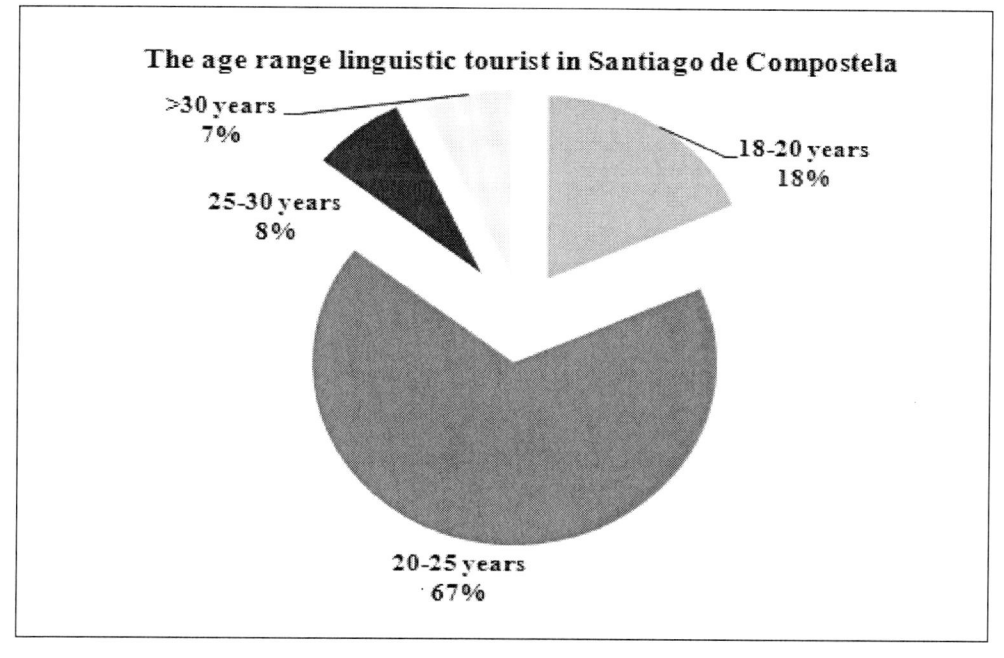

The age range linguistic tourist in Santiago de Compostela

- >30 years 7%
- 25-30 years 8%
- 18-20 years 18%
- 20-25 years 67%

Graphs 9 and 10. The gender of linguistic tourist in Santiago de Compostela and The age range linguistic tourist in Santiago de Compostela. Authors own.

In conclusion, the evolution of the demand of ELE courses in Santiago de Compostela presents a great concentration in public provision. In terms of the evolution of the demand, we can state that it is a positive evolution, and with great potential, although it does not reach the figures of other cities with similar characteristics such as Salamanca or Granada. To carry out an analysis of the demand in this sector of tourism, the methodology used was a survey of foreign students subscribed to the ELE centres in Santiago de Compostela, during the month of May 2009 The objective of this analysis was the modelling of the profile and the behaviour of a language tourist in the city, seeing as it is only through behavioural studies that we can achieve adequate and quality provision. The questionnaire was divided into various sections of study:

- Socio-demographic profile;
- Motivation and source of information on destination;
- Planning the trip. Method of reservation, transport and accommodation;
- Activities carried out during their stay and evaluation of the destination;
- The expenditure of linguistic tourists. The multiplier effect.

6.3.1. Socio-demographic Profile

The age range is very wide, although the most numerous group is found amongst young people between the ages of 21 and 25 (67.8%), showing a low presence of those under 18 years and older than 31 years (7%) of age. In regards to gender, there exists a clear majority of women (71.5%) – data very similar to the rest of Spain.

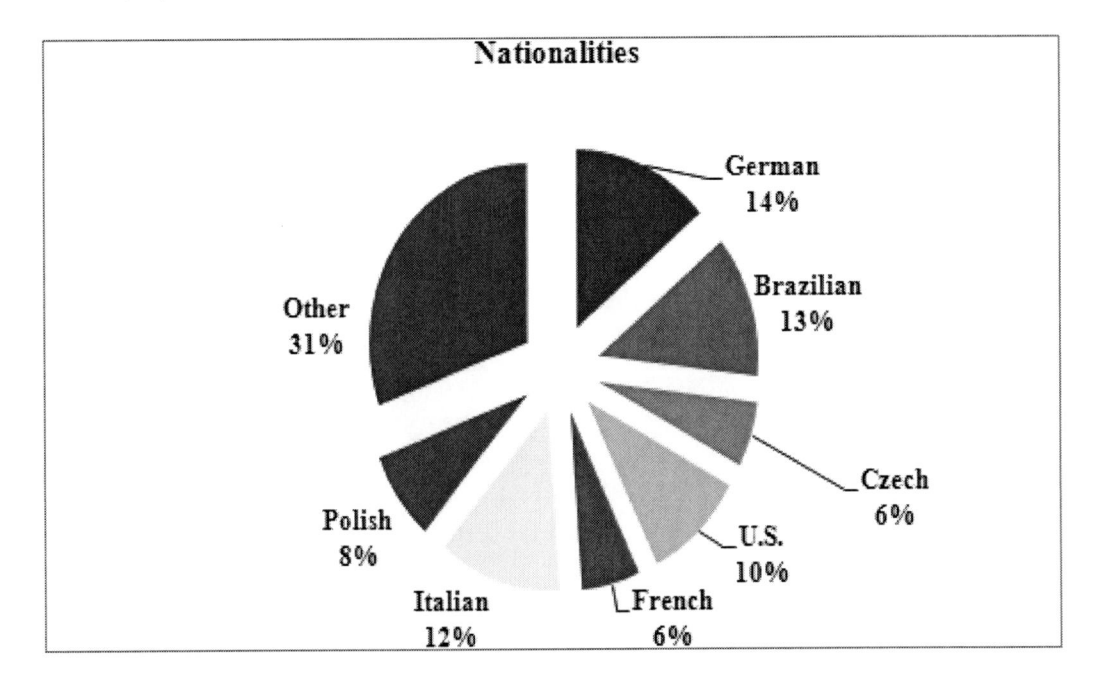

Graph 11. Nationality *of the tourists linguistic in* Santiago de Compostela. Authors own.

One of the aspects that most characterizes the demand of language tourist trips to Spain and where there exists a greater discrepancy with this type of trip to other countries such as the United Kingdom, Ireland, United States, France etc. is the lower presence of courses specifically for under 18s. In the case we are analyzing, there is only private provision of specific courses for under 18s. Another group which is also little developed is courses for over 50s, although we consider this to be a sector with great potential due to the changes in the habits of workers in the twenty-first century, amongst those which stand out are. Focusing on their places of origin, Europe is the main source (69%), the next is the American continent (24%) and some way behind are Asia and Oceania (6.5%). Within Europe the main places of origin are: Germany (14%), Italy (12.4%), Poland (8.1%), and France (6%). In the American continent, the United States (10.2%) and Brazil (12.9%) stand out.

6.3.2. The Motivation and Source of Information on Destination

In the first place students choose the study of Spanish as a foreign language for various reasons amongst which the following stand out: its significant importance for their working life, that the knowledge of Spanish allows them to travel in numerous countries, and also because they like studying languages. The motive for which they choose Santiago de Compostela as a language destination is for its attractive nature (70.97%). We find this data very interesting as with this information alone, on the importance of the image of Santiago de Compostela, appropriate marketing campaigns could be carried out to promote it as a destination.

6.3.3. Planning the Trip: Method of Reservation, Transport and Accommodation

In respect to finding out about the destination, means are varied: through their university (58%), through new technologies ie. the internet (13%) and through the medium of friends and family (12%). In relation to the manner of organising the trip we can see via the survey that new information channels like the internet have a high presence (35%) and also agreements between universities (29%), especially in the case of North Americans. Traditional channels such as travel agents are hardly noticeable (7%). This data is significant and should be taken into consideration in future promotional campaigns for the destination, namely that traditional channels such as fairs and travel agents are giving way to faster channels such as the internet. Also, the strengthening of links between universities is fundamental for an increase in this sector.

When we analyse how people travel to Santiago, the principal method of transport is by air (56%), followed by bus (12%) and train (10%). This data would increase in the case of air travel if we were to count all the language tourists that arrive by air to other cities such as Madrid or Porto and then take the bus or train to Santiago de Compostela, due to the city airport's lower level of international connections. In reference to the kind of accommodation chosen during their stay, we found that students who spend long periods in the city mostly opt to live in a shared flat (76.8%), followed by staying with local families (15.5%) and in third place, university residences (3.2%).

The average stay varies depending on nationality, the type of course, and the chosen teaching centre. 59% of ELE students surveyed have an average length of stay between 2 and 6 months los alumnus de CLM although students which study *International Courses Ldt.* stay for an average of 4 weeks.

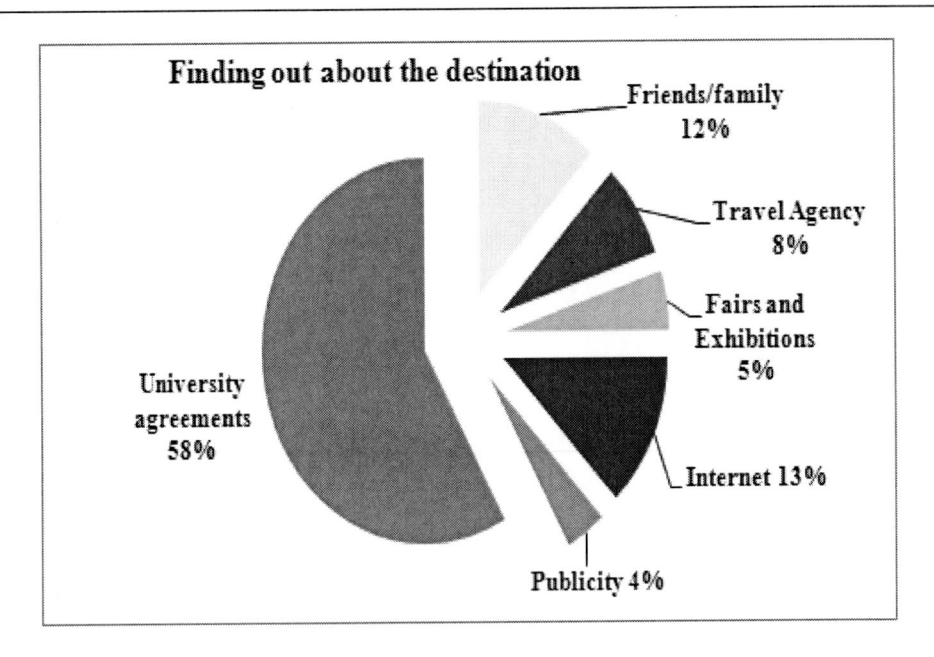

Finding out about the destination
- Friends/family 12%
- Travel Agency 8%
- Fairs and Exhibitions 5%
- Internet 13%
- Publicity 4%
- University agreements 58%

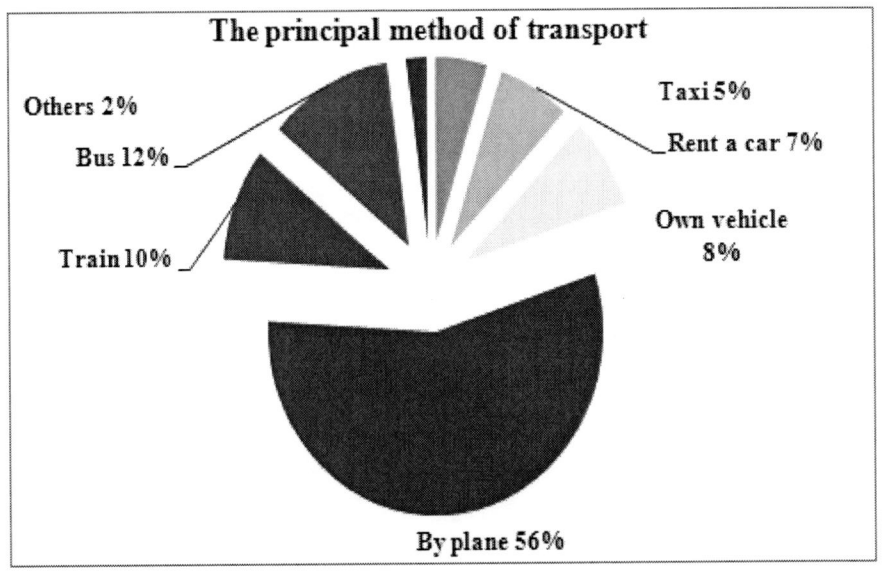

The principal method of transport
- Others 2%
- Bus 12%
- Train 10%
- Taxi 5%
- Rent a car 7%
- Own vehicle 8%
- By plane 56%

Graphs 12 and 13. Finding out about the destination and The principal method of transport *of the tourists linguistic in* Santiago de Compostela. Authors own.

6.3.4. Activities Carried out during their Stay. Evaluation of the Destination

As well as attending courses in Spanish as a foreign language, the language tourist carries out numerous tourist activities, from which arises the important economic impact which the city benefits from. Cultural activities are very present in the life of a language tourist, such as: going to the theatre, concerts, and cinema (11%), and visiting museums etc. (11%). They also enjoy the nightlife (12.5%), eat in restaurants (12.5%), and go shopping (12.1%), not only for souvenirs of the city, but also for their daily necessities, as well as buying gifts for families and friends.

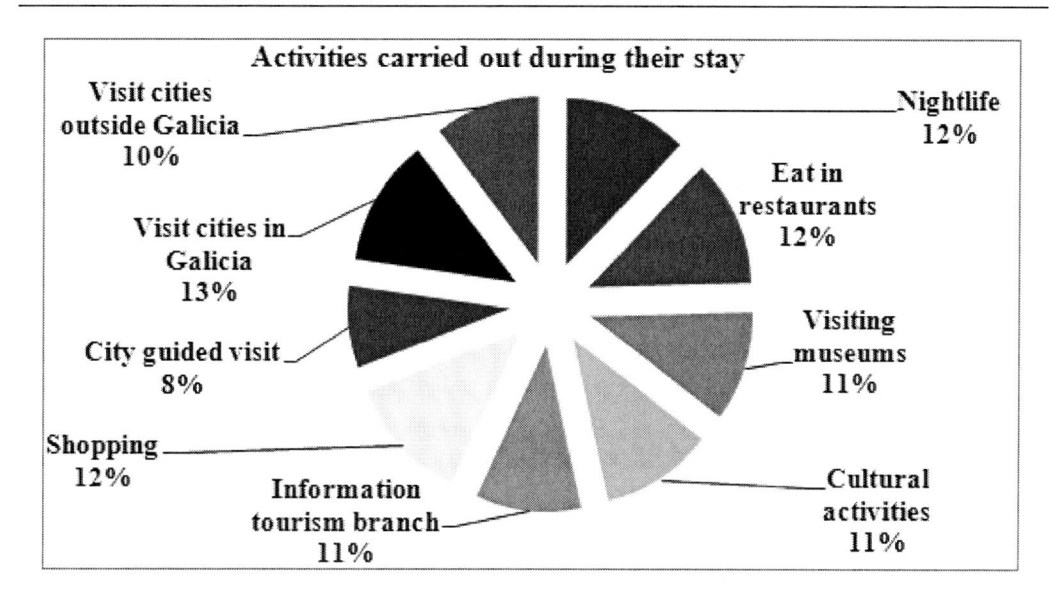

Graph 14. Activities carried out during their stay. Authors own.

Another characteristic of this tourist is that they like to travel not only to other Galician cities, with A Coruña and Vigo being the most popular, but also to other cities within Spain, mainly Madrid, Barcelona, Granada or Bilbao. On other occasions, due to the proximity to Portugal, they travel as far as Porto or Lisbon, or, taking advantage of cheap flights available from the city airport they make trips to Rome or Paris. The most travelled of the Europeans are the Germans (12%), followed by the Italians (11.3%) and from the American continent, the Brazilians (14%).

In reference to the value placed on the most important elements the city has to offer we can state that the gastronomy is very highly valued (57.9%) as much as the culture (53%), heritage (83%) and the University of Santiago de Compostela (65%). Other aspects of the destination are highly valued such as the significant cultural offerings, the peacefulness, and the safety of the city and the friendliness of the people; however, there are aspects, such as the climate and the lack of greater connections to the rest of Spain which are viewed poorly.

6.3.5. *Analysis of the Expenditure of Linguistic Tourists. The Multiplier Effect*

The analysis of expenditure has been the most difficult to quantify due to numerous difficulties that we have come across: credit given at the start of courses, currency exchange, and the reluctance to answer personal questions, and the suggested bias of students at C.L.M. where the survey was presented. Despite this, the results of the survey give us relevant data: the average monthly expenditure of the language tourist varies according to gender, in the case of women it is €835.60 whilst for men it is €1,061.60. If we compare them with the Salamancan figures, similarly the average is 1.058,38 € (*Observatorio Turístico de Salamanca*) during their visit to the city (excluding transport and complementary activities).

The most significant areas of expenditure, with some differences between men and women, are the following: accommodation (men 35% and women 22%), eating out (men 21% and women 22%), leisure (men 10% and women 13%) and shopping (men 12% and women 15%).

Something that is present in linguistic tourism that differentiates, characterises, and makes it more attractive, is its tourist *multiplier effect*. The more permanent nature of long stays at the destination of the language tourist means that many of them receive visits from other tourists (family, friends, and colleagues), a new phenomenon: the tourist is visited in turn by other tourists. The data obtained from our survey corroborates this hypothesis; the language tourist has an average of 3.6 visits with the number of visits by two people being the most frequent (27%).

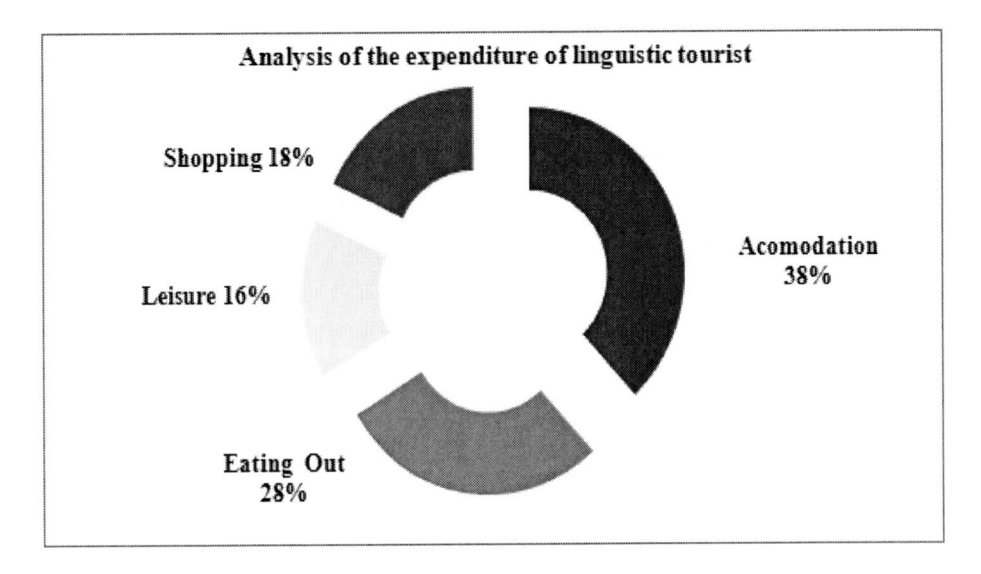

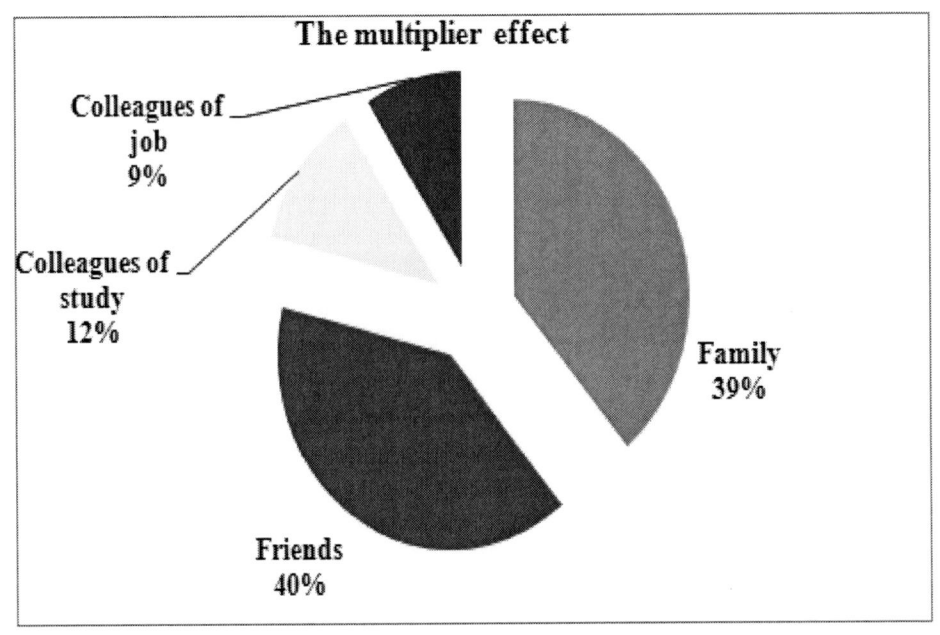

Analysis of the expenditure of linguistic tourist

- Shopping 18%
- Leisure 16%
- Acomodation 38%
- Eating Out 28%

The multiplier effect

- Colleagues of job 9%
- Colleagues of study 12%
- Family 39%
- Friends 40%

Graphs 15 and 16. Analysis of the expenditure of linguistic tourist in Santiago de Compostela. The multiplier effect of linguistic tourist in Santiago de Compostela. Authors own.

7. A STUDY OF THE DEMAND OF LANGUAGE TOURISM IN SALAMANCA (SPAIN)

The demand of language tourism in Salamanca is very positive, with figures situating it amongst the top language tourist destinations of Spain, with almost 30,000 students every year (in 2007 25,619 and in 2008 23,749) (*Ciudad del Español y Observatorio Turístico de Salamanca*), but these figures are down to the long-term work in the commercialisation of the product and policies coordinated as much by the regional government as at a local level and for a strong willingness on the part of all stakeholders to position themselves firmly in the cultural tourism sector. Proof of which are the numerous activities that have been carried out by the management and planning bodies of the city, amongst which we highlight: the creation of a special office for foreign students (*La ciudad del Español*), the creation of the brand "*Salamanca ciudad del español*", the commencement of areas of research by the *Observatorio Turístico de Salamanca* and activities coordinated on part of the organisms responsible for promotion both at a state and regional as well as local level.

The profile of a language tourist studying in Salamanca is very similar to other Spanish university cities, but with some slight variations. The following lists the main characteristics:

- Young, with an average age of 22;
- Mainly women (61.5%);
- Highly educated;
- The motive for studying Spanish is as a complement to their professional training/education;
- The main source countries are: France (15.8%), Germany (14%), and US (19.3%);
- The average stay in the city is around 4 weeks;
- The place of residence: host families, shared apartments, and university residence;
- During their stay they undertake numerous tourist activities;
- The seasonal concentration is July, August and September;
- The average expenditure per language tourist is €1,058.38 (*Observatorio Turístico de Salamanca y Ciudad del Español*).

CONCLUSION

In a globalised world, the learning of languages gathers greater importance every day. The objective of this report has been to present the teaching of Spanish as a new sector within tourism in which is combined the teaching of language along with economic exchange as a tourist resource. Language tourism is less invasive, with less marked seasonal concentration, a more culturally aware target public and consequently more respectful of the environment.

The case of Santiago de Compostela, the main object of our study, is one of a less developed position in the market due to the fact that until very recently linguistic tourism was not considered a worthwhile sector to develop, although it does appear in the Strategic Plan for the City (Plan de Marketing estratégico del Turismo de Santiago de Compostela, 2004). Currently, local authorities are beginning to understand that this sector of tourism is beneficial for the city, not only in terms of the economic impact that it brings about, but also because it is a kind of tourism that helps promote the image of the destination and is less invasive than other types of tourism such as group tours, which are so present in the city.

After analysing the market of the students that come to study Spanish in Santiago de Compostela we can confirm that they carry out numerous tourist activities in the city, within the region and also nationwide within Spain, such as: cultural visits, going to concerts, visiting museums, exhibitions and also shopping and leisure. Producing a great economic impact in the destination. But if just one characteristic deserves highlighting, which makes linguistic tourism so attractive and differentiates it from other types of tourism, it is that of the *multiplyer effect* – every language tourist generates 3, 6 visits from family members, friends or colleagues during their stay. A new phenomenon where the tourist, during their stay at the destination, is visited by other tourists.

Another characteristic that we consider of interest is the high degree of satisfaction of language tourists in the city. This satisfaction invokes plans to return to the city on future occasions, either alone or accompanied, which translates into an increase in tourists in the near future.

The interventions that we consider of most importance in order to develop the sector of linguistic tourism are:

- Better coordination by those responsible at the state, the autonomous regional, and the local level, with members of the sector.
- The creation of a homogeneous product, uniquely branding the destination in a way that drives and supports all the ELE centres, putting forward Santiago as a point of reference within Spain.
- The exploration of new market niches such as working with under 18s and over 50s as do other countries with a longer tradition of language teaching such as UK, France or Germany.
- The reduction in seasonal concentration through innovative products directed at a public with greater flexibility in their holiday arrangements.

We believe that it is very important to continue research into language tourism in the city. This report barely begins the task that should be the object of new and more ambitious approaches in the near future, to make possible sustained, professional management in this strategic sector.

In summary, we consider that language tourism has a strategic character – in the widest sense – because it helps to transmit local culture and reduce seasonality. The linguistic tourist has a high average expenditure, possesses a tourist *multiplier effect*, and develops a strong bond to their destination. For this reason Santiago de Compostela has an unequalled opportunity to develop this new sector, due to the characteristics of the city, to the presence of a university with more than 500 years of history and the critical mass that this implies, professionals of noted standing and, being already positioned within the landscape of global tourism, as a very prominent cultural destination, not just a religious one.

REFERENCES

Baralo, M. (2007) Enseñanza del español y turismo: las estancias lingüísticas available http://www.educacion.es/sgci/be/es/publicaciones/mosaico/mosaico20/mos20f.pdf.

Butler, R.W. (2005) the Tourism Area Life Cycle, Vol. 2 *Conceptual and Theorical Issues.* (Channel View Publications, Clevedon).

Cengiz, H., S. S., Eryilmaz, Y (2006) The importance of Cultural Tourism in the EU Integration Process. (*Paper presentd 42 nd ISoCaRP Congress 2006*) available http://www.isocarp.net/Data/case_studies/884.pdf

Coltman, M. M. (1989). *Introduction to Travel and Tourism an International Approach.* (York, Van Nostrand Reinhold).

Davó Cabra, J. Mª. (2002)El español como recurso económico en Francia (una aproximación desde el marketing). Available *El español en el mundo. Anuario del Instituto Cervantes*, 2002, p. 167-190.

El español en el mundo (2009). *Anuario del Instituto Cervantes.* (Madrid, BOE)

El mundo estudia español (2006). Ministerio de Educación y Ciencia. Subdirección General de cooperación Internacional. (Spain, Edt. Secretaría General Técnica) available http://www.mepsyd.es/redele/Biblioteca2007/elmundo/completo.pdf

El mundo estudia español (2007). Ministerio de Educación y Ciencia. Subdirección General de cooperación Internacional. (Spain, Edt. Secretaría General Técnica) available http://www.educacion.es/redele/elmundo/elmundo2007.shtml

Graddol, D. (2006) English Next. (United Kingdom: The English Company Ltd) available http://www.britishcouncil.org/learning-research-english-next.pdf

Herranz, G. (2006) El Turilingüísmo en España: actitudes y preferencias de los estudiantes universitarios estadounidenses de ELE. Available http://www.educacion.es/redele/Biblioteca2008/Ganfornina/MemoriaGanfornina.pdf

Internet World Stats available http://www.internetworldstats.com/stats7.htm

Instituto de Estudios Turísticos (2009) available http://www.iet.tourspain.es/informes/documentacion/FronturFamilitur/Balance%20del%20turismo%20en%20Espana%20en%202009.pdf

Lewis, M.P (Eds) (2009). *Ethnologue. Languages of the world.* (Dallas (USA), SIL International.

López Morales, H. (2008). *Enciclopedia del español en los Estados Unidos.* Anuario del Instituto Cervantes. (Madrid, Edit. Santillana)

Ministerio de Industria, Turismo y Comercio. RED.ES available http://www.nic.es/estadisticas/article/293

Kennett, B. (2002). Language Learners as Cultural Tourists. *Annals of Tourism Research,* Vol. 29, nª 2, pp. 557-559.

Montero, J.M; Móndejar, J., Vargas, M. Eds .Investigaciones, Métodos y Análisis del Turismo (Ferrari, G. Montero, J.M.; Mondéjar, J. y Vargas, M. Eds). *Septem Ediciones,* pp. 55-67. ISBN: 978-84-96491, 2010. DI: M-19104-2009.

Plan de impulso al turismo cultural e idiomático. *Serie Estudios de productos turísticos* (2002). (Madrid, Ed. Turespaña.)

Plan de Promoción del Turismo Cultural 2009-2012. (2009) available in www.mcu.es

Richards, G. (1996). *Cultural Tourism in Europe.* (Wallingford, CABI Publishing).

Smith, M.K. (2003) *Issues in Cultural Tourism Studies*. (London, Routledge).

Swarbrooke: Towards a sustainable future for cultural tourism: a European perspective. M. Robinson et al (ed) *Tourism and Culture: Managing Cultural Resources for the Tourist*, The Centre for Travel and Tourism in association with Business Education Publisher Ltd Great Britain pp.227-255(1996).

Taboada-de-Zúñiga, P. a. (2010) Language Learning as a Resource for Tourism. Case Study Santiago de Compostela. Prepare the Tourism Booming in China Proceedings. *The First International Conference on Tourism between China-Spain*. Editor Yuhua Luo, co-editors Marco A Robledo, Francisco Sastre, Xio ing, Teresa Palmer. Ed. Pearson. Palma de Mallorca, 2010. ISBN 978-84-8322-271-3.

Taboada-de-Zúñiga, P. b. (2010) Una aproximación al turismo idiomático: el caso particular de las ciudades históricas. Nuevas perspectivas del turismo para la próxima década. *III Jornadas de Investigación en Turismo. José Luis Jiménez Caballero and Pilar Fuentes Ruiz. Ed. Universidad de Sevilla.* ISBN- 978-84-693-2711-1.

Turismo idiomático. *Serie Estudios de productos turísticos* (2008) (Madrid, Ed. Egraf S.A.) 2008.

UNESCO (2010). Available http://whc.unesco.org/fr/etatsparties/es

Volumen y perfil de los estudiantes de español en la ciudad de Salamanca. (2008) *Observatorio Turístico Salamanca and Ciudad del español.* (Salamanca, Spain)

In: Research Studies on Tourism and Environment
Editors: J. Mondejar-Jimenez et al.

ISBN: 978-1-61209-946-0
© 2012 Nova Science Publishers, Inc.

Chapter 23

CONNECTING URBAN AND RURAL AREAS THROUGH A GREEN CORRIDOR: CASE STUDY OF TURIA FLUVIAL PARK (VALENCIA, SPAIN)

María José Viñals[*], Maryland Morant and Pau Alonso-Monasterio
Universidad Politécnica de Valencia, Spain

ABSTRACT

This chapter analyzes the connectivity effects afforded by a green corridor that links urban environments with natural and rural areas. This is the case of the Turia Fluvial Park in Valencia (Spain), a strategic project of a green corridor promoted by the Jucar River Basin Agency (*Confederación Hidrográfica del Júcar*). This corridor encompasses the Turia River as it flows through seven villages to the City of Valencia.

The Turia River Valley and the natural corridor have become a particular focus of recreation in the rural-urban fringe, positively affecting more than 1 million people. This green corridor includes a shared-use trail, providing a breathing space visually pleasurable, situated in a low-stress environment which is suitable for bicycling, horse riding and walking. It is an alternative setting that is separate from road traffic and concreted, where people can feel at ease and indulge their leisure needs in a natural area.

The most important outcome of this work is to highlight the vision of how this corridor can make this urban-rural fringe more attractive, accessible, and multi-functional, and to state the benefits of creating a connection both in people's mind as well as in the functioning of the environmental systems.

Keywords: Green Corridor, Urban-Rural Fringe, Recreation

---

[*] Contact author: mvinals@cgf.upv.es

1. INTRODUCTION

1.1. The Urban-Rural Fringes and Green Areas

The rural-urban fringe can be an attractive, accessible, diverse and multi-functional area that fulfils the needs of both urban and rural communities. By preserving existing corridors or creating additional linkages in the landscape, the enhancement of these values can be achieved.

Currently, some cities are trying to define community boundaries through green spaces, either to limit the urban growth (Monclús, 2006) or to preserve the integrity of the countryside surrounding the cities by designing this spaces to cover the minimum requirements of green areas in cities, that the World Health Organization estimated in $9m^2$/inhabitant (UN-Habitat, 2009). Green corridors represent an interesting alternative to manage the urban-rural fringe, even more when they are a popular mean for commuters to access the city or its surrounding towns, allowing, at the same time, a higher number of people to enjoy nature near their residential areas.

Several cities around the world had begun initiatives regarding this urban strategy. Among these it's worth to mention some that have already been executed, like the "Green U" in Stuttgart (Germany), the German Green Fringes in Munich (Pauleit & Oppermann, 2002) or the Botanic Spine in Dublin (Irland). This is the case also of the green rings of Vitoria-Gasteiz (De Juana, 2003) and Zaragoza both in Spain or the London Metropolitan Green Belt (United Kingdom), which encircles the greater London area (Kappler & Miller, 2009) and is completed with a set of green corridors. We can also find several examples of these strategic infrastructures in USA, as the Greater Grand Forks Greenway, a public park in North Dakota and Minnesota, the Ohlone Greenway in San Francisco, the Cherry Creek Greenway in Pennsylvania or the Rachel Carson Greenway in Maryland. In fact, green corridor planning has been the fastest among all planning and design activities in USA in the last years (Yun-Guo et al., 2005).

1.2. Green Corridors and Greenways

There are many definitions for the term 'green corridor' and many natural or rear/natural open spaces could be included in this concept. It is important to keep in mind that this is a technical conception based on the existence of linear-shaped open spaces which are characterized by attributes such as connectivity, access and their contribution to the creation of buffer areas.

Usually natural corridors can be found in remote geographical areas, thus fluvial valleys are one of the existent natural landforms that better fit into this definition. From the ecological point of view, especially when considering habitats, the existence of these spaces bears important biodiversity implications as conservation is an integrated key component in the environmental planning process. In fact, they could be assimilated as 'ecological corridors'.

Natural corridors normally don't need human interventions but sometimes it is needed to connect patches and linear strips of habitat. This connectivity would better maintain their environmental functions than if they were isolated (Girling & Kellet, 2005).

Rear/natural corridors are generalized and result especially important in the urban-urban fringe as they can be integrated with natural features of significance in order to adequately protect elements of greater biodiversity values and because of the connectivity functions that they can provide.

Beyond the environmental implications that green corridors have directly and indirectly in supporting biodiversity objectives; they provide and maintain connected open green spaces in areas of high density urban development.

For all these reasons, green corridors are currently a land-use and urban planning concern. What is particularly innovative is the advanced application of landscape ecological approaches to the landscape planning and management system.

In this school of thought, we can include Little's (1990) definition of green corridor. He defined them as open-spaces linking parks, nature reserves, cultural features, or historic sites with each other and with populated areas often through stream valleys and rivers. To a greater extent, this kind of approach is included in the discussion on how to 'rebuild', 'revitalize' and/or 'redevelop' the cities.

A similar emerging international strategic approach to land-use planning is the concept of 'green infrastructure'. Davies et al. (2009) and Baris et al. (2010) define green infrastructures as an interconnected network of green spaces that conserve natural ecosystem values and functions and provide associated benefits to human populations. These authors add that green infrastructure differs from conventional approaches to open space planning, where conservation values and actions are considered in concert with land use and social and economic development, growth management and built infrastructure planning. Green corridors are included in this framework as a linear green infrastructure.

Following Chiesura (2004), green corridors, proved environmental benefits as improving biodiversity, as well as its contribution to the quality of life by living in harmony with nature. The presence of natural areas provides important social and psychological benefits to human societies, which enrich human life with meanings and emotions.

On the other hand, it is common that green corridors include greenways as trails, towpaths or paths along riverbanks and canals, cycle ways through linear green spaces, and some rights of way. Greenways and green corridors are two concepts often treated effectively as one because they are both based on concepts of connectivity. But, usually greenways have a recreational emphasis while natural corridors make emphasis in ecological purposes as habitat and species conservation.

Greenways can be found especially in rear/natural corridors because they represent an important chance to link open green spaces within the urban areas (cities and country towns), providing aesthetic values, recreation opportunities and supporting wildlife conservation objectives. In fact, most of the times, greenways execution in this kind of corridors, involves ecological restoration projects. Nevertheless, they can also include social and cultural goals in order to promote the concept of sustainable land use. Then, the greenway planning has evolved from a single-objective paradigm of environmental protection or recreation to a multipurpose process that allocates greenways as a resource to satisfy the publics' demands of walking, cycling, horse riding, leisure and entertainment opportunities.

However, the access provided by greenways to nature, has its costs. Architect Ferguson (1995) noted that the best option is to stay out of nature's way. The major threats to biodiversity caused by greenway users and amenities setting (such a trails, interpretative facilities and streamside development) include disturbance or destruction of riparian forests along the river, infestation by invasive species and damage caused by storm water runoff. Therefore, greenways must be designed and executed with appropriate technologies in order to avoid these impacts.

1.3. Study Area: The Turia Fluvial Park

The geographic area targeted by the Turia Fluvial Park is the 26-kilometres along the Turia River, encompassing the final segment of the Turia River as it flows through Vilamarxant, Benaguasil, Riba-roja de Turia, L'Eliana, Paterna, Manises, Quart de Poblet, Mislata and the City of Valencia (Figure 1).

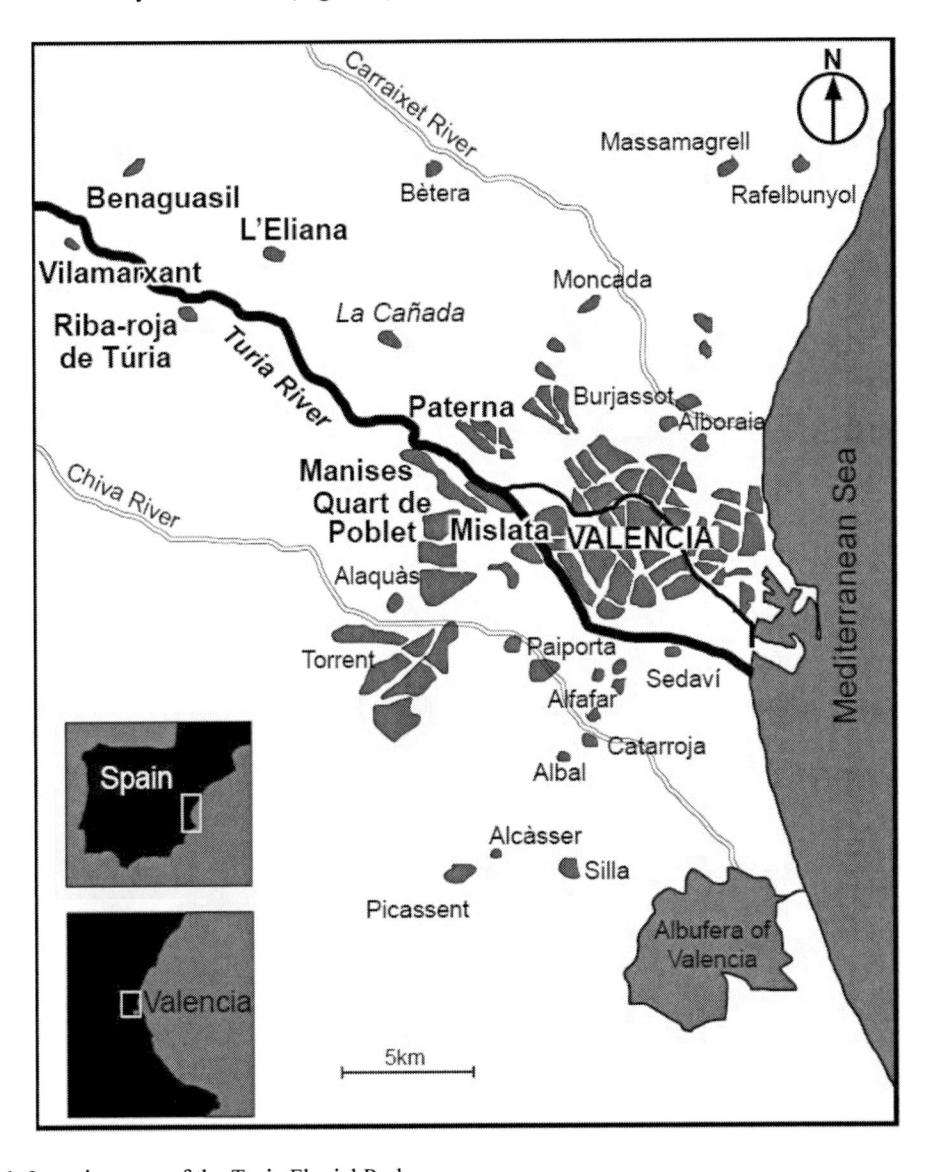

Figure 1. Location map of the Turia Fluvial Park.

The Turia River, located in the Valencia region (Spain), is a typical Mediterranean stream that harbors wildlife habitats with rare plants and animals as well as ethnographic elements and historic sites (Aguilella & Ríos, 2004). Following the report of the Conselleria de Medi Ambient, Aigua, Urbanisme i Habitatge (2009), a wide variety (about 166 species) of animals are commonly seen in and along the river, including: 166 vertebrate species, being 111 birds, 26 mammals, 14 reptiles, 6 amphibians and 9 fish. Among them, *Myotis capaccinii, Riparia riparia* and *Ardea purpurea* are species listed either endangered or threatened with extinction in the Valencian Regional Register, and there are about 20 protected species as for example the turtle *Mauremys leprosa* or *Arvicola sapidus*.

The riverine banks were dominated by *Arundo donax,* but we can find some patches of riparian forest and other restored areas with *Populus alba*, *Populus nigra*, *Salix alba*, *Salix elaeagnos*, *Salix purpurea*, *Tamarix gallica* and, less usual is *Ulmus minor*. A halophytic community of plants can be found in the riversides and some Potamogeton appear on the shallow and calm river water surface.

Some archaeological sites date from the Bronze Age (Lloma de Betxí) to post-roman visigothic period (Valencia la Vella) and historical sites date from the Spanish Civil War (military trenches) can be visible along the River.

For these reasons, this corridor is a valuable resource, forming a rich natural and cultural heritage and providing numerous benefits to the Valencian Region such as water for drinking, irrigation, recreational opportunities, sense of history, etc.

2. MATERIALS AND METHOD

This paper is the result of several field work campaigns based on the nature direct observation and getting empirical evidences about the users' attitudes and behaviors. A survey over 1,824 users was carried out by the Universidad Politécnica de Valencia research team during the spring of 2009 in order to identify the visitor's profile and the perceptions and expectations of the park users about the implementation of a new green corridor and related-recreational facilities.

Many in depth interviews, focus group meetings and stakeholders' consultations have been done to the promoters of the green corridor (Jucar River Basin Agency) in order to meet the expectations and interests of all the involved partners.

During the implementation period of the project, campaigns were held to inform and encourage citizens, landowners, developers, managers, and leaders of the community to take specific actions for a better stewardship of the natural and cultural resources of the river corridor.

A research work has been developed by the authors of this paper to design and set an accessibility plan and signage facilities to encourage sustainable visits. However, it has been observed that many visitors (direct estimations bring the number of more than 10.000 at week-ends) spontaneously indulged in activities even before the end of the restoration project.

3. RESULTS AND DISCUSSIONS

3.1. The Green Corridor Project

The Turia Fluvial Park is a strategic project of green corridor promoted and conducted by the Jucar River Basin Agency (Confederación Hidrográfica del Júcar, Ministerio Español de Medio Ambiente, Medio Rural y Marino) that started in 2007.

The vision of this project was conceived in order to respond to the European Water Framework Directive and to the Spanish Rivers Restoration Strategy goals in relation with river protection and restoration and with the aim to integrate fluvial ecosystems into the territorial planning.

While the interests and concerns addressed by this project are focused on the river corridor, they are not restricted to the corridor. Therefore, water quality concerns for the river require attention to the entire watershed. Interests in conservation and managing development in the river corridor also extend to the special character of the Turia River. Recreational use and access interests extend to potential connections with land and greenways that link the river to the greenway 'Garden of Turia' (ancient fluvial bed of the Turia River crossing the city of Valencia) (Viñals et al, 2004), to the protected area of the regional Turia Natural Park and recreation facilities and other natural and cultural resources beyond the defined corridor.

Therefore, the Jucar River Basin Agency has been conducting several hydraulic and environmental planning and economical efforts to create this park. The main goals of the plan concerned:

1. Reduce environmental degradation of the river's outstanding resources.
2. Enhance the highly valued natural, cultural, and scenic qualities of the river for the benefit and enjoyment of present and future generations.
3. Maintain and improve water quality in the Turia River to provide safe and healthy conditions for desired river uses.
4. Improve environmentally riverbanks and adjacent forested areas.
5. Balance recreational use and access with care, respect, and conservation of the river and its wildlife.

Actions afforded by this project were basically dealing with:

1. Water quality and flows management upstream.
2. Stability of the riverbanks.
3. Vegetation restoration works over the forested areas along the river.
4. Conservation of fluvial habitats and wildlife.
5. Cultural heritage elements restoration.
6. Creation of a greenway.
7. Access to the River improvements.
8. Mobilization of the natural and cultural resources for informal recreational purposes.
9. Provide recreational facilities and informative and directional signage.

One of the most outstanding facilities executed has been the greenway. This greenway is a linear shared-use path besides the stream linking the metropolitan area of Valencia with the reverie countryside towns and providing a breathing space that is visually pleasurable in a low-stress environment which is suitable for a wide variety of trip purposes such as bicycling, horse riding and walking. Definitively, it can be appointed as an alternative setting that is separate from motor vehicle traffic and concrete, where people can feel at ease and enjoy doing leisure activities in a natural area. Following Little's greenway classification (1990), this greenway can be basically included in the group of 'recreational greenways' defined as those featuring paths and trails of various kinds, often of relatively long distance, based on natural corridors (as it is our case) as well as canals, abandoned rail beds and other public right-of-way. But it is also recognized because it is an 'ecologically significant natural corridor' defined by the same author as found usually along rivers, streams and ridgelines, to provide for wildlife migration, nature study, and hiking.

It can be also highlighted that this green corridor links with the other above-mentioned open spaces found in the city of Valencia (Garden of Turia). This one can also be defined as an 'urban riverside greenway' following again Little (1990).

According to the survey results, this Turia Fluvial Park greenway has been recognized by users and citizens as an important recreation and townscape resource allowing park users to go bushwalking in close proximity to the city providing a sense of 'wilderness' and discovery.

This greenway is composed of a mixture of compacted gravel and earth road, wide enough for two-way travel with easy gradients. From a technical point of view the planning and design of this trail has posed many environmental challenges. The following issues are especially significant, as many of them have received extensive attention:

1. Positioning the greenway within the green corridor for avoiding land fragmentation.
2. Minimizing and managing environmental disturbance and impact, both during path construction and as the path sustains ongoing use.
3. Reducing storm water runoff and protecting against erosion.
4. Incorporating environmental restoration such as bioengineering and low-impact storm water management techniques.
5. Designing the trail to be compatible with or even reinforce the larger goals and recreational purposes of the corridor.

The Turia Fluvial Park shared-greenway does not permit the presence of motorized vehicles such as cars, motorcycles, etc. but is used by a diverse set of visitors representing different travel modes, using different types of equipment and traveling at different speeds. It is a management challenge to understand, even within the basic user categories of bicyclists, pedestrians, and horse riders, how diverse path users can be. This shared-use path is as we have seen, also appropriate for equestrian use within the same right-of-way but an additional bridleway has been implemented for horses, where appropriate, in order to facilitate transit in overcrowded points.

An additional emergency plan was designed in order to guarantee the users' safety which constituted of 25 access ways for emergency, police, and maintenance of the park vehicles.

Beyond this, the Turia Fluvial Park is benefiting from a coordinated plan to implement a signage system to assist the park users in finding their way through the park, providing safety feeling and to create identity by incorporating the park brand. Different signage types were set: directional signs, informative panels and points of interest signs following the same brand style used in the brochures, and in the self-guided map.

Information signs are located at main arrival points of the greenway, the directional signs are located at the key path intersections within the park as an aid for people to keep their bearing and know where they are. Point of interest signs are placed next to significant areas of vegetation or key features that have natural, historic or cultural importance. In a clear but imaginative form, these provide knowledge and insight to the area and the hidden treasures of the park.

The design of these materials is an invitation to explore and enjoy the park by using attractive graphic signs. All the signage elements include strategic themes, rules, safety recommendations and ethic and etiquette codes.

The two amenity areas of the Fluvial Turia Park contain a selection of furniture types consisting of pic-nic tables with seating benches, water fountains, rubbish bins, car parking, children's play facilities, an equestrian park, barbeques, bio-circuit area, etc.

3.2. Environmental and Social Implications

One of the main concerns of this work is to address the importance that this park has for natures' and citizens' well being.

Environmental implications of the project are evident because many improvements in habitats have taken place; especially in the riparian area thanks to the restoration works and autochthonous vegetation management (elimination of exotic species, over all *Arundo donax*; selective removal of cane, track clearance works and removal of weeds, intensify the number of plants; conversion to autochthonous riparian plants etc.). Therefore, 1,200,000 m^2 have been reforested with more than 200,000 plants of 26 native species.

Table 1. Targeted people directly involve in the Turia Fluvial Park Green Corridor Project

City/Town	Inhabitants
València	815,440
Mislata	43,756
Quart de Poblet	25,499
Manises	30,508
Paterna	64,023
L'Eliana	16,552
Riba Roja de Túria	18,938
Benaguasil	11,011
Vilamarxant	8,257
TOTAL	1,034,006

Source: From Population data referred to 01/01/2009 Royal Decree 1918/2009.

Regarding the fauna, particularly the greater variety of birds, fish and insects have increased their figures along the stream. What also enhances the biodiversity of bird life in the corridor is not just the diversity of the habitats themselves, but also the presence of fresh water.

About 20 heritage elements have been restored (water mills, small damps, traditional hydraulic devices, archaeological sites, etc.) in the framework of this project. Landscaping restoration works have improved river scenery in general terms.

On the other hand, the Turia Fluvial Park is a project affecting directly and positively 1,034,006 people (Table1). This green corridor project has mobilized 250 ha of open space for both conservation and recreational purposes, increasing access of urban dwellers to nature.

The target park users groups are basically pedestrians such as family groups, walkers, dog walkers, cyclists, horse riders, school groups and bird watchers. Moreover, people of all ages and abilities use and enjoy the greenway ranging from young to old people, from the novice cyclist to the professional runners as we can contrast from the surveys and from the weekly direct observations during two years.

Studying the answers of the open questions and final comments of the Turia Fluvial Park visitors' survey, and on-field evidences, some social benefits can be outlined. Therefore, easy access to nature and wildlife can be appointed as a source of positive feelings and beneficial services, which fulfill important immaterial and the non-consumptive human needs. It has been observed that the physiological benefits on residents' were: a healthier (not polluted) atmosphere, and the stimulation of pedestrianism and other outdoor related-activities. Psychologically speaking, it enhances stress-reduction benefits, mental health, a better quality of life because people have an easier access to wildlife, an increment of green space surface, and a landscapes' beautification have been recorded.

Residents and visitors social profit deal basically with recreational opportunities, and also with interaction among neighbors, civic pride, familiarity, community identity, sense of history, belonging and ownership. Beyond this, economic values have been identified in relation with the increasing properties value.

High level of satisfaction and fidelity with the park have been detected in the survey, in such a way that 94% of visitors have manifested their desire to come back again. This means that nearby inhabitants are satisfied with the corridor and probably feel that their life style has improved with this project, whereas foreign users have found an outdoor place where they can enjoy and spend a pleasant time.

The most current problems detected in the greenway deal with the conflict among different users. Conflicts on shared-greenways have been identified and sometimes interpreted as problems of success as an indication of the trail's popularity (Ryan, 1993). User conflicts can emerge when user goals and behaviors differ. Ethic and etiquette code, included in the signage system intend to minimize this kind of problems by showing the expected attitude and behavior from the different park users.

After analyzing the elements of the project, the discussion of the results has been centered on the question: *How well are human and wildlife sharing this green corridor?*

This corridor has been recently finished and it has resulted to be very popular. Probably the overcrowding of users can pose serious environmental problems in keeping natural and cultural resources in good conditions. A balance between public use and conservation is the key element to ensure functionality and the preservation of values.

Beyond technical considerations, at this point, it seems necessary to think about providing the park with a mid-term Management Action Plan. The Turia Fluvial Park project was conceived in the framework of a river basin management program even if it can be considered as an urban planning tool with important territorial implications in the short term. Therefore, a Management Action Plan can be very useful in order to provide guidance as we can observe in other USA green corridors such as the Ashley Scenic River Management Plan (Marshall, 2003) in South Carolina, or the Rouge Green Corridor in the Oakland County of Michigan.

Moreover, it is necessary to keep in mind that no direct statutory underpinning the protection status of green corridors usually exists but only indirect environmental laws provide legal support.

It is necessary to highlight that the management of green corridors is a complex task that needs to include many issues such as hydrology, hydraulics, fluvial geomorphology, biology, aquatics, ecology, socio-economics, politics, aesthetics, history, and culture, among others. In this point, it is important to take into account that corridor planning should ultimately be based not only on the biotic, abiotic and cultural factors peculiar to each local landscape, but also, on social basis and inhabitant's values and perceptions (Ahern, 1995).

This discussion is appropriate in our case study, because it is foreseen that the Turia Fluvial Park will record a successful arrival of visitors in the coming years. Thus, it will be necessary to balance recreational use level along with wildlife conservation, and this can only be achieved with specific related-regulations. It is important to outline that because not all the greenway's potentials have been developed until the moment because inter-town transport utilities using bicycles is a new challenge that offer many possibilities in a near future to the commuters.

CONCLUSION

We can finally conclude by outlining the following:

- The Turia Fluvial Park green corridor is playing an important role in the urban-rural fringe. It yields multiple benefits for all aspects of human life of its residents at an environmental, physiological, psychological, and economical level.
- This project has improved the valencian territorial planning system by linking the city of Valencia with the country towns along the river and with a Natural Protected Area (Parque Natural del Turia).
- The Turia greenway has the capability to protect and link natural, historical, cultural, ecological and economic resources into a type of system that has greater value and higher use as a hole than the sum of its parts.
- The analysis of the Turia Fluvial Park demonstrated that the planning and design of this green corridor had a significant effect on how people are using it.
- The corridors, like the case study presented, that consider equally recreational and ecological goals must have a very accurate and balanced planning, design and management in order to reach a healthy conservation status avoiding permanent threats.

Recommendations in order to guarantee the survival of ecological values of the green corridor and the success of the project in the long term are:

- Improve public attitudes and commitment to public participation.
- Reinforce green corridors' planning policies. This is an emerging issue but legal basis could be strengthened in order to protect the ecological values.
- The need to prepare an Action Management Plan in joint co-operation with all the involved partners and stakeholders.
- To be ready and flexible for new management expected challenges coming in the near future.

REFERENCES

Ahern, J. (1995). Greenways as a planning strategy. *Landscape Urban Planning,* No.33, pp.131–155.

Aguilella, A. & Ríos, S. (2004). Bosques, sotos y herbazales: quintaesencia de la ribera. (In Aguilella (coord.): *Monográfico Caminos de Plata* (pp. 92-101). Universitat de València)

Baris, M.E., Erdogan, E., Dilaver, Z. & Arslan, M. (2010). Greenways and the urban form: city of Ankara, Turkey. *Biotechnology & Biotechnological Equipment,* No. 24, pp.1657-1664.

Chiesura, A. (2004). The role of urban parks for the sustainable city. *Landscape and Urban Plan.,* No. 68, pp.129-138.

Conselleria de Medi Ambient, Aigua, Urbanisme i Habitatge. (2008). *Parc Natural del Turia. Memoria de gestión.* Report Generalitat Valenciana.

Davies, C., MacFarlane, R., McGloin, C. & Roe, M. (2008). *Green Infrastructure. Plan Guide.* Report National Infrastructure Plan, New Zeland Rayan.

De Juana, F. (2003, March). El Anillo Verde de Vitoria-Gasteiz: una propuesta para la integración armónica de la ciudad con el territorio *(Paper presented at the III Simposio Internacional sobre espacios naturales y rurales en áreas metropolitanas y periubanas: Los sistemas de espacios libres en la articulación de las áreas metropolitanas,* Barcelona)

Ferguson, B.K. (1996). Preventing the problems of urban runoff. *Renewable Resources Journal,* v.13, No. 14, pp.14-18.

Guirling, C. & Kellet, R. (2005). *Skinny street and green neighborhoods.* (Washington: Island Press)

Kappler, C. & Miller, L. (2009). Re-imagining the Urban Greenway: An Alternative Transportation Strategy and Vacant Land Use Plan for the Woodbridge Neighborhood of Detroit. Practicum submitted for the degree of Master of Landscape Architecture *(Natural Resources and Environment).* University of Michigan.

Little, C. E. (1990). *Greenways for America.* The John Hopkins University Press. London.

Marshall, B. (2003). *Ashley Scenic River Management Plan.* Report 25 South Carolina Department of Natural Resources.

Monclús, F.J. (2006). *Estratégias urbanísticas y crecimiento suburbano en las ciudades españolas:* el caso de Barcelona. Biblioteca Virtual Miguel de Cervantes. Retreived March 10, 2010, from http://www.cervantesvirtual.com

Pauleit, S. & Oppermann, B. (2002). *Case studies in green structure planning: The green structure of Munich, the need for and risk of regional cooperation.* (In: Duhem (ed). Report of Cost Action C11).

Ryan, K. L. (ed.) (1993). *Trails for the Twenty-First Century planning, design, and management manual for multi-use trails. Rails to Trails Conservancy.* Washington: Island Press.

Viñals, M.J., Cabrelles, G., Ferrer, C., Morant, M., Quintana, R., Ruiz, E. & Teruel, L. (2004). La ville de Valencia (espagne) et le fleuve Turia. *Un exemple de ville fluvial méditerranéenne. Gestions de fleuves Colloque, pp. 1-12.* Conservatoire regional des rives de la Loire et des ses affluents.

Yun-guo, L., Huang, Zhe, H., Gunan Ming, Z. & Xin, L. (2005). Greenways and their functions to eco-cities. *Ecological Economy,* v.1, No.4, pp.6-12.

In: Research Studies on Tourism and Environment
Editors: J. Mondejar-Jimenez et al.
ISBN: 978-1-61209-946-0
© 2012 Nova Science Publishers, Inc.

Chapter 24

SUSTAINABLE FOREST MANAGEMENT IN THE LAW 3/2008 OF CASTILLA LA MANCHA AND ADJUSTMENT TO STATE LAW AND THE EUROPEAN UNION ON THE MATTER

Jose Antonio Moreno Molina[*]
University of Castilla-La Mancha, Spain

ABSTRACT

The work carried out an analysis of several of the recent forest legislation of Castilla-La Mancha, in line with international forestry law, the law of the European Union and the State Basic Law on the matter, taking into account the many uses that our society demands of the mountain today, but above all, is based on the prioritization of environmental protection of the forest, in the protection of the biological dimensions of the forest. The object of study in particular the modern concept of forest as forest ecosystem.

1. THE KEY SOCIAL AND ENVIRONMENTAL FUNCTIONS OF FORESTS IN CASTILLA-LA MANCHA

Forests, welcoming a rich plant and animal life and balances of a typical natural habitat, are a natural resource whose contribution is crucial in maintaining the cycle of life and environmental conservation.

Social demands which are primarily faced today on forest resources, focus on what affects Castilla-la Mancha, are of three types:

1. Ecological function or regulating the dynamics of the biosphere, including soil protection against erosion, a major problem in large parts of the region, mainly in its

---

[*] Contact author: joseantonio.moreno@uclm.es

water mode. As recognized in the preamble of the Castilian-Manchego Law 3 / 2008, this erosion not only causes significant loss of soil fertility, as well as causing other side effects that diminish the effectiveness of certain infrastructure, especially roads and communication hydraulics. The existence of forests is essential, especially on sloping land, to mitigate the negative effects of erosion phenomenon, as well as containment of flooding and runoff regulation.

But we must also highlight the role of forests as a haven and refuge for wildlife, improving water quality, regulation of hydrological regime and its influence on the climate and atmosphere. In a time like that is in effect the Kyoto Protocol on reducing greenhouse gases in the United Nations Framework Convention on Climate Change (adopted in the European Union Council Decision of 25 April 2002), it is essential the role of forests as carbon sinks, reservoirs of greenhouse gases and also for biomass production and its potential in renewable energy (SARASÍBAR Iriarte, M., 2007).

2. Social Services at large. The forests, which represent an important cultural heritage, develop cultural functions, educational, recreational and improving the quality of life. Our citizens are increasingly demanding outdoor activities in contact with nature and the mountains and it highlights the significant development of rural tourism in Castilla-la Mancha. The region has gone from just over twenty-five thousand nights in 2001 to over sixty thousand in 2007, and in that same period has grown from 352 accommodation, which offers 2,442 beds, to 736, with 5,713 seats (Mondejar Jimenez, J ., Mondejar Jiménez, Ja And Vargas Vargas, M., 2008).

Keep in mind that ecotourism (UN Resolution 53/200 of 15 December 1998, proclaimed 2002 as International Year of Ecotourism) is currently the tourism sector has experienced a growth in the world according to reports from the World Tourism Organization. While the so-called "traditional tourism" registered a growth of 7.5% per year, ecotourism and rural tourism registered a 20% growth and now represent more than 5% of world tourism (source: http://www . fundacionglobalnature.org /, date accessed 29 November 2008).

But apart from that forests play a vital role in rural development, making up, properly managed, an important factor of stability of its population, to be a source of wealth and employment generating use of renewable resources they cherish. This crucial aspect of the forest has traditionally been emphasized by the European Union (see the above-mentioned resolution of 15 December 1998 on a forestry strategy for the European Union and the Plan of Action of the EU Forest 2006).

3. purely economic functions, by taking advantage of the many forest products. In the European Union's forestry and forest industries employ 3.4 million people (Communication from the Commission to the Council and the European Parliament of 15 June 2006, COM 2006 302 final). Furthermore, the EU is second largest producer after the United States, industrial roundwood and produces about 80% of the world's cork. According to the Spanish Forest Plan (2002), the comprehensive economic assessment of the Spanish forests, taking into account its production, recreational and environmental, amounts to about 136,000 million euros, very significant monetary value, but it leaves out what foresters call the value of internalization, such as carbon dioxide fixation mentioned, which has significant value in the market for emission rights, maintenance of biodiversity and landscape,

or the brake of the desertification process . Also left without assessing production aspects of forests such as hunting, fishing, grazing and wind potential, and environmental aspects as the protective functions of forest.

In many rural areas of Castilla-la Mancha forestry relevant manifests itself both in terms of employment and income generation.

But the significance of the functions of the forests are even more appreciated if we consider some statistics. Castilla-La Mancha has a forest area of just over 3,500,000 hectares, or no less than 44% of its territory, of which about three quarters of a wooded hill.

The land occupied by forests or forest vocation means about 28 percent of land area, according to the World Forestry Congress held in Paris in September 1991 (see Matthew Martin, R., 1992, p. 461). Forest area covers 37.8% of the territory of the European Union. As regards the Spanish forest land, according to the National Forest Inventory, covering 26 million hectares, equivalent to 51.93 of the national territory ", of which 14 million hectares in forest area, one of the most extensive in Europe (only surpassed by the 28,000,000 of Sweden, and Finland 23,299,000 France 16,242,000) and nearly 12 million forested area is treeless. The wooded area covering our territory is a 28 per cent of the total land area.

2. APPROVAL OF LAW 3 / 2008 OF JUNE 12, FORESTRY AND SUSTAINABLE FOREST MANAGEMENT OF CASTILLA-LA MANCHA, IN THE FRAMEWORK OF THE BASIC STATE LEGISLATION

The basic state law on forests is reflected today in the Law 43/2003 of 21 November. This rule, whose approval was instrumental in the Spanish Forest Strategy adopted on March 17, 1999 by the Sector Conference on the Environment and the Spanish Forest Plan approved by Cabinet in July 2002, came to repeal laws preconstitutional on March 10, 1941, Heritage State Forest, 8 June 1957, Montes, December 5, 1968, Wildland Fire and January 4, 1977, of Forest Production Development.

Law 43/2003, in line with international forestry law and the European Union law takes into account the multiple uses of our society demands of the mountain today, but above all, is based on the prioritization of environmental protection forest, in the protection of the biological dimensions of the forest. It defends our Constitution and the jurisprudence of the supreme interpreter of the Constitution, and moved in this direction Law 4 / 1989 of March 27, Conservation of Natural Areas (Article 9.2 which stated that the action of public administrations in Forest "will aim to achieve the protection, restoration, improvement and orderly utilization of forests, whatever their ownership, and its technical management should be consistent with its legal characteristics, ecological, and socio-economic forest, to prevail in any case the public interest over private ") and subsequent regional forest legislation, among which must be observed, but only partially affects forests, the Law of Castilla-La Mancha 9 / 1999 of 26 May, conservation of nature.

Also, the Law 42/2007 of December 13, Natural Heritage and Biodiversity, which repealed the Act 4 / 1989, focuses on the prevalence of environmental protection on the regional and town planning, and states that competent authorities shall ensure that the management of natural resources to produce the greatest benefits for current generations, without reducing their potential to meet the needs and aspirations of future generations, ensuring the maintenance and preservation of heritage, biodiversity and natural resources throughout the national territory, irrespective of ownership or legal status, according to their orderly use and restoration of their renewable resources. The principles underlying the Law 42/2007 was based, from the perspective of the account's own natural heritage, maintenance of essential ecological processes and basic life support systems, preservation of biodiversity, genetics, population and species, and the preservation of the variety, uniqueness and beauty of natural ecosystems, geological diversity and landscape.

In this sense, the actual grounds of the Law 43/2003 says that the provision is based on principles which are framed in the first and fundamental concept of sustainable forest management. Since he can deduct the other: the multifunctional, the integration of forestry planning in the planning, territorial cohesion and subsidiarity, the promotion of forest production and rural development, conservation of forest biodiversity, integration of forest policy in international environmental objectives, cooperation between administrations and the compulsory participation of all social and economic stakeholders in decision-making on the forest environment.

Well, how could it be otherwise, Law 3 / 2008 of June 12, Forestry and Sustainable Forest Management, contains the guiding principles of the State Law 43/2003 by proclaiming in Article 1 subject to " same principles and definitions contained therein, with the aim of preserving and protecting, promoting the restoration, improvement, sustainability and rational. "

Law 3 / 2008 is to collect in a single legislative body concerning matters all closely linked, but regulated earlier in Castilla-la Mancha in a dispersed manner: viz. the now repealed Law 2 / 1988 of 31 May, soil conservation and protection of natural vegetation cover, and the Decree 39/85 of 5 March, laying the groundwork for the establishment of agreements by the Junta de Comunidades de Castilla-La Mancha under Law 5 / 1977, to promote forest production, notwithstanding that they may continue to apply the agreements established with these bases, 73/1990 of 21 June laying approves the Regulations for the Implementation of the Law 2 / 1988, 61/1986 of 27 May on preventing and extinguishing forest fires, and 75/1986 of 24 June 1986 on the merits of investment in forest improvement the mountains of the public utility of local and provincial committees functioning of forests.

The Castilian-Manchego legislature has opted to transfer to the structure of the Autonomous State Law 43/2003, with the aim, says the preamble of the Law 3 / 2008, "to facilitate more accurate interpretation and application, as in the matters regulated by state regulations is interference that can not be subsumed under this statute, because such state of full competition, which, however, should remember those who have to ensure compliance with the regional standard ".

Thus, in parallel with the Basic Law of mountains, Law 3 / 2008 is divided into seven sections that deal with general provisions, classification of forests, sustainable forest management, conservation and forest protection, research, training, outreach, extension and forestry police, forestry development and the penalty system.

The Act also contains numerous transcripts of the basic provisions to prevent this way the constant referrals to state regulation.

3. THE CONCEPT OF FOREST AS ECOSYSTEM

Facing the limited concept of the mountain that provided the first article of the State Forestry Law of 8 June 1957, he meant by such a non-agricultural rural land, whether or not it populated by forest species (Esteve Pardo, J., 2005 , p. 85), Law 43/2003 came to enshrine in law the basic state the broad concept of the mountain that had gathered and regional forest legislation, as land that primarily meets or can meet environmental and protective functions.

The concept of forest of Article 5 of Law 43/2003 state incorporates the various functions of forest land and gives access to the regions in the range of regulation on abandoned agricultural land, urban and urbanizing soils and determining the size of the minimum unit to be considered in order to mount the law.

In line with the state provision, Article 3 of Law Mancha Castilian 3 / 2008 means all land in the forest species that grow trees, shrubs, shrub or herbaceous, either spontaneously or derived from sowing or planting, that meet or to meet environmental, protective, productive, cultural, scenic, or recreation. Forest species means any species of plant, whether tree, shrub, bush or grass that is not an exclusive feature of farming.

And for the Act are also considered as mountain,

a) The land barren, rocky and sandy.
b) The agricultural land for grazing abandonment of this activity for 10 consecutive years, provided they have been populated by forest vegetation and are capable of use or purpose forestry.
c) The permanent forest enclaves of agricultural land, they have a seating capacity of at least one area, provided that support thickets, lindazos, steep slopes or loose foot tree, shrub or bush of forestry, without prejudice to the provisions Article 49.
d) The land is devoted to temporary crops of forest into agricultural land with fast growing tree species for timber, firewood, fruit or twigs, in intensive regimen, or from other tree species, woody or herbaceous aromatics, seasoning or medicinal, which continue to be set mount at least for the duration of its use shifts. If the forest crop is within the margins of the hydraulic public domain, the forest condition is permanent.
e) The banks and thickets on the margins of public waterways that cut by streams, permanent or seasonal, continuous or discontinuous, and the margins of lakes and lagoons, to sustain or bodies as may be established trees, shrubs, shrub or herbaceous communities.
f) The agricultural enclaves and other areas covered by forests declared a public utility who have lost their ground cover, trees, shrubs and herbaceous communities on forestry, provided that the loss was due to administrative decision relapse prevalence file public interest or change of use and destination.
g) The grasslands installed on non-agricultural land.
h) In general, all land not qualify as described above, is ascribed to the purpose of being replanted or converted to forest use, as well as compensation from land use change forestry measures imposed in disciplinary proceedings, spaces concessions recovered mines, quarries, dumps, landfills and the like, or covered by the instruments of planning, management and forest management to be adopted under this Act

i) The buildings and infrastructure to service the mountain in which they are located.

However, no consideration of forest:

a) The cultivated agricultural land, unless the conditions are set out in paragraph 1 above.
b) The soils are classified as urban or developable developer programs of action adopted.
c) The linear plantings of trees or shrubs, whatever its purpose, when urban land to settle on or abutting public or private facilities and, in general, on land not affected by forest characteristics referred to in paragraph 1 above.
d) The areas for growing ornamentals and nurseries outside of the mountains, without prejudice to compliance with the laws that affect them.
e) The rural land with natural vegetation associated with agricultural practices, including the characteristic of the traditional fallow, herbaceous borders and own cousin colonizing abandoned crops, except as provided in paragraph 1.b).

In accordance with paragraph 3 of Article 3 of Law 3 / 2008, the concept of forest is independent of the affected area. However, plans for forest resources management may specify, as determined by regulation, minimum surface under which the land affected will not be considered forest administrative management purposes, subject to compliance with other provisions of this Act that are applicable.

The bush or forest land are difficult to sort or protected based on a residual concept, as did the State Forestry Act 1957. Therefore, the Law 3 / 2008 lays a positive conception to the time of cataloging the mountains or forest land, as is based on the intrinsic characteristics of different geographical areas, thus circumventing the residual conception that would result from the mere exclusion of areas intended for other uses while the concept of forest land are also added those that meet or can meet environmental, protective, productive, cultural, and recreational landscape, which not only improves the concept but becomes more in line with Article 45 of the Constitution, to take into account the following aspects of productivity, environmental. Establishing, finally, a fundamental idea for forest management, consisting of the mountains, and ecosystems that must be addressed in an integrated manner.

Prior to the adoption of state law 43/2003, various regional laws had endowed the concept of mountain of a more open and positive, also acknowledge explicitly the multiple social roles it plays. These rules, therefore, emphasized the functional aspects and finalists, who were integrated into the concept.

This line is clearly noticeable in Law 2 / 1988 of 31 May, Soil Conservation and Protection of Groundcovers and 9 / 1999 of 26 May, Nature Conservation of Castilla-La Mancha; 13/1990 of 31 December on the Protection and Development of the Heritage Forest of Navarra, 3 / 1993 of December 9, Forest Valencia, 8 / 1998 of June 26, Conservation of nature and natural areas in Extremadura; 5 / 1994 of 16 May, Building the wooded hills of Castilla y León, 16/1995, of 4 May, Forestry and nature protection of the Community of Madrid, 2 / 1992 of June 15 , Forest Andalusia 15/2006 of 28 December, Montes de Aragón and 3 / 2004 of 23 November, Forestry and Forest Management in the Principality of Asturias.

4. SUSTAINABLE FOREST MANAGEMENT AS A PRIME OBJECTIVE OF THE ACT 3 / 2008

Law 3 / 2008 ranks as the first for the legal and administrative management of forests is sustainable forest management, to highlight its importance entitles the standard itself.

This recognizes the protection of forests in Castilla-La Mancha, the principle of sustainable development, which have resulted in recent years in many international agreements, since the United Nations Conference on Environment and Development in Rio de Janeiro in 1992 will boost the process of raising awareness of the social and economic importance of proper management of forests (Agenda 21). The principle has also been given preferential attention by various instruments and also action in the framework of the European Union, from the successive declarations of the Ministerial Conferences on the Protection of Forests (Strasbourg 1990, Helsinki, 1993, Lisbon 1998, Vienna, 2003) to the Plan of Action of the EU Forest (Communication from the Commission to the Council and the European Parliament of 15 June 2006, COM 2006 302 final).

Sustainability criteria (on the structural principle of sustainability as environmental management, see Ortega Alvarez, L. 2005, pages 46 et seq.) Must therefore be present in the management of forest resources that Law 3 / 2008 aims to promote , since some of them, such as wood, represents an undeniable interest in regional economic life of the XXI century.

Prioritization is essential that the law makes the environmental protection of forest protection must not place the same level of attention to social needs, recreational and productive forests. Do not forget that the rule is, in short, a correlate of the biological dimensions of the forest. It defends our Constitution (art. 45.2) and the ultimate interpreter of constitutional law. The forest must be understood as well to keep in terms of their environmental values and, therefore, for all the services you are able to offer to the community. As a consequence of the profound cultural and legal developments around the environment, the forest should be defined and protected as a legally understood as an ecosystem that is as natural biological environment, understanding of all plant and animal life and also equilibria typical of a natural habitat.

Consideration of the mountains and forest ecosystems results in the treatment of such an integrated manner, which means the joint management of the flora, fauna and physical environment that is, to achieve a sustainable use natural resources, providing guarantees for the preservation of biological diversity and maintaining essential ecological processes.

Article 30 of Law 3 / 2008 defines sustainable forest management "organization, administration, development and use of forests, and intensity so as to maintain their biodiversity, productivity, vitality, potential and regeneration capacity, to meet now and in the future, their ecological, economic and social aspects at local, regional, national and global levels, without causing damage to other ecosystems. "

The forests are managed sustainably, integrating environmental considerations with economic, social and cultural rights, with the aim of conserving the natural environment while creating jobs and help to increase the quality of life and development expectations rural population.

Law 3 / 2008 forest management planning at two levels: at the top, setting out guidelines for sustainable forest management through the so-called management plans of forest resources (porphyry), county level or equivalent, and which are under to adjust management projects or working plan, on the lower level, which are the instruments of sustainable forest management at the direct application of forest or specific group of mountains.

The importance of the Law 3 / 2008 gives the exponent of sustainable forest management powers that it gives PORF because through them can be defined, for its territorial scope, characteristics to those mountains that meeting may be included in special protection schemes and, likewise, are empowered to modify, even within its territory, the minimum areas, in general, are set for agricultural land forest enclaves have the status of mountain and those areas for which the obligation is determined to have a sustainable forest management tool.

In terms of forest management projects, it is understood by those documents that summarize the organization in time and space of sustainable use of forest resources, timber and, on a hill or mountain group, for which shall include a detailed description of forest land in its ecological, legal, social and economic and, in particular, a forest inventory with a level of detail that allows making decisions about forestry to apply in each Mount units and estimates of his income.

A working plan and technical plan is that forest management project that by its singularity, prized for its small size, preference functions other than production of wood or cork-age masses treeless short or others set forth in instructions referred to below, require a simple adjustment of the management of its tree resources. Consistently, the forest inventory will be more simplified.

In accordance with paragraph 4 of Article 31 of Law 3 / 2008, the Board shall, in accordance with the basic guidelines common management and utilization of forests, adopted by the national government, the instructions for the management and utilization mountains of Castilla-La Mancha, whose approval for the Council of Government by decree. The development of management tools should be directed and supervised by professionals with a university forestry, and taking as a reference, where appropriate, porphyry within which is Mt. His approval rests with the Ministry competent in forestry, which have to do a period of six months, starting from its presentation. In the absence of an express decision, it is understood that the project has not been approved.

When a forest management tool to affect land included in the Regional Network of Protected Areas, will require a previous report of the managing body of the area in question, in order to ensure the project is compatible with the existence of the protected area.

REFERENCES

Esteve Pardo, J. (1995), *Reality and Prospects of the legal management of forests (ecological function and rational)*, Civitas, Madrid.

Garcia de Enterría, E. (1986), *The communal forms of forest ownership*, Santander.

Greenpeace (1991), *Buying Destruction: a Greenpeace report for Corporate Consumers of forest products*, Greenpeace International Publications.

Guaita, A. (1986), *Administrative Law. Water, forests and mines*, Civitas, Madrid.

Lázaro Benito, F. (1993), *The constitutional management of forest resources*, Tecnos, Madrid.

Marraco Solana, S. (1991), "The Spanish forest policy: recent developments and prospects", *Journal of Agro-Social Studies*, no. 158.

Martin Mateo, R. (1992), *Treatise on Environmental Law*, Trivium, Madrid, 2 vols.

Ministry of Environment and Rural Development (2006), *Catalogues of Public Utility Montes de Castilla-la Mancha* (monitoring and Marco Martinez Rivera Sanchez Palencia), Junta de Comunidades de Castilla-la Mancha, Toledo, 5 vols.

Mondejar Jiménez, J., Mondejar Jimenez, J.A. and Vargas Vargas, M. (2008), "Rural tourism in figures: Castilla-la Mancha", TURyDES, Vol 1, No 2.

Moreno Molina, J.A. (1998), *Environmental protection forests*, Marcial Pons, Madrid, 1998 and (2005) "Protection against forest fires" in *Systematic reviews of Law 43/2003 of November 21, Forestry. Studies of state and regional law* (coordinator Calvo Sanchez, L.), Thomson-Civitas, Madrid.

NIETO, A. (1991), *Communal property of the Montes de Toledo*, Madrid.

Oliván del Cacho, J. (1993), "The environmental protection in forest legislation," *Journal of Public Administration Aragonesa*, no. 2, (1995)

Ortega Alvarez, L. (2005), "The concept of environmental law" in VVAA, *Lessons of environmental law*, Lex Nova, Valladolid, 4 th ed.

Sarasíbar Iriarte, M. (2007), *The Right forestry to climate change: the environmental functions of forests*, Aranzadi, Pamplona.

Vicente Domingo, R. (1995), *Forest areas (The legal and natural resource management)*, Civitas, Madrid.

VVAA (2005), *Systematic Reviews to the Law 43/2003 of November 21, Forestry. Studies of state and regional law* (coordinator Calvo Sanchez, L.), Thomson-Civitas, Madrid.

In: Research Studies on Tourism and Environment
Editors: J. Mondejar-Jimenez et al.
ISBN: 978-1-61209-946-0
© 2012 Nova Science Publishers, Inc.

Chapter 25

RECOGNITION OF THE RESPONSIBILITY OF THE LOCAL SPANISH IN CASES OF NOISE

*A. Patricia Domínguez Alonso**
University of Castilla-La Mancha, Spain

ABSTRACT

The paper analyzes the requirements for recognition by the administrative litigation law the liability of local authorities in cases of failing to address noise pollution. It is a jurisprudence that has opened in recent years an important means of struggle of citizens against the serious problem of noise in Spain.

1. RECENT DEVELOPMENTS IN THE LEGAL TREATMENT OF NOISE POLLUTION

Jurisprudential recognition of the liability of the authorities in cases of failing to address noise pollution has opened in recent years an important means of struggle of citizens against this worrying phenomenon.

In fact, complaints to the councils have been and are now increasingly being force-one of the main means used by the public reaction to cases of noise nuisance.

Protection against noise pollution is covered in the constitutional mandate to protect health (article 43 of the Spanish Constitution -EC-) and the environment (Article 45 EC), and some fundamental rights enshrined in the Constitution, such as law rights to physical and moral integrity, to personal and family privacy and inviolability of the home (Articles 15 and 18 EC).

However, the sound lacked even the approval of Law 37/2003 of November 17, a general rule governing state level, and regulatory treatment unfolded, broadly speaking, among the provisions of the civil law as neighborly relationships and causation of damages, the rules on

---

* Contact author: patricia.dominguez@uclm.es

the limitation of noise in the workplace, the technical requirements for product approvals and local ordinances pertaining to the welfare city or urban planning.

The Law 38/1972 of 22 December on the protection of atmospheric environment, which was intended to prevent, monitor and correct the presence in air of material or energy sources that involve risk, harm, or serious problems for people and assets of any kind, did not express reference to noise pollution.

However, the Autonomous Communities themselves have enacted laws regulating protection against noise pollution.

Noteworthy in this regard, the Law 16/2002 of the Generalitat of Catalonia, June 28, protection against noise pollution, and the Law 7 / 1997, dated August 11, protection against noise pollution in Galicia, the Decree 78/1999 of 27 May, which regulates the protection against noise pollution of the Community of Madrid and the Law 7 / 2002, dated December 3, the Generalitat Valenciana, Protection Against Noise Pollution.

Directive 2002/49/EC of the European Parliament and the Council of 25 Jun. 2002 on the evaluation and management of environmental noise, states in its explanatory memorandum, which should achieve a high level of environmental protection and health and one of the objectives to be pursued is protection against noise.

The Environmental Noise Directive marks a new direction for the previous regulatory actions of the European Union in terms of noise. Previously, the regulations had focused on the sources of noise, which gave some results, but data show that the beneficial result of these measures on environmental noise has been reduced by the combination of other factors that have not yet been caught.

However, the scope and content of the Law 37/2003 is broader than the Directive by transposing it as the law does not end with the establishment of parameters and measures referred to by Directive concerning only the ambient noise, but has more ambitious goals. As stated in the preamble to the Spanish standard by attempting to give greater coherence to the management of noise pollution at the state level, the Act contains many provisions that go beyond the mere transposition of the directive and wants to promote actively, through proper distribution of administrative powers and the establishment of appropriate mechanisms, improving the sound quality of our environment.

2. THE REQUIREMENTS FOR JUDICIAL RECOGNITION OF THE RESPONSIBILITY FOR NOISY ACTIVITIES

2.1. Budget Needed. Complaint of Noise Pollution by the Affected Citizens

The ruling of the High Court of Valencia, Division of Administrative Litigation, section 3, of 1 June 1999 (JURIS LAW: 47893/1999) said on complaints from residents affected by noise and evidence reports:

"FOURTH .- When referring during most of the wording of the sentence some of the various complaints filed by residents of the building located in Plaza 13, Valencia Spanish Legion, now suffices to note that these are the parameter for the establishment the time period that extends, in the subjective understanding of the appellants, the inconvenience and damage caused to them by the development of the nightclub J.. It has to stand out from the fact and written complaints filed with the City Council both neighbors of the property specific - and, therefore, individually made - as representatives of the homeowners association. "

2.2. The Inactivity of the Administration

The first requirement for the recognition of responsibility is the operation of a public service. In cases of noise this budget largely concurs with the passivity is a city in dereliction of duty and responsibility in environmental matters is allocated to local councils in the state regulations.

Suffice it to recall in this connection that paragraphs f) and h) of Article 25.2 of Law 7 / 1985 of 2 April, the Local System, gives the municipality the exercise of competence in the field of environmental protection and public health, and that Article 42.3.a) of Law 14/1986 of 25 April, Surgeon General says the environmental health control, with specific reference to air pollution, as the responsibility of local councils and that Regulation inconvenient, unhealthy, harmful and dangerous states the general jurisdiction of the municipal bodies for the enforcement of the provisions on the matter (Article 6) and, more particularly, are recognized inspection functions on the activities that come to develop and authority to take action against the defects found (Articles 36 and 37).

For its part, the Noise Act (37/2003 of 17 November) states within the jurisdiction of Local Authorities, the City Council approved ordinances on noise and adapt the existing urban planning and the law, in relation to Article 6 of this same statute.

The ruling of the Supreme Court, Third Chamber, 7th Section, Judgement of May 29, 2003 (Rapporteur: Maurandi Guillén, Nicolás resource Antonio.N No: 7877/1999. JURIS LAW: 13323/2003) discusses the passivity of the City Council Sevilla follows:

"SIXTH. In the proceedings there is abundant evidence to form an opinion on the existence in the period from 1991 to 1997 claimed by the appellant, a noise pollution caused by noise and vibration of the club located at 20 Frederick Street Sánchez Bedoya de Sevilla, which directly and seriously affected the plaintiff's domicile (in no. 35 of the same street), and also a municipal inaction in relation to the problem that only ended in 1997. "

The ruling of the High Court of Valencia, Division of Administrative Litigation, section 3, of 1 June 1999 (JURIS LAW: 47893/1999) also discusses the insufficient measures taken by the City Council to view of complaints by neighbors to noise nuisance, concluding that "in the administrative record - and in this court - has not any information through which it can establish that the City Council adopted an activity implementation material requirements or legal requirements made by this administration against the owner of the business called J. disco. "

The important decision the Supreme Court, Third Chamber of Administrative Disputes Section 7th of April 10, 2003 (Rapporteur: Maurandi Guillén, Nicolás Antonio. Number of resource: 1516/1999. JURIS ACT: 2099 / 2003) resolves ultimately a matter arising from the appellant's claim that a municipality (Boiro) processing and initiate disciplinary proceedings as a precautionary shutdown of disco excess noise emissions:

"SIXTH. It is reprehensible passivity appreciate the City Council because, despite the proposal of the instructor, in that same file to which reference has been no evidence that the City Council adopt the precautionary measures that were requested. And let him Article 20 of

Law 7 / 1997, 11 Aug., protection against noise pollution in the Autonomous Community of Galicia "

Based on the foregoing, the Supreme Court finally declared that infringed the appellant's fundamental right to inviolability of his home on it by Article 18.2 EC.

However, this statement should be supplemented by a statement addressed to the full and effective restoration of fundamental rights violated, resulting in compensation for damages until the City Council does not take effective measures to do away the inconvenience caused the violation .

And the High Court referred the quantification of the indemnity to the stage of execution of sentence but setting some guidelines:

1) will take into account the cost of leasing a house with similar characteristics to that of the appellant in terms of extent and location, and
2) consider the time period between the date of the first applicant's request was not addressed and that other on which are put into place measures that will effectively do away the inconvenience of excessive noise (if already had taken while he has handled this process will be the date of such adoption).

In connection with this inactivity of the local administration for dereliction of their responsibilities in environmental matters include the Court of Justice of Andalusia from October 29, 2001 (Section One of the Chamber for Contentious Administrative Tribunal of Justice Andalusia based in Seville, appeal number 949/1998), which considers the appeal filed by the Neighborhood Association Torre del Oro, Monumental and Historical Center Barrio del Arenal against the City of Seville.

2.3. The Relationship of Causality

For a successful action against any liability Civil Service is essential to test the direct causal link between damage and normal or abnormal functioning of public services, such as direct driver of those,

Sheds light on the question the Court of Justice of Castilla-la Mancha, the Administrative Disputes Division, Section 1 (JURIS LAW: 1391967 / 2003) of 10 February 2003

Room. All legal issues of this claim equity flows in the proper scope of the requirement of causation and possibly in the termination of the compensation amount. In this regard, our Supreme Court has referred in general to a character direct, immediate or exclusive to particularize the causal link between the administrative and damage or injury that must be satisfied before liability can be assessed by public administrations, no excluded that the causal relation expressed - especially in cases of abnormal operating responsibility for public services - can appear in forms mediate, indirect or concurrent circumstances that may or may not result in a moderation of responsibility (Statements 8 Jan. 1967, 27 May. 1984 Apr 11. 1986 RA 2633 22 Jul. 1988, RA 6095, 25 Jan. 1997, RA 266, 26 Abr. 1997, RA 4307, 6 Oct. 1998, 6 Feb. 2001 , RA 653). In this way, will be the circumstances that define the reality and extent of potential harmful responsible in the event, with delimitation of the existence of causation. In this sense, this Court finds that there are, in this case, the objective circumstances that allow us to understand that there is an event or condition that may be considered relevant by itself to produce the final result as a precondition or "conditio sine qua non" that is, an act or fact without which it is inconceivable that other fact or event or outcome taking into account all the circumstances of the case (Judgement of the Supreme Court December 5, 1995, R. 9061, and 6 Feb. 2001, RA 653). "

The ruling of the High Court of Justice of the Principality of Asturias, Division of Administrative Litigation, Section 2, of September 10, 1999 (JURIS LAW: 410011 / 1999), analyzes the financial claim of a citizen of liability by damages generated to be deprived of rest is entitled during weekends, public holidays and eve as a result of the noise generated at the local pub called V. located in the basement of the building adjacent to yours:

"THIRD .- Well, not satisfied in the case under consideration the requirements for the birth of the duty of recovery from the defendant authority for damages claimed by the appellant, in particular, has not been established in this process that damages pain and suffering endured say that can be imputed, that is, legally attributed to a person other than the owner or manager or denounced local clients who come to it, by not prove the causal link between the activity of the City Gijon and inconvenience, because although these take place in an establishment subject to the necessary licenses and authorizations of municipal competition, not appear to have been dereliction of their duties by the defendant corporation, if comprised of numerous disciplinary proceedings that were initiated by excess opening hours and operating with the doors open, having demanded compliance with the regulations on sound insulation and ventilation measures, so the fact that the <<ruidos>> night originating from an establishment without further public can not derive a possible <<the responsibility of government because it would only be predicable of overseeing the activity and may be concluded that present in the case circumstances and strange or alien elements, the intervention of a third party, that are configured as causes in the production of damaging event for annoying and disrupting the causal link between the damage and the administrative act, which carries and prevents, not be a direct, immediate and exclusive cause and effect, the Administration should be obliged to compensate the damages requested by plaintiffs. "

However, as noted ALONSO GARCIA, C. ("The liability of the municipalities with their inaction in controlling noise," Administrative Justice No 23, 2004, p. 52), it should be noted that, in general, in cases of noise, liability imputed to the City Council does not have its origin in an injury caused directly by it, but it is derived from the conduct of a third party: the owner of the nuisance activity.

The ruling of the High Court of Valencia, Division of Administrative Litigation, section 3, of June 1, 1999 (LAW JURIS: 47893 / 1999) clearly argues the complaint of damage to the Administration through public powers of supervision of qualified activities, dismissing the argument that the intervention of a third party breaks the causal link idle time between the city and the damage caused to the appellants.

2.4. Damage Caused and Monetary Compensation

It is legal doctrine (for all of it, STS 10 July 1995) which states that: "in this procedure only claims for damages when it incorporated the restoration of the infringed right is imposed by necessary" or, in terms of STS 7 July 1995, RA 5762: ".. has taken a further step to a settled case law that supports the addition of claims for damages to the main claim when such compensation, by substitute or complementary, is the appropriate means for the full restoration of civil order perturbed by the wrongful act. .. Otherwise the statement obtained by the plaintiff would lack any real effectiveness with which the contested measure, despite its annulment by the decision, would have accomplished all its power against the actor harmful. "

Damage to the person highlighted in another paradigmatic ruling on noise, in this case from the criminal jurisdiction: the Supreme Court decision, Second, criminal, Judgement of 24 February 2003 (rapporteur: Granados Perez, Carlos, JURIS LAW: 1118/2003)

"FIRST .- (...) When it comes to noise pollution, both the Court of Human Rights and the jurisprudence of the Constitutional Court show the serious consequences of prolonged exposure to high levels of noise on the health of individuals, physical and moral, and social behavior in certain serious cases, while not endangering the health of people, can undermine their right to privacy and family in the home, to the extent that seriously impede or obstruct the free development of personality, noting that cases are particularly serious when it comes to continued exposure to intense noise levels. "

(...) For all that is left above, the appellant has created a serious danger to the physical, mental, personal and family privacy, comfort and quality of life for residents of the building that could be affected by inmissions <<ruido>> from the Party Room where he was responsible, having materialized in danger of serious harm to the health of these people. Is exceeded, then the threshold that separates the merely administrative unlawful criminal act.

The ruling of the High Court of Valencia, Division of Administrative Litigation, section 3, of 1 June 1999 (JURIS LAW: 47893/1999) means:

"(...) The most relevant parameter interpretation available to the Board for the purpose of establishing involvement by the noise caused by the harmful and unhealthy for the industry or commercial establishments in the health of citizens in respect for physical integrity and the inviolability of his home proprietary Case European Court of Human Rights López Ostra v. Spain recognizes that this venue is very difficult to determine the exact significance of the physical or mental injury caused by the emission inside the domicile of the claimants in some pain affecting the right under Article 8 of the Convention for the protection of human rights and fundamental freedoms and its translation into a monetary amount. "

Compensation for moral damage is recognized by the courts assessing the damages suffered by the victims. In the above statement we conclude that

".. The claimant has suffered an undeniable moral damage, apart from the nuisance caused by gas fumes, noise and smells from the factory, has suffered distress and anxiety endured watching the situation and how the health of his daughter was being degraded. The basis of the damages allowed are not provided an accurate calculation.

With this perspective, and analyzed all the parameters that have been reflected in the basis of law, the Court considers that the monetary amount requested by the actors in the statement of claim plead - 500,000 pesetas - fits or is proportional to the damage physical and mental generated by the violation of these fundamental rights of the person listed in articles 15 and 18 of the Spanish Constitution (...)".

For its part, the Court of Justice of Castilla-la Mancha, Administrative Litigation Division, Section 1 (JURIS LAW: 1391967 / 2003) of 10 February 2003, argues that

"It is evident that the actor and his family, for the development of the circumstances, excessive length in time (fifteen years), the struggle has had to develop to serve as the local authority has acted and only could correct the situation with your personal and financial effort (and thus leads to frustration, anxiety, insecurity, boredom ;...) be understood as equal (in the absence of other defined criteria and the importance that a declaration of responsibility has to the public purse), the recognition of a sum of three million pesetas, or its equivalent in euros, plus legal interest of the money from the date of its last claim dated May 4, 1999 until paid."

INDEX

#

20th century, 44, 91, 140, 229, 256
21st century, 24, 290
9/11, 146

A

access, xv, 58, 62, 63, 67, 90, 122, 140, 158, 160, 170, 172, 175, 176, 177, 178, 179, 205, 218, 219, 228, 260, 261, 262, 263, 264, 267, 270, 281, 282, 285, 314, 315, 318, 319, 321, 329
accessibility, xv, 148, 155, 281, 282, 283, 285, 286, 317
accommodation, xii, 30, 31, 34, 37, 39, 41, 71, 73, 113, 145, 147, 148, 149, 150, 154, 216, 217, 219, 261, 262, 265, 284, 293, 304, 305, 307, 326
accommodations, 248
accounting, 149, 270, 271, 273, 274, 275, 276, 277, 279
accounting standards, 279
accreditation, 122, 127, 128, 129, 295
acquisition of knowledge, 171, 178, 179, 180, 182
acquisitions, 182
active radicals, 7, 10
adaptability, 282
adaptation, 14, 281
adaptations, 283
adjustment, 332
adolescents, 55
adults, xv, 46, 207, 289
advertisements, 203
aesthetic, 239, 315
aesthetics, 322
affirming, 125
affluence, 62
Africa, 117, 250
age, x, xi, 46, 48, 57, 60, 66, 69, 70, 75, 77, 79, 80, 82, 83, 134, 136, 137, 139, 148, 158, 230, 282, 283, 294, 303, 304, 309, 332
agencies, 20, 91, 93, 102, 117, 284
AGFI, 153
aging population, 283
agriculture, 138, 140, 251
air carriers, 114
alternative treatments, 271
ambidexterity, 187
amphibians, 317
amplitude, 5, 214
anaerobic digestion, 12
analytical framework, 25
annual rate, 101
ANOVA, 163
anxiety, 340
arbitration, 201
Argentina, 259
aromatics, 329
Asia, 166, 167, 305
assessment, 47, 64, 120, 167, 242, 271, 273, 274, 277, 326
assets, 64, 174, 179, 336
assimilation, 14
asylum, 212
atmosphere, 248, 321, 326
attribution, xi, 119, 120, 123, 127, 128
audit, 275
audits, 17, 20
Austria, 59
authenticity, 185
authorities, 16, 208, 260, 327, 335
authority, 212, 219, 337, 339, 340
automation, 274
autonomous communities, 295, 298
autonomy, 214
awareness, xv, 21, 64, 241, 249, 281, 283, 285, 289, 331

B

badminton, 161
baggage, 116
banking, 97
banks, 148, 317, 329
Barbados, 165
barriers, 23, 207, 239, 284
base, 29, 52, 53, 95, 137, 139, 246, 247, 248, 249, 275
beautification, 321
behaviors, x, 57, 59, 60, 191, 317, 321
Belarus, 95, 100
Belgium, 44, 66
benchmarking, 270
beneficiaries, 286
benefits, xv, 15, 25, 63, 67, 90, 114, 115, 146, 157, 159, 163, 173, 174, 177, 237, 240, 241, 247, 249, 306, 313, 315, 317, 321, 322, 328
beverages, 29, 89, 147
bias, 293, 307
biodegradation, 3
biodiversity, xi, 119, 125, 129, 135, 159, 160, 163, 240, 249, 250, 274, 314, 315, 321, 326, 328, 331
biological treatment., ix, 1, 2, 8
biomass, 2, 12, 326
biosphere, 325
biotechnology, 183
biotic, 322
birds, xiv, 126, 237, 242, 246, 317, 321
border control, 93
border crossing, 93
boredom, 340
bounds, 126, 128
Brazil, 186, 291, 298, 305
breakdown, 10, 11, 39
breathing, xv, 313, 319
breeding, 141
Britain, 142, 184
Bulgaria, 98
business management, 18, 271
business strategy, 114
businesses, xiii, 97, 172, 201, 202, 227
buyers, 84

C

cabinet, 327
campaigns, 53, 64, 294, 305, 317
canals, 315, 319
capital markets, 97
carbon, 326
carbon dioxide, 326
Caribbean, 85, 166
carob, 141
case law, 123, 124, 127, 339
case study, 142, 186, 187, 322
cash, 271
cash flow, 271
casinos, 100, 104
category a, 113, 204
cattle, 136
causal relationship, 150
causality, 128
causation, 335, 338
Central Europe, 91, 98, 104
certification, 14, 17, 18, 20, 21, 22, 24, 25, 160, 163, 165, 262, 295
CFI, 153
challenges, xiv, 2, 14, 247, 281, 282, 319, 323
chemical, 2, 11, 12, 64, 97, 239
chemical reactions, 2
Chicago, 54
children, 72, 73, 202, 203, 205, 207, 213, 283, 320
Chile, 259
China, 185, 312
circulation, 121
CIS, 16
cities, xi, xii, xv, 29, 47, 49, 51, 70, 73, 92, 100, 133, 134, 142, 154, 155, 169, 170, 172, 187, 212, 289, 298, 299, 304, 305, 307, 309, 314, 315, 324
citizens, xvi, 90, 91, 140, 230, 282, 283, 319, 326, 335, 340
city, xii, xv, 85, 134, 138, 140, 145, 147, 148, 265, 278, 293, 301, 303, 309, 313, 316, 320, 337, 338, 339
civil action, 232
civil law, 335
Civil War, 317
civilization, 91
clarity, 225
classes, 108
classification, 14, 34, 149, 242, 247, 251, 254, 255, 256, 257, 319, 328
cleaning, 20, 218
clients, 18, 91, 98, 106, 109, 112, 114, 115, 117, 230, 339
climate, 91, 134, 158, 159, 160, 164, 254, 258, 307, 326, 333
climate change, 158, 159, 160, 164, 333
climates, 260, 266
clothing, 116, 147, 239
cluster analysis, xi, 69, 70, 71, 75, 161
clustering, 173
clusters, 75, 79, 80, 170, 184, 185, 186, 187

coercion, 207, 213, 219
coffee, 212
coherence, 152, 336
collaboration, xii, 136, 137, 169, 170, 172, 266, 282
collisions, 119, 122, 124, 129
commerce, 240, 284
commercial, x, xi, xiv, 27, 111, 114, 115, 147, 202, 205, 206, 207, 208, 215, 216, 223, 225, 230, 253, 255, 257, 259, 264, 340
common rule, 116
communication, 17, 19, 29, 98, 156, 276, 282, 285, 290, 291, 326
communication strategies, 29
communities, xiv, 157, 159, 163, 237, 238, 240, 241, 249, 250, 298, 314, 329
community, xii, xiv, 28, 113, 140, 169, 170, 172, 173, 179, 237, 239, 241, 248, 249, 250, 282, 314, 317, 321, 331
comparative analysis, xi, 69, 70, 77, 89, 90
compatibility, 59, 187
compensation, 112, 113, 114, 115, 116, 117, 120, 208, 209, 225, 249, 329, 338, 339
competition, xii, xiii, 169, 170, 172, 178, 179, 201, 202, 203, 208, 209, 238, 261, 328, 339
competitive advantage, 20, 21, 170, 171, 173, 174, 175, 176, 177, 178, 182, 183, 184, 186, 187
competitive conditions, 175
competitiveness, x, 17, 27, 171, 173, 175, 178, 273
competitors, 203, 296, 300
complement, 84, 177, 187, 259, 300, 309
complementarity, 173
complementary products, 172
complexity, 274
compliance, xi, 119, 120, 121, 123, 124, 125, 126, 128, 129, 139, 202, 246, 274, 276, 328, 330, 339
compost, 12
compounds, 2, 10, 11
compression, 2
computing, 275
conception, 45, 213, 215, 286, 314, 330
conceptual model, 186
conceptualization, ix, 13
conditioning, 260
conference, 59
configuration, 177, 215
conflict, x, 57, 64, 124, 153, 321
conformity, 55
congress, 250, 311, 327
connectivity, xv, 313, 314, 315
consensus, 213
consent, 213, 217, 218
conservation, xi, xii, xiv, 16, 25, 119, 120, 124, 125, 126, 128, 129, 133, 134, 135, 136, 137, 138, 157, 158, 218, 237, 238, 240, 241, 242, 248, 249, 273, 277, 314, 315, 318, 321, 322, 325, 327, 328
consolidation, 205
constitution, 121, 202, 203, 214, 215, 218, 327, 330, 331, 335
constitutional law, 331
construction, 2, 11, 64, 89, 97, 161, 163, 216, 228, 319
consulting, 46
consumer protection, 202, 205
consumers, xi, xiii, xiv, 13, 18, 89, 111, 112, 114, 116, 201, 202, 204, 205, 206, 207, 208, 209, 224, 225, 253, 255, 258, 259, 264
consumption, 7, 14, 19, 28, 29, 82, 87, 89, 147, 156, 160, 204, 255, 258, 278
consumption habits, 87
consumption patterns, 14
contact time, ix, 1, 6, 7, 10, 11
content analysis, xii, 157, 160, 162, 163, 165
control group, 195
controversial, 228
controversies, 123
convention, 239
convergence, 270, 273, 276, 277
cooperation, 170, 171, 172, 175, 178, 179, 328
coordination, 17, 136, 173, 202, 292, 310
copper, 44, 166
coral reefs, 159, 160, 161, 164, 166
correlation, 153, 154
cost, ix, 1, 2, 4, 11, 112, 113, 114, 115, 190, 196, 198, 225, 273, 338
cost-benefit analysis, 190, 196, 198
cotton, 141
Council of Europe, 283
Council of Ministers, 283
covering, 46, 58, 242, 327
creativity, 172
crimes, 213, 215
crises, 109
criticism, 249
crops, 329, 330
crown, 137, 139
crystallization, 44
cultivation, 239, 245, 246, 254
cultural heritage, xi, 133, 135, 136, 138, 141, 149, 153, 154, 155, 170, 173, 258, 294, 299, 317, 326
cultural values, 258, 294
culture, 16, 29, 30, 43, 79, 91, 134, 135, 136, 138, 139, 141, 142, 146, 170, 173, 175, 179, 182, 192, 240, 254, 260, 261, 286, 292, 294, 295, 298, 307, 310, 322
currency, 91, 307
customers, 115, 116, 146, 160, 230, 231, 232

cycles, 2, 240
cycling, 62, 315
Czech Republic, 95, 98, 99

D

damages, xi, 117, 119, 120, 121, 122, 123, 124, 126, 127, 128, 138, 208, 209, 227, 231, 232, 233, 335, 338, 339, 340
dance, 40, 296
danger, 136, 138, 230, 340
data gathering, 275
decision-making process, 276
deconcentration, 295
defects, 337
defence, 127, 140, 209, 210
deficiencies, 277
deficit, xiii, 226, 237, 245
degradation, 12, 136, 240, 249, 251
democratization, 112
demographic data, 59
denial, 113
Denmark, 59
Department of Agriculture, 140
dependent variable, 150
deposits, 44, 232
depth, ix, xii, 1, 6, 169, 239, 292, 317
derivatives, 138
destiny, 140, 216
destruction, 2, 160, 238, 240, 250, 315
detection, 192, 193, 195
developed countries, 13
developing countries, 90
deviation, 6, 49, 55, 107, 164
dichotomy, 53
diffusion, 174, 175, 184
digestion, 12
dignity, 202, 203, 284
direct cost, 273
direct observation, 317, 321
directors, 292
disability, 282, 283, 284, 285, 286
discrimination, 54, 283
dissatisfaction, 117
distress, 340
distribution, 3, 20, 30, 32, 34, 37, 75, 79, 80, 147, 277, 298, 336
diversification, 175, 295
diversity, 186, 188, 242, 281, 282, 321, 328, 331
division of labor, 177
Dominican Republic, 116
drainage, 140
drinking water, 160

drug trafficking, 217
dyes, 12
dynamism, 14, 70

E

earnings, 89, 240, 285
Easter, xi, 29, 32, 34, 60, 69, 70, 73, 74, 75, 76, 77, 79, 80, 83, 84
Eastern Europe, 98
eco-innovation, ix, 13, 14, 15, 16, 17, 18, 19, 20, 21, 22, 25
ecological restoration, 315
ecological systems, xiv, 70, 237
ecology, 322
economic activity, x, 27, 28, 46, 159
economic behaviour, 271, 277
economic crisis, 90, 296
economic development, xii, 157, 166, 167, 171, 174, 187, 315
economic growth, 17, 90, 92, 137, 140, 173
economic incentives, 251
economic performance, 97, 108
economic policy, 137
economic progress, 282
economic resources, 137, 322
economic transformation, 91
economic values, 238, 321
economic well-being, 13
economics, 183, 322
ecosystem, xiv, 67, 158, 159, 160, 161, 163, 164, 237, 238, 242, 247, 249, 315, 331
ecotourism, vi, xiv, 16, 23, 158, 166, 237, 240, 241, 247, 248, 249, 250, 251, 326
editors, 199, 312
education, xv, 19, 64, 72, 73, 136, 141, 173, 192, 196, 198, 240, 258, 289, 291, 292, 309
educational institutions, 47, 49, 51
elaboration, 15, 16, 19, 20, 21, 22, 58, 92, 94, 98, 102, 103, 105, 149, 213, 254
election, 141
electricity, 278
emergency, 319
emission, 166, 278, 326, 340
emotional state, 150
empirical studies, 180
employees, 89, 93, 98, 172, 217, 233
employers, 73, 114
employment, 14, 23, 28, 47, 48, 49, 51, 73, 75, 115, 146, 172, 249, 294, 296, 326, 327
employment status, 75
empowerment, 241, 250
EMS, 19, 20

endangered species, 129
energy, 2, 3, 5, 6, 11, 14, 19, 21, 22, 23, 25, 97, 161, 239, 336
energy consumption, 19, 21
enforcement, 240, 337
entrepreneurs, 21, 114, 134, 173, 226, 227, 230, 231, 232, 233, 255, 266
environment, ix, xii, xv, 14, 17, 19, 21, 23, 45, 46, 47, 49, 50, 51, 52, 53, 54, 59, 65, 66, 70, 93, 119, 129, 134, 137, 138, 139, 157, 158, 159, 160, 164, 170, 174, 179, 184, 190, 192, 195, 198, 239, 240, 241, 247, 248, 249, 250, 262, 269, 273, 284, 289, 295, 309, 313, 319, 328, 331, 335, 336
environmental aspects, 16, 157, 163, 273, 327
environmental awareness, ix, 13
environmental care, 148, 154
environmental change, 246
environmental characteristics, 82
environmental conditions, 121
environmental degradation, 158, 318
environmental effects, 90
environmental factors, 71, 274, 277
environmental impact, 13, 15, 16, 17, 159, 167, 273, 277
environmental management, 17, 19, 20, 24, 166, 273, 331
environmental policy, 161, 163
environmental protection, xiv, xvi, 11, 14, 70, 121, 237, 273, 315, 325, 327, 331, 333, 336, 337
Environmental Protection Agency, 242
environmental quality, 13, 14, 22
environmental resources, ix, 273
environmental standards, 18, 277
EPS, 106
equipment, ix, 1, 5, 6, 21, 178, 203, 226, 283, 319
equity, xii, 145, 173, 285, 286, 338
erosion, 160, 238, 239, 240, 319, 325
Estonia, 97, 100
ethics, x, 57, 60, 64, 65
ethnic groups, 67
etiquette, x, 57, 60, 62, 64, 65, 320, 321
eucalyptus, 139
Europe, 12, 17, 25, 71, 91, 116, 156, 184, 254, 282, 283, 291, 305, 311, 327
European Commission, 16, 18, 19, 20, 21, 22, 23, 283
European Community, 165
European market, 22
European Parliament, 113, 116, 326, 331, 336
European Union (EU), 16, 18, 19, 20, 23, 25, 71, 85, 92, 97, 111, 113, 209, 256, 260, 274, 283, 311, 325, 326, 327, 331, 336
EUROSTAT, 15, 16, 71

evidence, xv, 59, 66, 122, 124, 128, 289, 336, 337
evolution, 14, 20, 33, 136, 140, 173, 193, 195, 212, 302, 304
exaggeration, 234
exchange rate, 90
exclusion, 114, 213, 215, 218, 224, 227, 286, 330
execution, 206, 218, 315, 338
exercise, 62, 63, 122, 129, 209, 212, 214, 219, 224, 232, 233, 282, 337
expenditures, 36, 89, 93, 96, 108, 147
experimental design, 197
expertise, 171, 286
exploitation, 14, 123, 125, 177, 179, 182, 188, 238, 244, 245, 246, 247
exports, 90, 258
exposure, 264, 340
external environment, 273, 277
external relations, xii, 169, 171, 175, 176, 177, 179, 180, 181, 182
externalities, 170, 172, 187, 189
extinction, 129, 317
extraction, 46, 138, 240

F

factor analysis, 149
factories, 185
faith, 227, 231
families, 65, 72, 95, 242, 305, 306, 309
family income, 48
family life, 216
family members, 283, 295, 310
farms, 129, 136
fauna, xi, 119, 123, 129, 131, 250, 321, 331
fear, 61, 234
feelings, 45, 60, 61, 63, 321
fencing, xi, 119, 121, 125, 126, 127, 128, 129
fertility, 326
fidelity, x, 27, 321
fights, 286
financial, xiv, 23, 97, 108, 109, 204, 205, 241, 269, 270, 271, 273, 274, 275, 276, 277, 279, 339, 340
financial condition, 109
financial institutions, 97
financial reports, 270, 273, 275
financial sector, 205
Finland, 91, 142, 327
firm size, 178
firm value, 185
fish, 239, 317, 321
fishing, 63, 90, 246, 248, 327
fixation, 326
flexibility, 115, 172, 256, 310

flight, 115, 116, 117
flights, 113, 114, 115, 116, 307
flooding, 326
flora, xi, 91, 119, 123, 139, 250, 331
fluid, 3
focus groups, x, 43, 44, 45, 46
food, 28, 43, 89, 97, 116, 147, 173, 238, 239, 258, 259, 261, 262, 264, 265, 266
footwear, 29, 140, 183
force, xiii, 53, 114, 127, 146, 207, 215, 223, 226, 227, 231, 233, 335
Ford, 14
Fordism, 24
foreign companies, 97
foreign language, xv, 289, 291, 295, 298, 300, 305, 306
forest ecosystem, xvi, 325, 331
forest fire, 328, 333
forest management, 328, 329, 330, 331, 332
forest resources, 325, 330, 331, 332
formation, xii, 2, 145
formula, 102
foundations, 241, 262
fragility, 238
France, xv, 19, 23, 72, 91, 254, 256, 259, 261, 266, 289, 291, 295, 296, 305, 309, 310, 327
fraud, 206, 208
freedom, 197, 204, 207, 212, 213, 248, 284
freedom of choice, 207
freshwater, 167
funding, 160
funds, 97

G

gambling, 203
genetics, 328
geographical origin, 124
geography, 136, 141
Germany, 12, 17, 20, 23, 24, 25, 91, 93, 95, 99, 261, 291, 296, 305, 309, 310, 314
glasses, xiii, 189, 190
global warming, 159, 160
globalised world, 309
globalization, 177, 294
goods and services, 15, 29, 91, 202, 207, 238
grants, 121, 140, 213, 214, 218
graph, 50, 52, 302
grass, 329
grasslands, 329
gravity, 248
grazing, 246, 327, 329
Great Britain, 312

greenhouse, 326
greenhouse gases, 326
Gross Domestic Product (GDP), 90, 92, 93, 97, 108, 146, 258, 295
grouping, 125
growth, xv, 8, 15, 19, 20, 23, 70, 72, 136, 140, 146, 147, 170, 185, 244, 246, 257, 258, 289, 295, 302, 314, 315, 326
growth rate, 244
guesthouses, 217
guidance, 173, 322
guidelines, 48, 159, 173, 254, 273, 274, 275, 283, 331, 332, 338
guiding principles, 262, 328
guilty, 192

H

habitat, 125, 127, 242, 314, 315, 325, 331
habitats, 129, 237, 242, 314, 317, 318, 320, 321
harassment, 207
harbors, 317
harmonization, 137, 269, 271
harmony, 65, 315
Hawaii, 86
health, 47, 48, 49, 50, 51, 77, 79, 82, 100, 136, 192, 203, 335, 336, 337, 340
health condition, 136
health services, 47, 48, 49, 50, 51
height, 137, 139, 242
heterogeneity, xii, 169, 170, 175, 176, 181, 185, 186
high school, 60, 65
highways, 123, 147
historical character, xi, 133, 134, 135
history, 91, 134, 146, 290, 310, 317, 321, 322
homeowners, 337
homes, 30, 212
homogeneity, 170
Hong Kong, 84, 278
horses, 230, 231, 319
hospitality, 22, 25, 54, 148, 155, 258, 259, 271, 279
host, 241, 309
hotel, xiii, 18, 19, 20, 21, 24, 72, 73, 85, 89, 98, 112, 113, 114, 116, 117, 140, 204, 211, 212, 217, 218, 219, 231, 270, 271, 272, 273, 274, 276, 277, 278
hotels, xi, xiii, xiv, 19, 30, 72, 93, 97, 98, 100, 102, 106, 108, 111, 113, 114, 148, 160, 211, 217, 218, 261, 269, 270, 272, 276, 282, 283, 284
house, 85, 215, 217
household income, 244
housing, 216, 247
human, xiii, 135, 136, 137, 161, 176, 178, 189, 190, 214, 239, 281, 282, 285, 314, 315, 321, 322, 340

human behavior, xiii, 189, 190
human capital, 178
human development, 282
human resources, 176
human right, 214, 340
humidity, 260
Hungary, 98, 99
Hunter, 14, 24
hunting, xi, xiv, 119, 120, 121, 122, 123, 124, 125, 126, 127, 128, 129, 237, 238, 242, 246, 248, 259, 262, 263, 265, 327
hybrid, 12
hydrolysis, 7, 11
hygiene, 148, 154, 274, 277
hypothesis, 108, 191, 197, 308

I

ideal, 51, 53, 54, 177, 212, 247
identification, 28, 251, 255, 273
identity, xii, 55, 133, 135, 214, 320, 321
ideology, 28
illumination, 154
image, 21, 29, 44, 45, 46, 54, 134, 158, 203, 209, 213, 258, 273, 296, 298, 299, 305, 310
images, 142
immersion, ix, 1
immigration, 247
Impact Assessment, 250
imprisonment, 213
improvements, 19, 22, 141, 318, 320
inauguration, x, 57, 60, 64
incidence, 120, 125, 128
income, xiv, 22, 28, 32, 37, 39, 40, 72, 75, 93, 159, 160, 164, 192, 195, 196, 198, 237, 238, 240, 241, 244, 245, 246, 247, 249, 272, 273, 327, 332
incompatibility, 121, 125
independence, 75, 80
indirect effect, 93, 181
individuals, 30, 32, 37, 45, 46, 47, 48, 53, 70, 77, 141, 153, 175, 190, 191, 193, 195, 198, 202, 212, 219, 263, 282, 340
industrial sectors, 20
industrialization, 141
industries, ix, 1, 2, 11, 13, 16, 89, 173, 174, 179, 186, 249, 258, 326
industry, ix, xi, xiv, 1, 2, 13, 14, 15, 16, 18, 19, 20, 22, 23, 24, 44, 69, 70, 85, 89, 90, 91, 93, 96, 108, 111, 112, 114, 140, 146, 158, 165, 170, 171, 173, 176, 183, 185, 241, 253, 255, 256, 259, 266, 271, 273, 274, 276, 277, 278, 279, 282, 340
inequality, 202

information technology, 178
infrastructure, 97, 134, 171, 172, 179, 248, 283, 315, 326, 330
injuries, 124
injury, 338, 339, 340
insects, 321
insecurity, 61, 340
insertion, 224
institutions, x, xv, 21, 27, 49, 113, 170, 172, 173, 182, 205, 248, 281, 283, 285
insulation, 339
integration, 173, 177, 180, 282, 283, 285, 328
integrity, 230, 240, 314, 335, 340
intellectual capital, 186
interdependence, xii, 169, 170, 172, 174, 179, 199, 241
interface, 3
interference, 328
internalization, 326
international communication, 290
international trade, 205
internationalization, xiv, 269, 270
intervention, 136, 179, 232, 339
intrusions, 140
inventors, 174
investment, x, 27, 204, 205, 241, 328
investments, x, 27, 97, 175, 249, 263
investors, 97, 101
Iowa, 199
Iran, vi, xiv, 1, 237, 241, 242, 251
Iranian dairy producer, around, ix, 1, 2
Ireland, 142, 170, 188, 305
Iron Curtain, 91, 98
irradiation, ix, 1, 2, 4, 6, 12
irrigation, 238, 242, 246, 317
islands, xii, 157, 158, 159, 160, 161, 163, 164, 165, 166, 186
isolation, 129, 174
issues, xiii, xiv, 59, 61, 64, 116, 117, 120, 121, 124, 165, 211, 241, 250, 269, 319, 322
Italy, 17, 19, 20, 71, 72, 85, 183, 185, 186, 254, 256, 259, 266, 291, 305

J

Japan, 25
joint stock company, 99, 100
jurisdiction, 209, 337, 340
justification, 230, 232, 233

K

knowledge acquisition, xii, 169, 177, 178, 179, 180, 181, 188

L

laboratory tests, 191
labour market, 291
lakes, 53, 91, 100, 262, 329
landfills, 329
landscape, xiv, 62, 141, 237, 241, 248, 262, 310, 314, 315, 322, 326, 328, 330
landscapes, 65, 141, 321
language skills, 292
languages, xv, 289, 290, 291, 292, 294, 305, 309
Latvia, 100
laws, xiii, 135, 140, 201, 202, 256, 322, 327, 330, 336
lead, 43, 175, 178, 180, 214, 232, 233, 238, 246, 249
leadership, 19, 175
learning, xv, 158, 170, 174, 175, 177, 178, 179, 185, 186, 187, 188, 289, 292, 296, 309, 311
legal issues, 338
legal protection, 211, 218, 250
legislation, xi, xiv, xv, 21, 23, 113, 119, 120, 121, 122, 123, 124, 125, 127, 129, 134, 202, 219, 249, 253, 274, 277, 325, 327, 329, 333
leisure, xv, 30, 34, 35, 37, 39, 43, 60, 62, 95, 100, 112, 160, 248, 282, 285, 286, 291, 307, 310, 313, 315, 319
leisure time, 62, 248
level of education, 46, 75
life cycle, 14, 166
light, xiii, 189, 190, 338
limited company, 300
limited liability, 100, 227
linguistic tourism, xv, 289, 293, 294
Lithuania, 93, 95, 98, 100
litigation, xvi, 335
livestock, xiv, 121, 237, 245, 246, 249
living environment, 46, 47, 48, 50, 51, 52, 53, 54
loans, 205
local authorities, xvi, 209, 309, 335
local community, 28, 173, 238, 240, 248, 250
local government, 173
longitudinal study, 182
love, 141
loyalty, 84, 113, 134, 295
LTD, ix, 1
Luo, 312
lying, 239

M

magazines, 266
majority, x, xiii, 17, 19, 57, 60, 61, 62, 63, 89, 95, 97, 136, 211, 213, 215, 291, 295, 298, 304
Malaysia, 156
mammals, 317
man, 65, 122
management, xi, xiv, 2, 14, 15, 18, 21, 22, 24, 59, 65, 66, 114, 117, 119, 129, 136, 137, 139, 140, 141, 147, 160, 161, 163, 164, 167, 169, 173, 176, 178, 183, 186, 249, 257, 266, 269, 270, 271, 273, 276, 277, 279, 292, 309, 315, 318, 319, 320, 322, 323, 324, 327, 328, 329, 330, 331, 332, 336
manufacturing, 25
mapping, 44, 251
marital status, xi, 69, 70, 75, 79, 80, 83
market capitalization, 102, 108
market economy, 91, 202
market segment, 71
marketing, xi, 29, 41, 44, 54, 69, 70, 71, 82, 146, 187, 254, 256, 258, 294, 300, 305, 311
marketing strategy, 44
Maryland, vii, 313, 314
MAS, 14
mass, xiv, 3, 139, 140, 158, 240, 269, 270, 276, 310
mass media, 139
materials, 62, 238, 320
matrix, 103
matter, xvi, 63, 64, 112, 117, 120, 124, 137, 151, 217, 226, 228, 231, 232, 274, 277, 325, 337
measurement, ix, 13, 14, 15, 16, 18, 47, 96, 150, 224, 226
measurements, 154, 155, 277
media, 97, 173, 205
mediation, 201
Mediterranean, v, 57, 134, 141, 283, 317
membership, 170
memory, 28, 187
mental health, 63, 321
messages, 65
meta-analysis, 250
metaphor, 177
meter, ix, 1, 2, 141
methodology, xii, 17, 45, 89, 157, 160, 304
Mexico, 186, 190
microclimate, 259, 263, 266
Microsoft, 274
migrants, 247
migration, 90, 319
military, 218, 317
Ministry of Education, 291
minors, 203, 207, 213

misconceptions, 206
mission, 166, 296
mixing, 3
modelling, 145, 304
models, xii, 22, 102, 104, 129, 145, 146, 166, 169, 181, 191
modern economies, 170, 174
modern society, 140
modernization, 140
modifications, 212, 275
modus operandi, 218
mother tongue, 291
motivation, 15, 293
multiculturalism, 290
multidimensional, 45, 171
multinational firms, 170
multiplier, 293, 295, 304, 308, 310
multiplier effect, 293, 295, 304, 308, 310
municipal solid waste, 158
museums, 173, 306, 310
music, x, 27, 28, 29, 41, 296
mutual respect, 282

N

narcotic, 204
nationality, 32, 34, 35, 148, 305
native species, 246, 320
natural habitats, 240, 242
natural resource management, 333
natural resources, 19, 44, 90, 137, 157, 171, 241, 247, 249, 328, 331
nature conservation, 248
negative consequences, 249
negative effects, 115, 159, 175, 326
Netherlands, 97, 251, 279
neutral, 190, 191, 192, 193, 198
new market niches, 310
New Zealand, 166
NGOs, 18
nitrogen, ix, 1, 2, 7, 11
nitrogen compounds, 7
normal distribution, 48
North America, 298, 305
Norway, 91
nuisance, 335, 337, 339, 340
null, 75
null hypothesis, 75
nutrients, ix, 1, 2, 11
nutrients balancing, ix, 1, 2, 11

O

obstacles, 122
Oceania, 305
officials, 46, 216
oil, 97, 138, 140
olive oil, 260, 264, 265
omission, 206
one dimension, 52
open spaces, 59, 215, 314, 319
operational frequency, ix, 1, 5
operations, 25, 249, 271
opinion polls, 22
opportunities, xiv, 71, 115, 172, 237, 241, 270, 315, 317, 321
optimization, 4
oral tradition, 138
organ, 263
organic compounds, 12
organism, ix, 1, 2
organize, 181, 264
outreach, 328
outsourcing, 13, 178
overlap, 202, 256
ownership, 321, 327, 328, 332
ozone, 11

P

Pacific, 166
pain, 339, 340
Pakistan, 12
PAN, 109
parallel, x, 29, 43, 97, 282, 328
parallelism, 218
parameter estimates, 103, 104, 105
Parc Fluvial del Túria, v, x, 57, 58, 62, 64
parents, 203, 207
parliament, 113
participants, 46, 137, 240
pastures, 246
patents, 15, 16, 25
penalties, 190, 192, 194, 195, 206
people with reduced mobility (PRM), xiv, 281
permit, 120, 126, 218, 260, 319
personal contact, 177
personality, 255, 257, 260, 264, 266, 340
phenolic compounds, 12
Philippines, 282
phosphors, ix, 1, 2, 11
physical activity, 67
physical environment, 45, 241, 331

plants, ix, xiv, 1, 2, 11, 136, 237, 238, 239, 242, 246, 317, 320
plastics, xiii, 189, 190
platform, 28, 172
playing, 322
pleasure, 261, 262
Poland, xi, 89, 90, 91, 92, 93, 94, 95, 96, 97, 98, 99, 100, 108, 305
police, 217, 319, 328
policy, xii, xv, 23, 25, 145, 146, 161, 164, 165, 204, 207, 219, 283, 289, 295, 296, 300, 328, 332
policymakers, 163
political system, 90
politics, 322
pollution, xvi, 14, 15, 20, 239, 335, 336, 337, 338, 340
population, 11, 16, 44, 46, 47, 48, 72, 73, 91, 112, 147, 148, 198, 199, 238, 243, 244, 249, 251, 261, 283, 284, 290, 294, 301, 326, 328
population growth, 238, 243, 244
portfolio, 97
Portugal, 86, 259, 264, 266, 279, 307
positive relationship, 34
poverty, 90
poverty reduction, 90
precedents, 211
predictability, 154
prejudice, 116, 121, 203, 210, 329, 330
preparation, xiv, 240, 253, 273, 274, 276
preservation, 147, 249, 321, 328, 331
prestige, 115, 260, 298
prevention, 121, 125, 128, 226, 239
primacy, 212
principles, 126, 137, 203, 262, 271, 272, 275, 328
private schools, 300
privatization, 91
probability, 191, 193, 197
probability distribution, 191
probe, ix, 1, 6, 11
process control, 21
producers, xiv, 13, 175, 253, 254, 255, 257, 258, 259, 266
production function, 247
professional management, 310
professionals, 270, 271, 310, 332
profit, 209, 257, 275, 321
profitability, ix, 13, 25, 114, 170, 271
project, xv, 23, 24, 58, 69, 165, 263, 274, 276, 313, 317, 318, 320, 321, 322, 323, 332
proliferation, 275
promote innovation, 19, 175
promoter, 260, 261
property rights, 238

proposition, 176, 177, 178, 180
protection, xi, xii, xiii, xiv, xvi, 70, 113, 119, 121, 129, 133, 134, 135, 136, 137, 138, 139, 140, 141, 159, 160, 163, 201, 202, 203, 205, 208, 209, 211, 212, 213, 214, 215, 216, 217, 218, 219, 223, 224, 225, 226, 231, 234, 237, 238, 249, 250, 318, 322, 325, 327, 328, 330, 331, 332, 333, 336, 338, 340
prototype, 275
pruning, 138
public administration, 137, 172, 327, 338
public domain, 329
public goods, 192, 196, 198
public health, 337
public interest, 135, 327, 329
public safety, 92
public service, 337, 338

Q

qualitative research, 53
quality of life, 59, 134, 137, 241, 315, 321, 326, 331, 340
quality standards, 14
quantification, 338
quantitative research, 53
quantitative technique, 44
Queensland, 184
questioning, 175
questionnaire, xii, 46, 47, 145, 148, 149, 293, 304
quotas, 47

R

race, 66
radicals, 2
radio, 228
radius, 2, 3
rainfall, 242
rangeland, 246
rate of return, 101, 108
rationality, 191
raw materials, 239
RDP, 130
reactions, 2
real property, 215
real time, 225, 226, 277
reality, xiv, 154, 214, 281, 282, 285, 286, 293, 338
recall, 64, 337
reception, 217, 276
recognition, xiv, xvi, 140, 201, 253, 254, 255, 256, 257, 259, 266, 335, 337, 341
recombination, 175

Index

recommendations, 39, 182, 282, 283, 320
recovery, 138, 146, 205, 339
recreation, xv, 62, 93, 166, 239, 284, 285, 313, 315, 318, 319, 329
recreational, xii, 47, 48, 49, 50, 51, 58, 59, 60, 61, 67, 72, 133, 134, 135, 136, 137, 148, 246, 315, 317, 318, 319, 321, 322, 326, 330, 331
rectification, 208, 209
recycling, xiii, 189, 190
red wine, 260, 262, 263
redundancy, 171, 173, 175
reform, 17, 120, 225
regeneration, 28, 172, 183, 331
regional cooperation, 324
regression, 149, 152, 153
regression weights, 153
regulations, xii, 97, 113, 116, 123, 125, 128, 134, 135, 142, 157, 158, 202, 203, 208, 219, 254, 263, 270, 271, 322, 328, 336, 337, 339
rejection, 125
relatives, 73, 76, 77, 78, 81, 82, 93, 95
relaxation, 63, 65
relevance, xiv, xv, 120, 176, 178, 254, 281, 282, 283, 295
reliability, 75, 80, 152, 232
Religous Music Week, x, 27
renewable energy, 160, 326
rent, xiii, 223, 229, 230
repetitions, 46, 197
replication, 185
repression, 232
reprocessing, 232
reproduction, 28
reputation, xii, 19, 97, 133, 134, 135, 141, 208, 230, 254, 256
requirements, xiv, xvi, 19, 125, 129, 135, 140, 202, 205, 207, 209, 218, 225, 234, 257, 269, 275, 314, 335, 336, 337, 339
researchers, x, 46, 57
reserves, 70, 112, 315
residuals, 153
resistance, 146, 282
resolution, 113, 125, 151, 213, 218, 283, 326
resource management, 71
resources, ix, xi, xiv, 14, 15, 16, 20, 22, 25, 70, 111, 112, 114, 115, 116, 117, 134, 158, 160, 166, 167, 170, 171, 176, 178, 183, 237, 238, 239, 240, 242, 244, 246, 247, 249, 254, 255, 259, 274, 277, 282, 283, 284, 286, 294, 295, 317, 318, 321, 326, 328, 331, 332
response, 48, 148, 190, 191, 270
response time, 48

restaurants, 17, 29, 30, 73, 93, 97, 98, 99, 100, 102, 104, 106, 108, 112, 148, 255, 259, 266, 282, 283, 306
restitution, xiii, 223, 226, 227, 228, 229, 230, 232
restoration, 58, 136, 241, 317, 318, 319, 320, 321, 327, 328, 338, 339
restrictions, 92
revaluation, 136, 141
revenue, 102, 104, 114, 115, 117, 164, 192, 193, 196, 270
rights, 113, 114, 115, 116, 122, 137, 202, 203, 206, 207, 208, 209, 212, 214, 217, 218, 238, 282, 286, 315, 326, 331, 335, 338, 340
rings, 314
risk, xiv, 14, 21, 115, 125, 126, 129, 140, 146, 190, 191, 192, 193, 195, 198, 232, 233, 237, 266, 275, 324, 336
risk management, 146
risks, 15, 54, 126, 203, 206
RMSEA, 153
robberies, 227, 233, 234
Romania, 100
roots, 139, 153
routes, 136, 140
routines, 178
rubber, 139
rules, xiii, 97, 113, 114, 116, 120, 121, 125, 176, 182, 201, 202, 203, 208, 219, 254, 257, 272, 273, 276, 277, 320, 330, 335
runoff, 242, 316, 319, 323, 326
rural areas, xv, 90, 313, 327
rural development, 158, 163, 326, 328
rural population, 244, 331
Russia, 91, 95, 98

S

safety, 11, 121, 122, 123, 124, 146, 203, 226, 274, 277, 307, 319, 320
salinity, 160
sample design, 196
sample survey, 48
SAP, 113, 225, 228
savings, 19, 22, 204
scale economies, 14
schema, 275
school, 140, 172, 296, 300, 301, 315, 321
science, 146
Scientz-IID with 20 kHz, ix, 1
scope, xiii, 2, 113, 117, 120, 136, 155, 215, 217, 223, 225, 226, 227, 233, 275, 332, 336, 338
sea level, 242, 261, 262, 264
seasonal changes, 59

seasonality, 89, 102, 103, 104, 105, 294, 310
second generation, 14
second language, 290, 291
Second World, 98
secondary education, 48
secondary schools, 291
security, 61, 146, 148, 155, 192, 212, 213
sediment, 239
seed, 139
Semana de Música Religiosa (SMR), x, 27
semiconductor, 186
sensation, 66
sensation seeking, 66
sensitivity, 193, 195, 196, 198
Serbia, 98
service provider, 230
services, xii, 15, 17, 19, 20, 28, 34, 41, 70, 71, 84, 89, 92, 93, 98, 100, 104, 106, 109, 112, 113, 116, 117, 145, 146, 147, 148, 149, 150, 151, 154, 155, 163, 171, 172, 173, 178, 192, 196, 198, 202, 203, 205, 206, 207, 219, 225, 238, 248, 249, 259, 271, 282, 283, 285, 296, 321, 331
sewage, 11, 274, 277
sex, 48, 75
Seychelles, 165
shape, 3, 16, 173
shear, 2
sheep, 246
shellfish, 239
shortage, 249
showing, 64, 79, 158, 277, 304, 321
shrubs, 135, 136, 329, 330
side effects, 326
SIDS, 165
significance level, 75, 80, 103, 104, 105
signs, 58, 213, 215, 320
silver, 261
Singapore, 59, 65
skewness, 48
Slovakia, 93, 95, 100
sludge, ix, 1, 2, 3, 4, 5, 6, 7, 8, 9, 10, 11, 12
small firms, 181
social behavior, 340
social behaviour, 271, 277
social benefits, 321
social capital, 174, 185, 186, 240
social development, 165
social environment, 48, 49, 199
social identity, 28
social image, 273
social interactions, 170, 174
social life, 50, 52, 53, 54
social network, 170, 173, 174, 175, 187

social programs, 141
social relationships, 175
social responsibility, 274
social roles, 330
social sciences, 28, 151, 173
social structure, 174
society, xvi, 13, 28, 135, 172, 178, 205, 223, 285, 325, 327
Socrates, 300
software, 30, 276
sole proprietor, 99
solidarity, 114
solution, 2, 64, 137, 160, 163
South Africa, 259
South America, 298
South Asia, 167
South Pacific, 165
Soviet Union, 92
sowing, 329
Spain, v, vi, vii, x, xi, xii, xiv, xv, xvi, 13, 17, 19, 20, 21, 27, 29, 39, 41, 57, 58, 67, 69, 70, 71, 72, 73, 83, 84, 85, 91, 111, 112, 119, 120, 127, 133, 134, 135, 142, 145, 147, 157, 165, 169, 185, 189, 190, 201, 205, 209, 211, 212, 213, 219, 223, 224, 253, 254, 255, 256, 258, 259, 260, 261, 262, 264, 266, 267, 269, 272, 274, 275, 277, 281, 289, 291, 295, 296, 297, 298, 300, 302, 304, 305, 307, 309, 310, 311, 312, 313, 314, 317, 325, 335, 340
Spanish Constitution, 121, 137, 213, 214, 217, 218, 335, 340
spatial location, 171
specialisation, 300
specialists, xiv, 73, 253
specialization, 171, 173
species, xi, 91, 119, 120, 121, 122, 126, 129, 135, 136, 137, 138, 139, 141, 160, 238, 242, 315, 316, 317, 320, 328, 329
specific tax, 275
spending, x, 27, 29, 30, 31, 32, 33, 34, 35, 36, 37, 39, 41, 216, 249
spillovers, 174, 179, 187
SPSS software, 163
stability, 326
stabilization, 12
stakeholders, ix, xii, 53, 169, 170, 174, 292, 309, 317, 323, 328
standard deviation, 107
standardization, 22, 273, 275, 276
stars, 160
state, xv, 11, 19, 63, 80, 90, 91, 96, 97, 98, 120, 123, 124, 127, 150, 226, 227, 228, 254, 255, 257, 259, 260, 261, 262, 264, 265, 266, 283, 292, 295, 296,

300, 304, 307, 309, 310, 313, 327, 328, 329, 330, 333, 335, 336, 337
state enterprises, 91
Statement of Financial Accounting Standards-SFAS 131, xiv, 269
states, 92, 135, 138, 158, 160, 165, 167, 204, 205, 207, 208, 260, 283, 327, 336, 337, 339
Statistical Package for the Social Sciences, 293
statistics, x, 13, 18, 57, 71, 73, 103, 197, 244, 246, 271, 327
steel, 186
stockholders, 238
storage, 3, 239, 242
strategic management, 170, 173, 187
stress, xv, 63, 64, 313, 319, 321
structural equation modeling, 54
structure, 30, 45, 90, 93, 94, 95, 98, 140, 141, 148, 183, 190, 193, 195, 198, 242, 255, 295, 324, 328
style, 264, 320, 321
styles, 85
subscribers, 33, 39
substitutes, 175
substitution, 121, 123, 238
substrate, x, 43
substrates, 7, 8, 11
sugar beet, 246
Sun, 23
supervision, 136, 232, 339
support staff, 284
Supreme Court, 124, 127, 204, 216, 217, 227, 337, 338, 340
surface area, 91, 254, 255, 261, 262
surveillance, 125, 126, 129
survival, 14, 129, 136, 137, 138, 323
sustainability, ix, 13, 15, 70, 137, 146, 247, 274, 294, 328, 331
sustainable development, 16, 67, 136, 137, 140, 141, 160, 239, 249, 274, 331
Sweden, 23, 66, 91, 187, 327
Switzerland, 250, 279, 291
symbiosis, 294

T

talent, 173
tangible resources, 295
target, 158, 172, 203, 309, 321
tax collection, 193, 194, 195
taxes, 190, 192, 193, 195, 196, 198, 206
taxonomy, 271, 275, 276
TDI, 272
teachers, 291
teams, 186

techniques, 30, 203, 260, 271, 286, 319
technological advances, 254
technologies, 14, 15, 17, 266, 290, 291, 305, 316
technology, ix, 1, 11, 14, 22, 178, 185, 188, 240, 254, 259, 261
Tehran Pegah Dairy Complex (TPDC), ix, 1, 2
telecommunications, 173
telephone, 48, 155, 207
telephones, 226
temperature, 2, 6, 260, 261, 264
tension, 153
terminals, 284
territorial, xiii, 29, 113, 137, 169, 171, 174, 178, 180, 185, 318, 322, 328, 332
territory, 91, 120, 123, 125, 171, 172, 174, 327, 328, 332
testing, 254, 293
Thailand, 167
theatre, 28, 261, 294, 306
theft, 228
Third World, 249
thoughts, x, 57, 60
threats, 213, 247, 315, 322
time series, 104
Title I, 203, 208
Title IV, 203, 208
tobacco, 29
total product, 255
trade, 20, 92, 206, 207, 232
traditions, 135, 141, 299
training, xiv, 176, 179, 237, 241, 249, 284, 300, 309, 328
transaction costs, 172
transactions, 101, 201, 272
transcripts, 328
transducer, 5
transformation, 2, 90, 257
transition period, 91
translation, 23, 142, 340
transmission, 137, 176, 270
transparency, 90
transport, xi, xiii, 28, 30, 31, 39, 89, 111, 112, 113, 116, 117, 148, 173, 216, 223, 230, 231, 232, 284, 293, 304, 305, 306, 307, 322
transportation, 61, 115, 117, 148, 155, 172, 284
travel agency, 98, 100, 117, 266
treasury, 98
treatment, ix, xiii, 1, 2, 7, 8, 11, 12, 120, 206, 211, 216, 224, 230, 238, 283, 331, 335
trial, 229, 230
Turia Fluvial Park, vii, xv, 313, 316, 318, 319, 320, 321, 322
Turia River Valley, xv, 58, 313

Turkey, 67, 86, 323
turnover, 99, 100, 108, 146, 176
turtle, 317

U

Ukraine, 91, 95
ultrasound, 3, 5, 11, 12
uncertain outcomes, 122
UNESCO, xii, 70, 145, 147, 148, 290, 295, 312
uniform, xiv, 3, 129, 269, 270, 273
unions, 18
united, xv, 23, 54, 72, 90, 112, 117, 199, 259, 279, 283, 289, 291, 296, 298, 305, 311, 314, 326, 331
United Kingdom (UK), xv, 12, 23, 25, 54, 65, 72, 184, 186, 199, 279, 289, 291, 295, 296, 305, 310, 311, 314
United Nations (UN), 23, 158, 159, 283, 285, 290, 314, 326, 331
United States, 112, 117, 259, 291, 298, 305, 326
Universal Declaration of Human Rights, 285
universe, 282
universities, 190, 297, 298, 299, 300, 305
updating, 175
urban, xii, xv, 28, 29, 47, 49, 50, 51, 59, 61, 63, 65, 66, 67, 70, 90, 133, 134, 137, 140, 141, 142, 169, 170, 172, 173, 284, 294, 298, 313, 314, 315, 319, 321, 322, 323, 329, 330, 336, 337
urban areas, 63, 90, 134, 284, 298, 315
urban life, 47, 49, 51, 65
urbanization, 66
USA, 12, 41, 59, 86, 311, 314, 322
UWI, ix, 1, 2

V

Valencia, v, vi, vii, x, xi, xv, 13, 23, 57, 58, 67, 72, 73, 76, 77, 78, 79, 81, 82, 125, 130, 133, 134, 135, 136, 138, 139, 140, 141, 142, 157, 169, 219, 220, 298, 313, 316, 317, 318, 319, 322, 324, 330, 336, 337, 339, 340
validation, 275
valuation, 115, 116, 238, 250
vandalism, 61
vapor, 2
variables, x, xi, xii, xiii, 30, 43, 48, 59, 61, 63, 69, 70, 71, 75, 79, 80, 83, 102, 103, 106, 107, 108, 145, 147, 149, 150, 151, 152, 153, 155, 161, 181, 189, 190, 192
variations, 121, 309
varieties, xiv, 253, 254, 259, 261, 263, 264
VAT, 205

vegetation, 135, 239, 320, 328, 329, 330
vehicles, 121, 205, 224, 225, 226, 228, 319
vein, 45
velocity, 3
Venezuela, 250
ventilation, 339
venue, 142, 340
vibration, 5, 337
victims, 340
violence, 203
vision, xv, 24, 82, 111, 225, 313, 318
visual field, 121
visualization, 52
vocabulary, 146
vulnerability, 160

W

walking, x, xv, 57, 62, 313, 315, 319
Washington, 66, 250, 323, 324
waste, ix, 1, 2, 3, 6, 7, 8, 10, 11, 12, 15, 19, 158, 160, 161, 163, 164, 192, 274, 277
waste management, 160, 161, 163, 164
wastewater, ix, 1, 2, 11, 12
wastewater treatment plants (WTPs), ix, 1, 2
water, 3, 12, 15, 19, 140, 141, 159, 160, 161, 163, 164, 238, 239, 242, 244, 246, 316, 317, 318, 319, 320, 321, 326
water quality, 318, 326
water supplies, 238
watershed, 318
waterways, 329
wavelengths, 3
wealth, 28, 259, 282, 296, 298, 326
weapons, 122
websites, 105, 160, 161, 163, 164, 165
welfare, 336
well-being, 137
wetlands, xiii, xiv, 237, 238, 239, 242, 244, 246, 247, 249, 250
wilderness, 53, 319
wildlife, x, 43, 54, 65, 120, 124, 125, 129, 315, 317, 318, 319, 321, 322, 326
wildlife conservation, 315, 322
withdrawal, 206, 246
wood, 123, 260, 331, 332
work environment, 48, 49
workers, 140, 141, 247, 291, 305
working hours, 285
World Bank, 90
World Health Organization, 314
worldwide, xiv, 59, 89, 91, 93, 97, 177, 269, 272, 290

X

Xinzhi Bio-Instrument LTD, ix, 1
XML, 274, 275

Y

Yale University, 156
yield, 114, 117, 263, 264, 270
young people, xv, 202, 289, 304